# Australian Signpost MATHS

AF746579

# 3


Alan McSeveny
Diane McSeveny-Foster
Bob Collard
Alan Parker
Rachel McSeveny

# Short Contents



**Sections**

**Pearson Australia**
(a division of Pearson Australia Group Pty Ltd)
459–471 Church Street, Level 1, Building B, Richmond, Victoria 3121
PO Box 23360, Melbourne, Victoria 8012
www.pearson.com.au





First published 2018 by Pearson Australia
2025 2024 2023 2022
10 9 8 7 6 5 4 3 2 1





Content and Learning Specialist: Julian Lumb
Project Manager: Shelly Wang
Production Manager: Elizabeth Gosman & Aptara
Editor: Aptara
Designer: Anne Donald & Jennifer Johnston
Cover Designer: Jennifer Johnston
Desktop operator: Aptara
Rights & Permissions Editor: Samantha Russell-Tulip
Senior Publishing Services Analyst: Rob Curulli
Cover art: Michael Barter
Cover background: rf_vector/shutterstock.com
Illustrators: Michael Barter, Wendy Gorton, Nives Porcellato, Andrew Craig
Printed in Malaysia (CTP-VVP)

ISBN 978 1 4886 2184 0
Pearson Australia Group Pty Ltd ABN 40 004 245 943

**Acknowledgements**
All material identified by ACARA is material subject to copyright under the *Copyright Act 1968* and is owned by the Australian Curriculum, Assessment and Reporting Authority 2018. ACARA neither endorses nor verifies the accuracy of the information provided and accepts no responsibility for incomplete or inaccurate information. In particular, ACARA does not endorse or verify that:
- The content descriptions are solely for a particular year and subject;
- All the content descriptions for that year and subject have been used; and
- The author's material aligns with the Australian Curriculum content descriptions for the relevant year and subject.

You can find the unaltered and most up to date version of this material at http://www.australiancurriculum.edu.au/ This material is reproduced with the permission of ACARA.

We thank the following for their contributions to our text book:
**Shutterstock**: Jane Kelly, p. 125.

Every effort has been made to trace and acknowledge copyright. However, if any infringement has occurred, the publishers tender their apologies and invite the copyright holders to contact them.

# What is Australian Signpost Maths?

Australian Signpost Maths is a mathematics activity book series for students from Foundation to Year 6. The series has been written to meet the requirements of the Australian Curriculum.

The components of the series include Student Books, Teacher's Books, Mentals Books and an interactive Website. Teachers can select an appropriate program for every student from the rich and varied material provided.

The content has been carefully sequenced within each year level and across the series to take into account students' likely mathematical development.

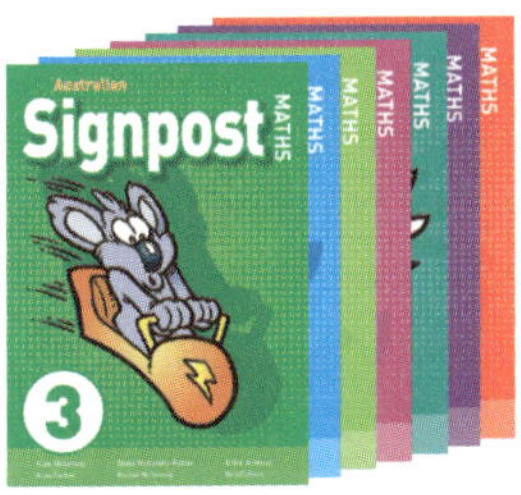

Student Books

Teacher's Books

Mentals Books

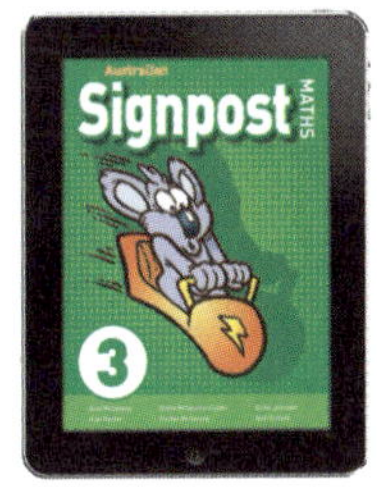

Website

# Structure of Australian Signpost Maths

Australian Signpost Maths emphasises the curriculum's syllabus content as well as problem-solving strategies, language development and the use of technology.

The syllabus is organised into three content strands and four proficiency strands:

**Content Strands**

- Number and Algebra
- Measurement and Geometry
- Statistics and Probability

**Proficiency Strands** (see page iv)

- Understanding
- Fluency
- Problem Solving
- Reasoning

The curriculum's **general capabilities** are developed throughout the Australian Signpost Maths program. These are:

- literacy
- numeracy
- information and communication technologies (ICT)
- critical and creative thinking.

Australian Signpost Maths also provides opportunities to develop other general capabilities, such as personal and social competence and intercultural understanding.

The cross-curriculum dimensions of the syllabus – 'Aboriginal and Torres Strait Islander histories and cultures', 'Asia and Australia's engagement with Asia' and 'Sustainability' – are embedded in the program.

To maximise the benefits of the program, the Student Book, Teacher's Book, Mentals Book and Website should be used together.

The structure of the **Student Book** allows teachers to determine both the order and the extent of content covered. Strands are organised separately so that the teacher, not the Student Book, decides the content of the next lesson. However, a suggested term program (see page x of this book) and a detailed program (see the Teacher's Book and Website) are also provided.

The **Teacher's Book** also provides lesson plans for each page of the Student Book and blackline masters to assist teachers in implementing the program.

The **Mentals Book** mixes examples from all strands. It revises the content covered in the Student Book. Each content strand is thoroughly covered, with the proficiency strands incorporated within each section. A special feature woven throughout the Mentals Book is the tables program in the four operations.

The innovative **Website** help teachers to bring mathematics alive with technology. The website provides interactive maths tools, games and practice opportunities as well as relevant resource masters and worksheets for all year levels. These can be used for whole-class, small-group and individual learning. The website also includes **Concept Check-In** a new diagnostic screener.

Student Book pages are colour-coded by section.

| | | |
|---|---|---|
| Number and Algebra A | Measurement and Geometry A | Statistics and Probability |
| Number and Algebra B | Measurement and Geometry B | Answers |

# Australian Curriculum Proficiency Strands

The proficiency strands of the Australian Curriculum describe how content is explored or developed – that is, the 'thinking and doing' of mathematics.

## Understanding

**Learning the concepts**

*Students build a robust knowledge of adaptable and transferable mathematical concepts. They make connections between related concepts and progressively apply the familiar to develop new ideas. They develop an understanding of the relationship between the 'why' and the 'how' of mathematics.**

Conceptual understanding of maths ideas includes the explanation of a concept using text and diagrams. This occurs throughout Australian Signpost Maths at the top of many pages and is indicated by the Concepts icon.

## Fluency

**Using the concepts**

*Students develop skills in choosing appropriate procedures, carrying out procedures flexibly, accurately, efficiently and appropriately, and recalling factual knowledge and concepts readily.**

The practice of maths skills to build fluency occurs on every page of Australian Signpost Maths.

## Problem Solving

**Applying concepts and strategies to develop solutions to problems**

*Students develop the ability to make choices, interpret, formulate, model and investigate problem situations, and communicate solutions effectively.**

Problem solving provides opportunities for students to use strategies and skills such as investigating and questioning, to collaborate with others and to communicate their findings to different audiences. Such activities are often indicated throughout Australian Signpost Maths by the Activity and Investigation icons.

## Reasoning

**Coherent and logical thought**

*Students develop an increasingly sophisticated capacity for logical thought and actions, such as analysing, proving, evaluating, explaining, inferring, justifying and generalising.**

Students require opportunities to explain their mathematical thinking and can do so through both diagrams and written explanations. Reasoning questions are located throughout Australian Signpost Maths.

* The Australian Curriculum: Mathematics, v8.3 – Content structure

## Special Features of Australian Signpost Maths

- **Traffic Light** system allows students to reflect on their work and highlight any units that they are having trouble understanding. They tick the red for units they feel they still don't understand, and green for those they feel they understand fully.
- Exercises are **well graded**. New work is reinforced in the Mentals Book.
- **Answers** are supplied in the back of this book as well as in the Teacher's Book.
- **Concept Check-In** diagnostic screener (on the Website) provides a snapshot of the class' conceptual understandings to aid in classroom management. It also allows teachers to measure progress over time.
- The eight **Diagnostic Tests** (now also in the back of this book) allow the teacher to discover each student's strengths and weaknesses, and the cross-references direct students to the pages where that work is introduced. Answers are supplied in the Teacher's Book.
- The **Dictionary** at the beginning of this Student Book will help students to learn the language of mathematics.
- **ID Cards** (in the Mentals Book, Teacher's Book and Website) review the language of mathematics by asking students to identify common terms, shapes and symbols.
- Important **rules and concepts** are clearly highlighted.
- **Worked examples** and explanations are given throughout the Student Book where new ideas are introduced.
- The use of **colour** makes emphasis clear and is highly motivating.
- **Cartoons** give instruction and friendly advice.
- **Interactive activities** are provided on the website for whole-class, small-group and individual learning.

## Australian Signpost Icons

Signpost icons are used throughout the book as cues to the essential nature of exercises and activities, and as a guide to ways of engaging with them. These icons often indicate alternative or more concrete approaches to dealing with concepts.

This icon highlights **important rules and concepts** occurring throughout the book. It often appears with worked examples.

Activities provide **applications and enrichment**. These activities usually involve the use of concrete materials and partner or group work.

These enjoyable activities are used to **motivate and involve** students in mathematical pursuits. They usually involve games and puzzles.

Investigations allow students to **explore and discover** maths concepts.

This icon indicates the use of computers, calculators or other **information and communications technology**.

# 3 Contents and Syllabus Overview



**KEY**

- Number and Algebra
- Measurement and Geometry
- Statistics and Probability

## Number and Algebra A

| Page | Unit | Title | Sub-strand | Number and place value | Fractions and decimals | Patterns and algebra | Content | Counting and numeration | Place value | Fractions | Number patterns | Suggested progress |
|---|---|---|---|---|---|---|---|---|---|---|---|---|
| 1 | 1:01 | Skip Counting | | | | ■ | | | | | ● | Term 1 |
| 2 | 1:02 | Odd and Even Numbers | | ■ | | ■ | | ● | | | ● | |
| 3 | 1:03 | Odd and Even Numbers | | ■ | | ■ | | ● | | | ● | |
| 4 | 1:04 | Numbers to 1 000 | | ■ | | | | ● | ● | | | |
| 5 | 1:05 | Numbers to 1 000 | | ■ | | | | ● | ● | | | |
| 6 | 1:06 | Counting | | ■ | | ■ | | ● | | | ● | |
| 7 | 1:07 | Counting | | ■ | | ■ | | ● | | | ● | |
| 8 | 1:08 | Numbers to 1 000 | | ■ | | | | ● | ● | | | T1, T2* |
| 9 | 1:09 | Numbers to 1 000 | | ■ | | | | ● | ● | | | |
| 10 | 1:10 | Fractions of a Whole | | | ■ | | | | | ● | | Term 2 |
| 11 | 1:11 | Fractions of a Collection | | | ■ | | | | | ● | | |
| 12 | 1:12 | Numbers to 10 000 | | ■ | | | | ● | ● | | | |
| 13 | 1:13 | Numbers to 10 000 | | ■ | | | | ● | ● | | | |
| 14 | 1:14 | Fractions | | | ■ | | | | | ● | | |
| 15 | 1:15 | Comparing Fractions | | | ■ | | | | | ● | | |
| 16 | 1:16 | Number Patterns | | | | ■ | | | | | ● | T3, T4* |
| 17 | 1:17 | Numbers to 10 000 | | ■ | | | | ● | ● | | ● | |
| 18 | 1:18 | Ordering Numbers | | ■ | | | | ● | | | | |
| 19 | 1:19 | Rounding | | ■ | | | | ● | ● | | | |
| 20 | 1:20 | Fractions | | | ■ | | | | | ● | | Term 3 |
| 21 | 1:21 | Fractions in Our World | | | ■ | | | | | ● | | |
| 22 | 1:22 | Numbers to 10 000 | | ■ | | | | ● | ● | | | |
| 23 | 1:23 | Place Value to 10 000 | | ■ | | | | ● | ● | | | |
| 24 | 1:24 | Number Patterns | | | | ■ | | | | | ● | |
| 25 | 1:25 | What's the Rule? | | | | ■ | | | | | ● | |
| 26 | 1:26 | Expanded Notation | | ■ | | | | | ● | | | T5, T6* |
| 27 | 1:27 | Numbers to 10 000 | | ■ | | ■ | | ● | | | ● | |
| 28 | 1:28 | Number Patterns | | ■ | | ■ | | ● | | | ● | |
| 29 | 1:29 | Number Patterns | | | | ■ | | | | | ● | |
| 30 | 1:30 | Numbers to 10 000 | | ■ | | | | ● | ● | | | |
| 31 | 1:31 | Expanded Notation | | ■ | | | | ● | ● | | | |
| 32 | 1:32 | Numbers to 10 000 | | ■ | | | | ● | ● | | | Term 4 |
| 33 | 1:33 | Place Value to 10 000 | | ■ | | | | ● | ● | | | |
| 34 | 1:34 | Numbers to 10 000 | | ■ | | | | ● | ● | | | T7, T8* |
| 35 | 1:35 | Making Number Patterns | | | | ■ | | | | | ● | |
| 36 | 1:36 | Rounding | | ■ | | | | ● | ● | | | |

* Suggested placement for Diagnostic Tests 1 to 8 (see the Teacher's Book).

## Number and Algebra B

| | | | Sub-strand | | | Content | | | | | | |
|---|---|---|---|---|---|---|---|---|---|---|---|---|
| **Page** | **Unit** | **Title** | **Number and place value** | **Money and financial mathematics** | **Patterns and algebra** | **Addition** | **Subtraction** | **Multiplication** | **Division** | **Place value** | **Number patterns** | **Suggested progress** |
| 37 | 2:01 | Australian Money | | ■ | | ● | | | | | | Term 1 |
| 38 | 2:02 | Addition and Subtraction Facts | ■ | | ■ | ● | ● | | | | ● | |
| 39 | 2:03 | Strategies | ■ | | | ● | ● | | | | | |
| 40 | 2:04 | Number Facts, × 2 | ■ | | | | | ● | | | | |
| 41 | 2:05 | Addition by Looking for 10s | ■ | | | ● | | | | | | |
| 42 | 2:06 | Patterns in Adding and Subtracting | ■ | | ■ | ● | ● | | | | ● | |
| 43 | 2:07 | Patterns Involving Subtraction | ■ | | ■ | | ● | | | | ● | |
| 44 | 2:08 | Number Facts, × 5, × 10 | ■ | | | | | ● | | | | |
| 45 | 2:09 | Number Facts, × 1, × 0 | ■ | | | | | ● | | | | |
| 46 | 2:10 | Mental Strategies | ■ | | | ● | | | | | | T1, T2* |
| 47 | 2:11 | Subtraction from 2-Digit Numbers | ■ | | | | ● | | | | | |
| 48 | 2:12 | Multiplication Facts | ■ | | | | | ● | | | | |
| 49 | 2:13 | Number Facts, × 3 | ■ | | | | | ● | | | | Term 2 |
| 50 | 2:14 | Addition to 99, No Trading | ■ | | | ● | | | | | | |
| 51 | 2:15 | Addition to 99, No Trading | ■ | | | ● | | | | ● | | |
| 52 | 2:16 | Multiplication | ■ | | ■ | | | ● | | | ● | |
| 53 | 2:17 | Number Facts, Multiplication | ■ | | | | | ● | | | | |
| 54 | 2:18 | Multiplication Facts | ■ | | | | | ● | | | | |
| 55 | 2:19 | Relating × and ÷ | ■ | | | | | ● | ● | | | |
| 56 | 2:20 | Money | ■ | ■ | | ● | | | | | | |
| 57 | 2:21 | Shopping | ■ | ■ | | ● | | | | | | T3, T4* |
| 58 | 2:22 | Relating Addition and Subtraction | ■ | | | ● | ● | | | | | |
| 59 | 2:23 | Inverse Operations, + and – | ■ | | | ● | ● | | | | | |
| 60 | 2:24 | Sharing and Grouping | ■ | | | | | | ● | | | |
| 61 | 2:25 | Modelling Division | ■ | | | | | | ● | | | |
| 62 | 2:26 | Change from $2 | ■ | ■ | | ● | | | | | | |
| 63 | 2:27 | Money | ■ | ■ | | ● | | | | | | Term 3 |
| 64 | 2:28 | Addition to 99, No Trading | ■ | | | ● | | | | ● | | |
| 65 | 2:29 | Subtraction, No Trading | ■ | | | | ● | | | ● | | |
| 66 | 2:30 | Multiplication | ■ | | | | | ● | | | | |
| 67 | 2:31 | Linking Multiplication and Division | ■ | | | | | ● | ● | | | |
| 68 | 2:32 | Multiples | ■ | | ■ | | | ● | | | ● | |
| 69 | 2:33 | Division Facts from Multiplication Facts | ■ | | | | | ● | ● | | | |
| 70 | 2:34 | Addition to 99 with Trading | ■ | | | ● | | | | | | |
| 71 | 2:35 | Addition to 99 with Trading | ■ | | | ● | | | | | | T5, T6* |
| 72 | 2:36 | Addition to 99 with Trading | ■ | | | ● | | | | ● | | |
| 73 | 2:37 | Addition to 99 with Trading | ■ | | | ● | | | | ● | | |
| 74 | 2:38 | Problem Solving | ■ | | | ● | ● | ● | ● | | | |
| 75 | 2:39 | Problem Solving | ■ | | | ● | ● | ● | ● | | | |
| 76 | 2:40 | Addition Involving Hundreds | ■ | | | ● | | | | ● | | Term 4 |
| 77 | 2:41 | Addition Involving Hundreds | ■ | | | ● | | | | ● | | |
| 78 | 2:42 | Subtraction to 999 | ■ | | | | ● | | | ● | | |
| 79 | 2:43 | Addition to 999 | ■ | | | ● | | | | ● | | |
| 80 | 2:44 | Checking Subtraction by Addition | ■ | | | ● | ● | | | | | |
| 81 | 2:45 | Inverse Operations, × and ÷ | ■ | | | | | ● | ● | | | |
| 82 | 2:46 | Counting Change | ■ | ■ | | ● | | | | | | T7, T8* |
| 83 | 2:47 | Rounding to the Nearest 5c | ■ | ■ | | ● | | | | | | |
| 84 | 2:48 | Problem Solving | ■ | | | | | ● | ● | | | |
| 85 | 2:49 | Subtraction with Trading | ■ | | | | ● | | | ● | | |
| 86 | 2:50 | Subtraction with Trading | ■ | | | | ● | | | ● | | |

* Suggested placement for Diagnostic Tests 1 to 8 (see the Teacher's Book).

| Measurement and Geometry A | | | | | | | | | | | | |
|---|---|---|---|---|---|---|---|---|---|---|---|---|
| Page | Unit | Title | Sub-strand | Using units of measurement | Content | Length | Area | Volume | Capacity | Mass | Time | Suggested progress |
| 87 | 3:01 | Direct Comparison of Length | | | | ● | | | | | | Term 1 |
| 88 | 3:02 | Time Revision | | | | | | | | | ● | |
| 89 | 3:03 | Analogue Time | | | | | | | | | ● | |
| 90 | 3:04 | Analogue and Digital Time | | | | | | | | | ● | |
| 91 | 3:05 | Analogue and Digital Time | | | | | | | | | ● | |
| 92 | 3:06 | The Metre | | | | ● | | | | | | |
| 93 | 3:07 | Using the Metre | | | | ● | | | | | | T1, T2* |
| 94 | 3:08 | Centimetres | | | | ● | | | | | | Term 2 |
| 95 | 3:09 | Measuring with Centimetres | | | | ● | | | | | | |
| 96 | 3:10 | Comparing Areas | | | | | ● | | | | | |
| 97 | 3:11 | Area | | | | | ● | | | | | |
| 98 | 3:12 | Estimating the Litre | | | | | | | ● | | | |
| 99 | 3:13 | The Litre | | | | | | ● | ● | | | |
| 100 | 3:14 | The Kilogram | | | | | | | | ● | | T3, T4* |
| 101 | 3:15 | Using the Kilogram | | | | | | | | ● | | |
| 102 | 3:16 | Analogue Time | | | | | | | | | ● | |
| 103 | 3:17 | Analogue and Digital Time | | | | | | | | | ● | |
| 104 | 3:18 | Recording Length | | | | ● | | | | | | Term 3 |
| 105 | 3:19 | Finding Length | | | | ● | | | | | | |
| 106 | 3:20 | The Calendar | | | | | | | | | ● | |
| 107 | 3:21 | Analogue and Digital Time | | | | | | | | | ● | |
| 108 | 3:22 | The Millimetre | | | | ● | | | | | | |
| 109 | 3:23 | Using Millimetres | | | | ● | | | | | | T5, T6* |
| 110 | 3:24 | Area | | | | | ● | | | | | |
| 111 | 3:25 | Comparing Areas | | | | | ● | | | | | |
| 112 | 3:26 | The Gram | | | | | | | | ● | | |
| 113 | 3:27 | Using Grams | | | | | | | | ● | | |
| 114 | 3:28 | Using a Ruler | | | | ● | | | | | | Term 4 |
| 115 | 3:29 | Comparing Masses | | | | | | | | ● | | |
| 116 | 3:30 | The Millilitre | | | | | | ● | ● | | | T7, T8* |
| 117 | 3:31 | Using Litres | | | | | | | ● | | | |
| 118 | 3:32 | The Calendar | | | | | | | | | ● | |
| 119 | 3:33 | Standard Metric Units | | | | ● | | | ● | ● | | |

*Suggested placement for Diagnostic Tests 1 to 8 (see the Teacher's Book).

## Measurement and Geometry B

| Page | Unit | Title | Sub-strand | Shape | Location and transformation | Geometric reasoning | Content | 2D Space | 3D Space | Location | Transformations | Angles | Suggested progress |
|---|---|---|---|---|---|---|---|---|---|---|---|---|---|
| 120 | 4:01 | Properties of 3D Objects | | ● | | | | | ● | | | | Term 1 |
| 121 | 4:02 | Symmetry | | | ● | | | ● | | | ● | | |
| 122 | 4:03 | Properties of 3D Objects | | ● | | | | | ● | | | | |
| 123 | 4:04 | Symmetry in Our World | | | ● | | | ● | | | ● | | |
| 124 | 4:05 | Parallel and Perpendicular Lines | | ● | | ● | | ● | ● | | | ● | |
| 125 | 4:06 | Regular and Irregular Shapes | | ● | | ● | | ● | | | | ● | T1, T2* |
| 126 | 4:07 | Position and Giving Directions | | | ● | | | | | ● | | | |
| 127 | 4:08 | Position and Giving Directions | | | ● | | | | | ● | | | Term 2 |
| 128 | 4:09 | Shapes Revision | | ● | | | | ● | | | | | |
| 129 | 4:10 | Properties of 2D Shapes | | ● | | | | ● | | | | | |
| 130 | 4:11 | Investigating Angles | | | | ● | | | | | | ● | T3, T4* |
| 131 | 4:12 | Angles | | | | ● | | | | | | ● | |
| 132 | 4:13 | The Trapezium and Parallelogram | | ● | | | | ● | | | | | |
| 133 | 4:14 | Features of 2D Shapes | | ● | | | | ● | | | | | |
| 134 | 4:15 | Right Angles | | | | ● | | | | | | ● | Term 3 |
| 135 | 4:16 | Angles | | | | ● | | | | | | ● | |
| 136 | 4:17 | Describing Position | | | ● | | | | | ● | | | |
| 137 | 4:18 | Pathways Between Places | | | ● | | | | | ● | | | T5, T6* |
| 138 | 4:19 | Investigating Prisms and Cylinders | | ● | | | | | ● | | | | Term 4 |
| 139 | 4:20 | Investigating Pyramids | | ● | | | | | ● | | | | |
| 140 | 4:21 | Creating Maps | | | ● | | | | | ● | | | |
| 141 | 4:22 | Mazes | | | ● | | | | | ● | | | |
| 142 | 4:23 | Investigating Spheres | | ● | | | | | ● | | | | |
| 143 | 4:24 | 3D Objects | | ● | | | | | ● | | | | |
| 144 | 4:25 | Properties of 2D Shapes | | | ● | | | ● | | | ● | | T7, T8* |
| 145 | 4:26 | Properties of 2D Shapes | | | ● | | | ● | | | ● | | |
| 146 | 4:27 | The Net of a Cube | | ● | | | | | ● | | | | |

*Suggested placement for Diagnostic Tests 1 to 8 (see the Teacher's Book).

## Statistics and Probability

| Page | Unit | Title | Sub-strand | Chance | Data representation and interpretation | Content | Chance | Data investigation | Data representation | Suggested progress |
|---|---|---|---|---|---|---|---|---|---|---|
| 147 | 5:01 | Using Blocks in Graphs | | | ● | | | | ● | Term 1 |
| 148 | 5:02 | Tables and Graphs | | | ● | | | ● | ● | |
| 149 | 5:03 | Tables and Graphs | | | ● | | | | ● | T1, T2* |
| 150 | 5:04 | Chance | | ● | ● | | ● | ● | | |
| 151 | 5:05 | Predicting Outcomes | | ● | ● | | ● | ● | | Term 2 |
| 152 | 5:06 | Picture Graphs | | | ● | | | ● | ● | T3, T4* |
| 153 | 5:07 | Making Graphs | | | ● | | | ● | ● | |
| 154 | 5:08 | Reading Tables and Graphs | | | ● | | | | ● | Term 3 |
| 155 | 5:09 | Reading for Number | | | ● | | | | ● | |
| 156 | 5:10 | Reading Picture Graphs | | | ● | | | | ● | |
| 157 | 5:11 | Drawing Graphs | | | ● | | | ● | ● | T5, T6* |
| 158 | 5:12 | Ordering Events | | ● | | | ● | | | T7, T8* |
| 159 | 5:13 | Repeating an Experiment | | ● | ● | | ● | ● | ● | Term 4 |

*Suggested placement for Diagnostic Tests 1 to 8 (see the Teacher's Book).

# Suggested Program

| | Weeks 1–10 | Weeks 11–20 | Weeks 21–30 | Weeks 31–end |
|---|---|---|---|---|
| Number and Algebra A | 1:01–1:09 | 1:10–1:19 | 1:20–1:31 | 1:32–1:36 |
| Number and Algebra B | 2:01–2:12 | 2:13–2:26 | 2:27–2:39 | 2:40–2:50 |
| Measurement and Geometry A | 3:01–3:07 | 3:08–3:17 | 3:18–3:27 | 3:28–3:33 |
| Measurement and Geometry B | 4:01–4:07 | 4:08–4:14 | 4:15–4:18 | 4:19–4:27 |
| Statistics and Probability | 5:01–5:04 | 5:05–5:07 | 5:08–5:12 | 5:13 |

It is assumed that there are 10 weeks in each term.

The eight Diagnostic Tests are found in the Teacher's Book.

See the Contents and Syllabus Overview on pages vi–ix for suggested placement of each test.

# Contents Cross-reference

## Numbers and Algebra

| 1 Whole numbers | Pages | Australian Curriculum Reference |
|---|---|---|
| Three-, four- and five-digit numbers and place value | 4, 5, 6, 7, 8, 9, 12, 13, 17, 18, 22, 23, 26, 27, 30, 31, 32, 33, 34 | Recognise, model, represent and order numbers to at least 10 000 (ACMNA052); Apply place value to partition, rearrange and regroup numbers to at least 10 000 to assist calculations and solve problems (ACMNA053) |
| Odd and even numbers | 2, 3, 42 | Investigate the conditions required for a number to be odd or even and identify odd and even numbers (ACMNA051); Describe, continue, and create number patterns resulting from performing addition or subtraction (ACMNA060) |
| Rounding off | 9, 19, 23, 36, 83 | Recognise, model, represent and order numbers to at least 10 000 (ACMNA052); Apply place value to partition, rearrange and regroup numbers to at least 10 000 to assist calculations and solve problems (ACMNA053); Represent money values in multiple ways and count the change required for simple transactions to the nearest five cents (ACMNA059) |

| 2 Addition | Pages | Australian Curriculum Reference |
|---|---|---|
| Mental strategies | 36, 39, 41, 42, 46, 50, 58, 59, 63, 70, 71, 77, 79, 83 | Apply place value to partition, rearrange and regroup numbers to at least 10 000 to assist calculations and solve problems (ACMNA053); Recognise and explain the connection between addition and subtraction (ACMNA054); Recall addition facts for single-digit numbers and related subtraction facts to develop increasingly efficient mental strategies for computation (ACMNA055); Represent money values in multiple ways and count the change required for simple transactions to the nearest five cents (ACMNA059); Describe, continue, and create number patterns resulting from performing addition or subtraction (ACMNA060) |

| | | |
|---|---|---|
| Written strategies | 38, 41, 51, 64, 72, 73, 76, 77, 79, 80 | Recognise and explain the connection between addition and subtraction (ACMNA054); Recall addition facts for single-digit numbers and related subtraction facts to develop increasingly efficient mental strategies for computation (ACMNA055) |
| Problem solving | 35, 51, 59, 64, 72, 74, 75, 83 | Recall addition facts for single-digit numbers and related subtraction facts to develop increasingly efficient mental strategies for computation (ACMNA055); Represent money values in multiple ways and count the change required for simple transactions to the nearest five cents (ACMNA059); Describe, continue, and create number patterns resulting from performing addition or subtraction (ACMNA060) |

| 3 Subtraction | Pages | Australian Curriculum Reference |
|---|---|---|
| Mental strategies | 30, 36, 38, 39, 42, 43, 47, 58, 59, 74, 78 | Apply place value to partition, rearrange and regroup numbers to at least 10 000 to assist calculations and solve problems (ACMNA053); Recognise and explain the connection between addition and subtraction (ACMNA054); Recall addition facts for single-digit numbers and related subtraction facts to develop increasingly efficient mental strategies for computation (ACMNA055) |
| Written strategies | 65, 78, 80, 85, 86 | Apply place value to partition, rearrange and regroup numbers to at least 10 000 to assist calculations and solve problems (ACMNA053); Recognise and explain the connection between addition and subtraction (ACMNA054); Recall addition facts for single-digit numbers and related subtraction facts to develop increasingly efficient mental strategies for computation (ACMNA055); Describe, continue, and create number patterns resulting from performing addition or subtraction (ACMNA060) |
| Problem solving | 30, 35, 59, 65, 74, 75, 80 | Recognise and explain the connection between addition and subtraction (ACMNA054); Recall addition facts for single-digit numbers and related subtraction facts to develop increasingly efficient mental strategies for computation (ACMNA055); Describe, continue, and create number patterns resulting from performing addition or subtraction (ACMNA060) |
| Relating addition and subtraction | 58, 59, 80 | Recognise and explain the connection between addition and subtraction (ACMNA054); Recall addition facts for single-digit numbers and related subtraction facts to develop increasingly efficient mental strategies for computation (ACMNA055) |

| 4 Multiplication | Pages | Australian Curriculum Reference |
|---|---|---|
| Multiplication facts: 2, 3, 5 and 10 | 1, 40, 44, 45, 48, 49, 52, 53, 54, 55, 66, 67, 69 | Recall multiplication facts of two, three, five and ten and related division facts (ACMNA056); Represent and solve problems involving multiplication using efficient mental and written strategies and appropriate digital technologies (ACMNA057); Describe, continue, and create number patterns resulting from performing addition or subtraction (ACMNA060) |
| Mental and written strategies | 40, 49, 52, 66, 69, 81, 84 | Recall multiplication facts of two, three, five and ten and related division facts (ACMNA056); Represent and solve problems involving multiplication using efficient mental and written strategies and appropriate digital technologies (ACMNA057); Describe, continue, and create number patterns resulting from performing addition or subtraction (ACMNA060) |

| | | |
|---|---|---|
| Multiples, factors and square numbers | 68, 84 | Recall multiplication facts of two, three, five and ten and related division facts (ACMNA056); Represent and solve problems involving multiplication using efficient mental and written strategies and appropriate digital technologies (ACMNA057); Describe, continue, and create number patterns resulting from performing addition or subtraction (ACMNA060) |
| Problem solving | 53, 74, 75, 84 | Recall multiplication facts of two, three, five and ten and related division facts (ACMNA056); Represent and solve problems involving multiplication using efficient mental and written strategies and appropriate digital technologies (ACMNA057) |

| 5 Division | Pages | Australian Curriculum Reference |
|---|---|---|
| Mental and written strategies | 60, 61, 69, 81, 84 | Recall multiplication facts of two, three, five and ten and related division facts (ACMNA056) |
| Division facts | 55, 61, 67, 69 | Recall multiplication facts of two, three, five and ten and related division facts (ACMNA056) |
| Problem solving | 60, 61, 74, 75 | Recall multiplication facts of two, three, five and ten and related division facts (ACMNA056) |
| Linking multiplication and division | 55, 67, 69, 81 | Recall multiplication facts of two, three, five and ten and related division facts (ACMNA056) |

| 6 Fractions | Pages | Australian Curriculum Reference |
|---|---|---|
| Common fractions | 10, 11, 14, 15, 20, 21 | Model and represent unit fractions including $\frac{1}{2}$, $\frac{1}{4}$, $\frac{1}{3}$, $\frac{1}{5}$ and their multiples to a complete whole (ACMNA058) |
| Comparing fractions | 14, 15, 20 | Model and represent unit fractions including $\frac{1}{2}$, $\frac{1}{4}$, $\frac{1}{3}$, $\frac{1}{5}$ and their multiples to a complete whole (ACMNA058) |
| Count using fractions | 21, 28 | Model and represent unit fractions including $\frac{1}{2}$, $\frac{1}{4}$, $\frac{1}{3}$, $\frac{1}{5}$ and their multiples to a complete whole (ACMNA058); Describe, continue, and create number patterns resulting from performing addition or subtraction (ACMNA060) |

| 7 Money | Pages | Australian Curriculum Reference |
|---|---|---|
| Representing money values | 37, 56, 57, 63, 83 | Recall addition facts for single-digit numbers and related subtraction facts to develop increasingly efficient mental strategies for computation (ACMNA055); Represent money values in multiple ways and count the change required for simple transactions to the nearest five cents (ACMNA059) |
| Giving change | 62, 82 | Recall addition facts for single-digit numbers and related subtraction facts to develop increasingly efficient mental strategies for computation (ACMNA055); Represent money values in multiple ways and count the change required for simple transactions to the nearest five cents (ACMNA059) |

| 8 Number patterns | Pages | Australian Curriculum Reference |
|---|---|---|
| Patterns | 1, 6, 7, 16, 22, 24, 25, 27, 28, 29, 35, 38, 42, 43, 53 | Investigate the conditions required for a number to be odd or even and identify odd and even numbers (ACMNA051); Recall addition facts for single-digit numbers and related subtraction facts to develop increasingly efficient mental strategies for computation (ACMNA055); Describe, continue, and create number patterns resulting from performing addition or subtraction (ACMNA060) |

| | | |
|---|---|---|
| Pattern rules | 6, 7, 25, 28, 29, 35 | Investigate the conditions required for a number to be odd or even and identify odd and even numbers (ACMNA051); Recall addition facts for single-digit numbers and related subtraction facts to develop increasingly efficient mental strategies for computation (ACMNA055); Recall multiplication facts of two, three, five and ten and related division facts (ACMNA056); Describe, continue, and create number patterns resulting from performing addition or subtraction (ACMNA060) |
| Number sentences using 'is equal to', 'is less than', 'is greater than' | 9, 18, 27, 28, 34, 43, 51, 53, 70, 71, 72, 76, 79 | Recognise, model, represent and order numbers to at least 10 000 (ACMNA052); Apply place value to partition, rearrange and regroup numbers to at least 10 000 to assist calculations and solve problems (ACMNA053); Recall addition facts for single-digit numbers and related subtraction facts to develop increasingly efficient mental strategies for computation (ACMNA055); Recall multiplication facts of two, three, five and ten and related division facts (ACMNA056) |

| 9 Calculators | Pages | Australian Curriculum Reference |
|---|---|---|
| Using a calculator | 16, 24, 30, 36, 70, 71 | Apply place value to partition, rearrange and regroup numbers to at least 10 000 to assist calculations and solve problems (ACMNA053); Represent and solve problems involving multiplication using efficient mental and written strategies and appropriate digital technologies (ACMNA057) |

# Measurement and Geometry

| 1 Using units of measurement | Pages | Australian Curriculum Reference |
|---|---|---|
| Reading scales | 94, 95, 108, 109, 114 | Measure, order and compare objects using familiar metric units of length, mass and capacity (ACMMG061); |
| Length | 58, 87, 92, 93, 94, 95, 104, 105, 108, 109, 114, 119 | Measure, order and compare objects using familiar metric units of length, mass and capacity (ACMMG061); |
| Area | 96, 97, 110, 111 | Compare and order several shapes and objects based on length, area, volume and capacity using appropriate uniform informal units (ACMMG037) |
| Capacity and volume | 98, 99, 116, 117, 119 | Measure, order and compare objects using familiar metric units of length, mass and capacity (ACMMG061); |
| Mass | 100, 101, 112, 113, 115, 119 | Measure, order and compare objects using familiar metric units of length, mass and capacity (ACMMG061); |
| Telling the time | 88, 89, 90, 91, 102, 103, 107 | Tell time to the minute and investigate the relationship between units of time (ACMMG062) |
| Units and duration of time | 106, 107, 118 | Tell time to the minute and investigate the relationship between units of time (ACMMG062) |
| Calendars | 106, 118 | Tell time to the minute and investigate the relationship between units of time (ACMMG062) |
| Time lines, timetables | 106, 107 | Tell time to the minute and investigate the relationship between units of time (ACMMG062) |

| 2 Shape | Pages | Australian Curriculum Reference |
|---|---|---|
| 2D shapes | 124, 125, 128, 129, 132, 133, 144 | Make models of three-dimensional objects and describe key features (ACMMG063); Identify angles as measures of turn and compare angle sizes in everyday situations (ACMMG064); Identify symmetry in the environment (ACMMG066); Describe and draw two-dimensional shapes, with and without digital technologies (ACMMG042) |
| 3D objects | 120, 122, 124, 138, 139, 142, 143, 146 | Make models of three-dimensional objects and describe key features (ACMMG063); Identify angles as measures of turn and compare angle sizes in everyday situations (ACMMG064) |
| Cones, cylinders and spheres | 138, 142, 143 | Make models of three-dimensional objects and describe key features (ACMMG063) |
| Prisms and pyramids | 138, 139, 143 | Make models of three-dimensional objects and describe key features (ACMMG063) |
| Angles | 124, 130, 131, 134, 135 | Make models of three-dimensional objects and describe key features (ACMMG063); Identify angles as measures of turn and compare angle sizes in everyday situations (ACMMG064) |

| 3 Location and transformation | Pages | Australian Curriculum Reference |
|---|---|---|
| Maps, scales and keys | 126, 127, 137, 140, 141 | Create and interpret simple grid maps to show position and pathways (ACMMG065) |
| Describing position | 136 | Create and interpret simple grid maps to show position and pathways (ACMMG065) |
| Symmetry | 121, 123 | Identify symmetry in the environment (ACMMG066) |
| Flip, slide, turn and tessellations | 144, 145 | Identify symmetry in the environment (ACMMG066); Investigate the effect of one-step slides and flips with and without digital technologies (ACMMG045) |

## Statistics and Probability

| 1 | Data representation and interpretation | Pages | Australian Curriculum Reference |
|---|---|---|---|
| | Data collection | 41, 148, 152, 153, 157 | Conduct chance experiments, identify and describe possible outcomes and recognise variation in results (ACMSP067); Identify questions or issues for categorical variables. Identify data sources and plan methods of data collection and recording (ACMSP068); Collect data, organise into categories and create displays using lists, tables, picture graphs and simple column graphs, with and without the use of digital technologies (ACMSP069); Interpret and compare data displays (ACMSP070) |
| | Graphs | 147, 148, 149, 152, 153, 154, 155, 156, 157, 159 | Identify questions or issues for categorical variables. Identify data sources and plan methods of data collection and recording (ACMSP068); Collect data, organise into categories and create displays using lists, tables, picture graphs and simple column graphs, with and without the use of digital technologies (ACMSP069); Interpret and compare data displays (ACMSP070) |
| | Tables | 147, 148, 149, 154, 157 | Identify questions or issues for categorical variables. Identify data sources and plan methods of data collection and recording (ACMSP068); Collect data, organise into categories and create displays using lists, tables, picture graphs and simple column graphs, with and without the use of digital technologies (ACMSP069); Interpret and compare data displays (ACMSP070) |

| 2 | Chance | Pages | Australian Curriculum Reference |
|---|---|---|---|
| | Chance experiments | 150, 151, 159 | Conduct chance experiments, identify and describe possible outcomes and recognise variation in results (ACMSP067); Collect data, organise into categories and create displays using lists, tables, picture graphs and simple column graphs, with and without the use of digital technologies (ACMSP069); Interpret and compare data displays (ACMSP070) |
| | Describing likelihood | 150, 151, 158, 159 | Conduct chance experiments, identify and describe possible outcomes and recognise variation in results (ACMSP067) |

**abacus**

An instrument used for counting and calculating.

- This abacus shows 546.

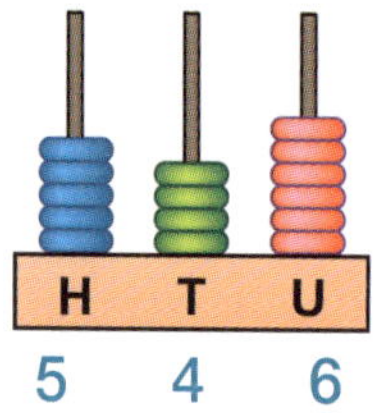


**analogue time**

The time shown on a clock face.

- 13 minutes to 6 is the time on this analogue clock.

**angle**

The amount of turning between two lines about a common point.

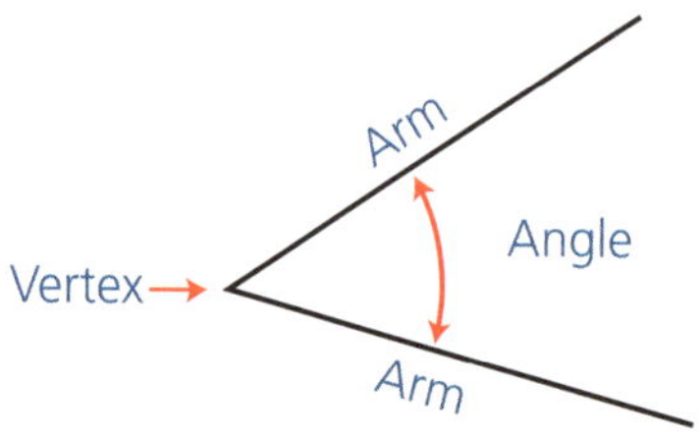


- Right Angle (Quarter Turn)

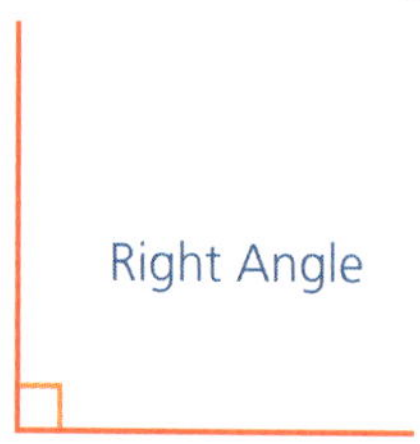


- Straight Angle (Half Turn)

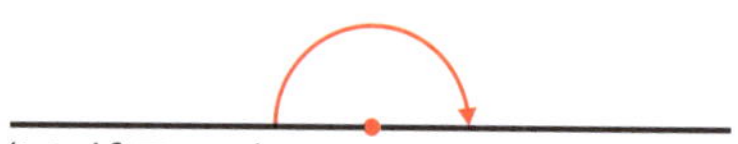

**anticlockwise** and **clockwise**

The direction of a turn.

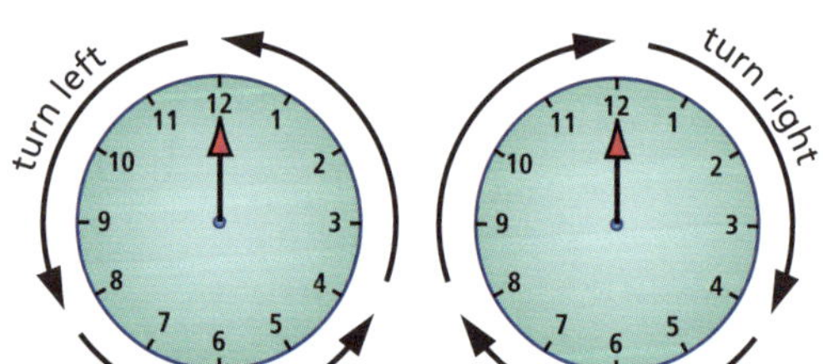


Anticlockwise

Clockwise (as on a clock)

**area**

The measure of the amount of surface.

**axis of symmetry**

See *line of symmetry*.

A line that divides a picture in half so that each half is the mirror image of the other part.

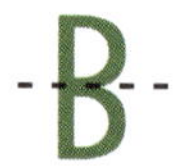


- One Axis of Symmetry

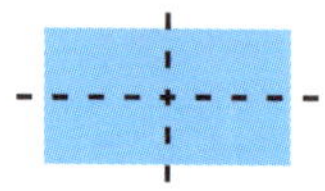

- Two Axes of Symmetry

The plural of *axis* is *axes*.

**bar graph**

A graph that uses horizontal bars to compare the size of groups.

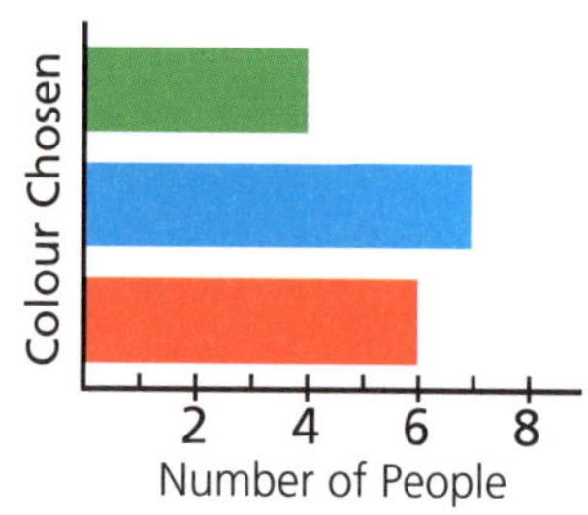


**calculator**

A machine used to perform mathematical operations.

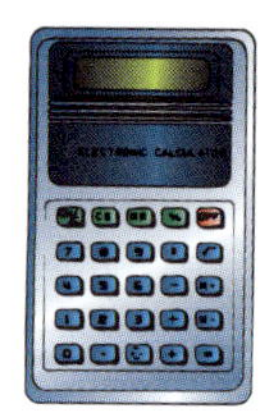

**capacity**

The amount that a container can hold.

- less than 1 litre

- more than 1 litre

**centimetre (cm)**

A unit of length equal to one hundredth of a metre.

- 100 cm = 1 m

## chance

The chance (or probability) of something happening is its likelihood of happening.

- If you toss a coin, there is an even chance of tossing a head.

See also *probability*.

## circle

A two-dimensional shape bounded by a single curved line on which every point is the same distance from the centre.

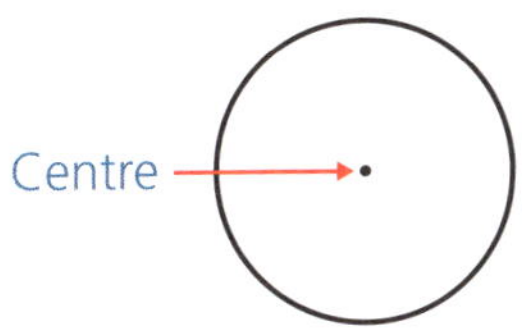


## column graph

A graph that uses columns to compare the size of groups.

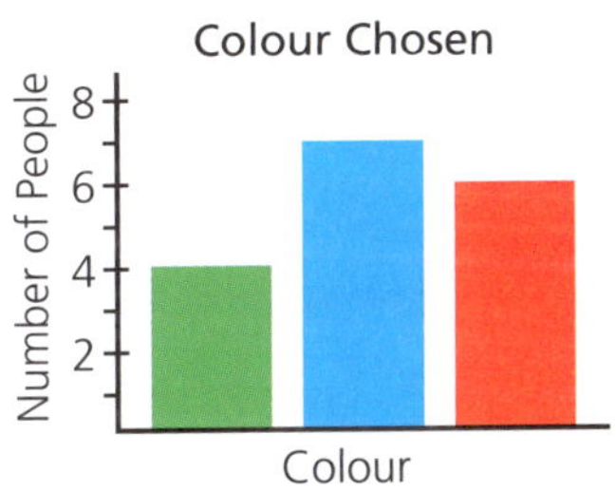


## cone

A three-dimensional object with a circular base that tapers to a point.

## counting number

Any number used in counting.

- 1, 2, 3, 4, 5, …

  Zero is not a counting number.

## cube

A three-dimensional object that has six equal square faces, eight vertices and twelve equal edges.

## cylinder

A three-dimensional object with two equal circular faces and one curved surface.

## difference

The amount by which two numbers are separated.

- 9 − 6 = 3

  The difference between 9 and 6 is 3.

## digital time

Time expressed using digits.

- This digital clock shows 24 minutes past 10.

## digits

Symbols used to write a number.

- 6

  Six is a 1-digit number.

- 47

  Forty-seven is a 2-digit number.

## division (÷)

Breaking up groups into equal parts.

- 6 ÷ 3

  **a** Sharing 2 each

  **b** Grouping 2 groups

## double

Twice as much.

- Double 6 = 6 + 6 or 2 × 6

  12 is double 6.

## edge

The intersection of two surfaces.

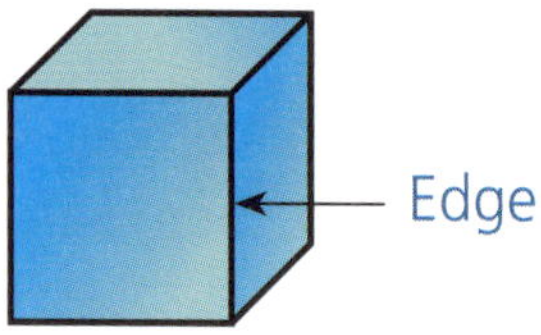


## even number

Any number that is a multiple of two and can be grouped in twos.

- 16, 300, 4394

## expanded notation

A way of writing numerals to show the place value of each digit.

- 137 = (1 × 100) + (3 × 10) + 7

## face

A flat surface of a three-dimensional object.

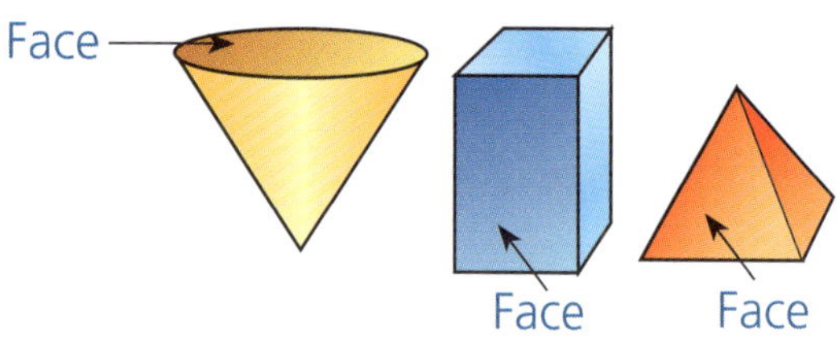


## flip (reflection)

To turn over.

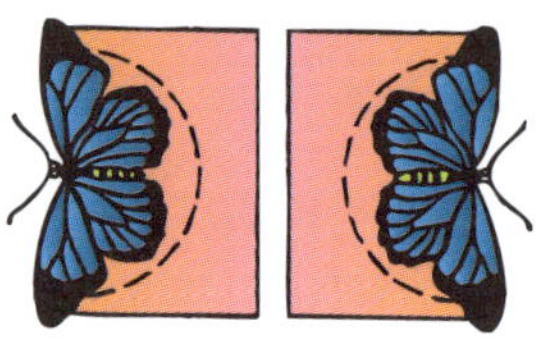

## fraction

Any part of a whole, group or object.

- 2 out of 6 shaded 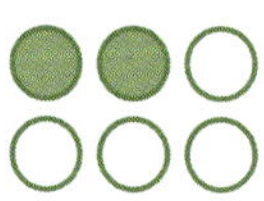
- $\frac{1}{4}$ is shaded 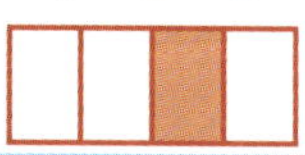

## gram (g)

A unit of mass used to measure how heavy something is.

- 1 kilogram = 1 000 grams

## graph

A diagram or drawing used to record a collection of data.

Bar Graph

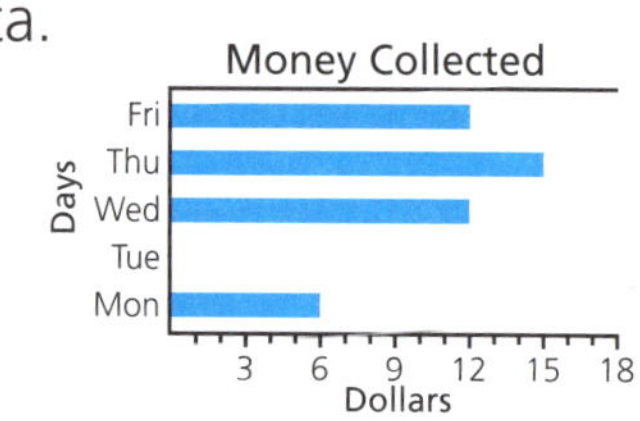


Column Graph

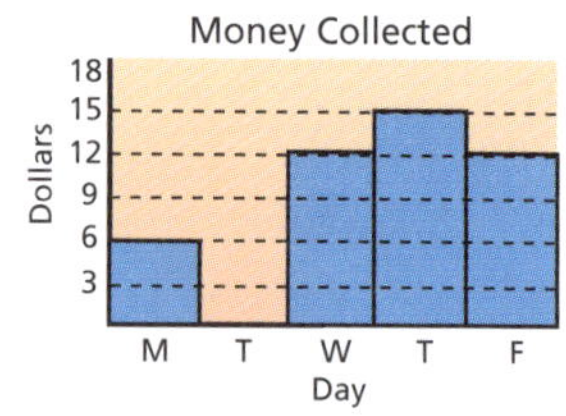


Picture Graph

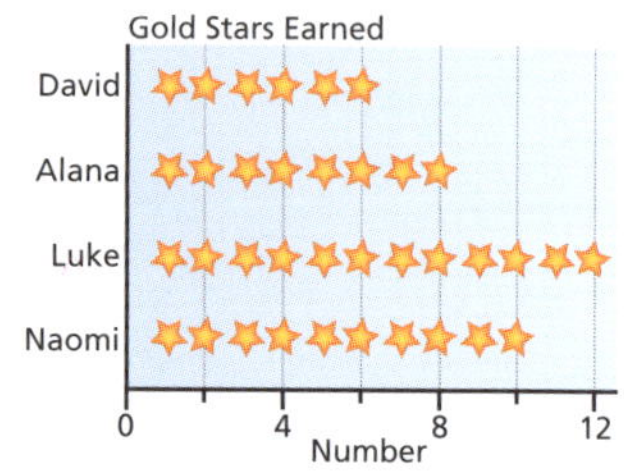


## greater than (>)

A way of showing that a number is larger than another number.

- 7 > 3

See also *less than (<)*.

## heft

To judge weights by lifting them with your hands.

## hexagon

A polygon with six sides.

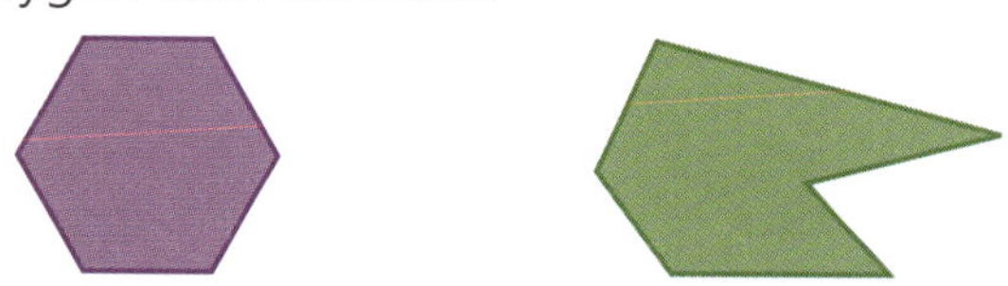

Regular Hexagon    Irregular Hexagon

See also *polygon*.

## inverse operations

Adding 8 is the opposite (the inverse) of subtracting 8.

- 100 + 8 − 8 = 100

Multiplying by 2 is the opposite (the inverse) of dividing by 2.

- 4 × 2 ÷ 2 = 4

## jump strategy

Adding or subtracting numbers, jumping by tens or ones.

- 52 − 14 = 38

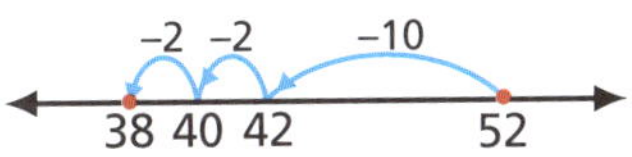


## kilogram (kg)

The basic unit of mass, equal to 1000 grams.

- 1 kg = 1 000 g

## less than (<)

A way of showing that a number is smaller than another number.

- 3 < 7

See also *greater than (>)*.

## line

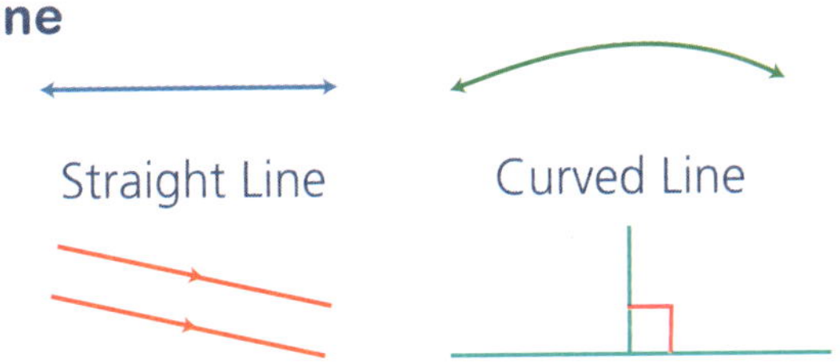


## line of symmetry

A line that divides something in half so that each half is a mirror image of the other part.

Line of Symmetry

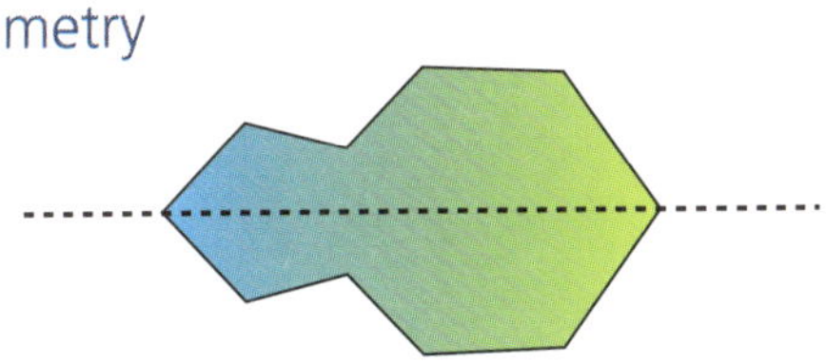

## litre (L)

A unit of capacity (or volume) used for the measurement of liquids.

- 1 L = 1 000 mL

## map or plan

A picture of an area viewed from above.

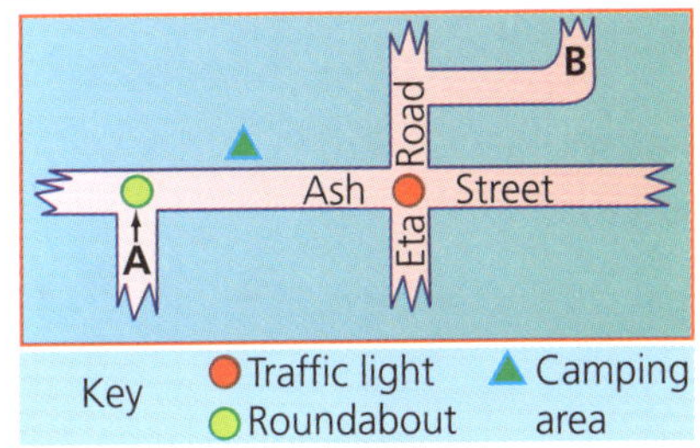


## mass

The amount of matter in an object, a measure of how heavy it is.

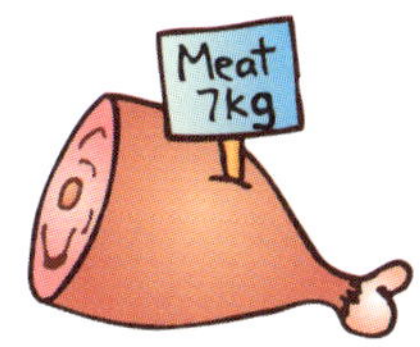


## metre (m)

The basic unit of length, equal to 100 centimetres.

- 1 m = 100 cm

## millilitre (mL)

A unit of capacity (or volume) equal to one thousandth of a litre.

- 1 000 mL = 1 L

## millimetre (mm)

A unit of length equal to one tenth of a centimetre, or one thousandth of a metre.

- 10 mm = 1 cm
- 1 000 mm = 1 m

## money

## multiple

The result of multiplying a counting number by another counting number.

- The multiples of 5 are 5, 10, 15, 20, …

## multiplication (×)

Combining equal groups.

- 3 × 5 = 15

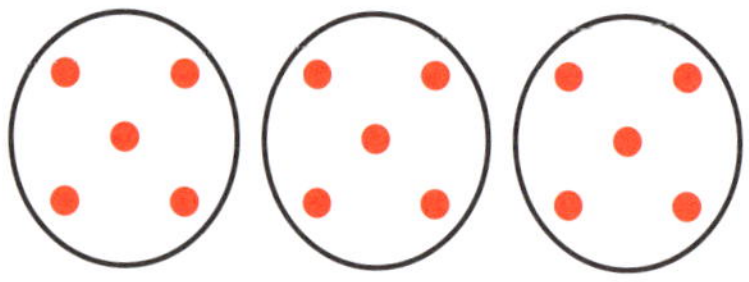

## net

A flat shape that can be folded to make a three-dimensional object.

Net of a Cube

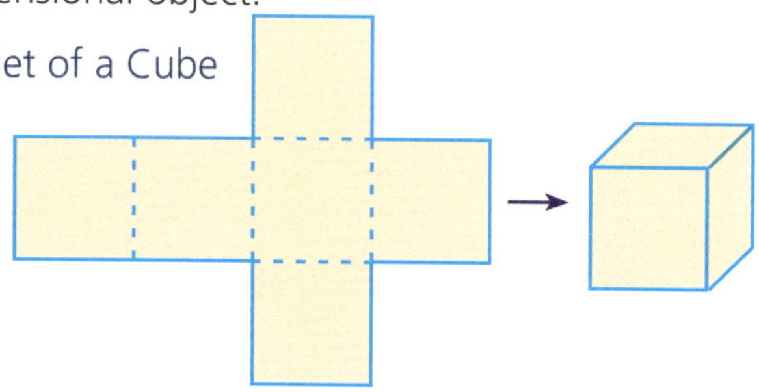

## number line

A line on which numbers are marked.

Number lines can be used to show operations.

- 2 + 2 + 2 = 6

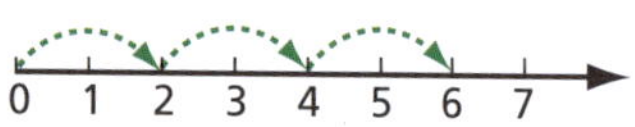


## numeral expander

Breaks up a number into hundreds, tens and ones.

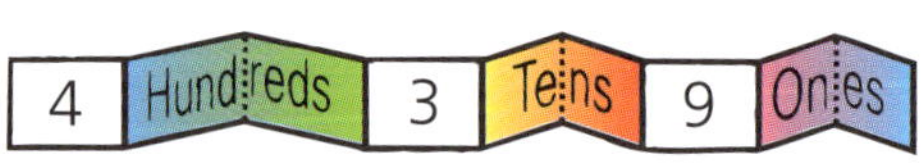


- This numeral expander shows 439.

## object

The term used to describe a three-dimensional shape.

Hexagonal Prism

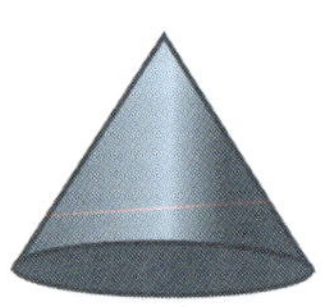

Cone

## octagon

A polygon with eight sides.

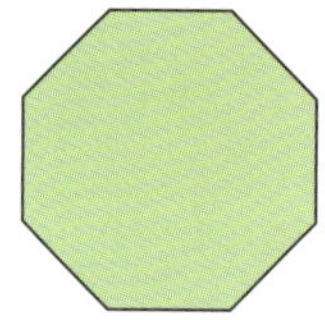

Regular Octagon

Irregular Octagon

See also *polygon*.

## ordinal number

A number that shows the position or place of an object.

1st 2nd 3rd 4th

Ordinal Numbers

## oval

A squashed circle.

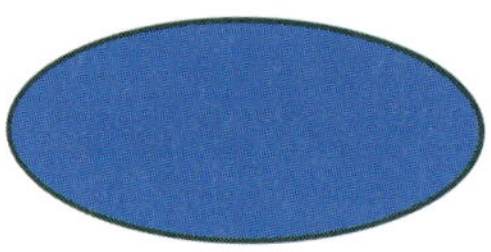

## parallel lines

Straight lines on the same flat surface that do not meet.

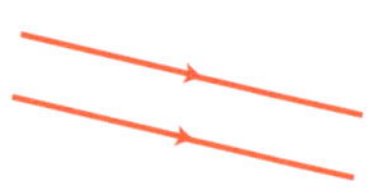

## parallelogram

A shape with 4 sides such that the pairs of opposite sides are parallel and equal.

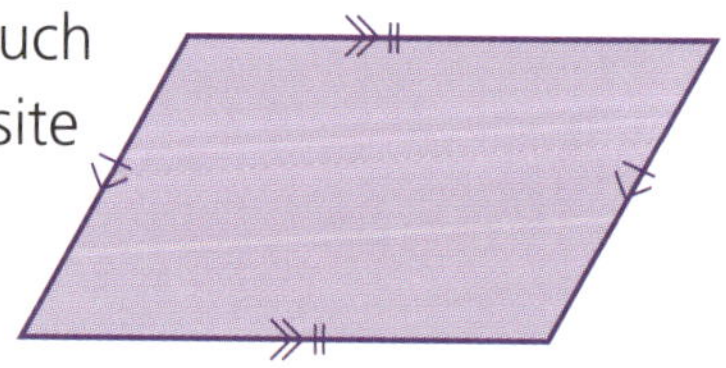

## **patterns** and **rules**

Sets of numbers that can be made by using a rule.

- The rule for 5, 9, 13, 17 … is **Add 4**.

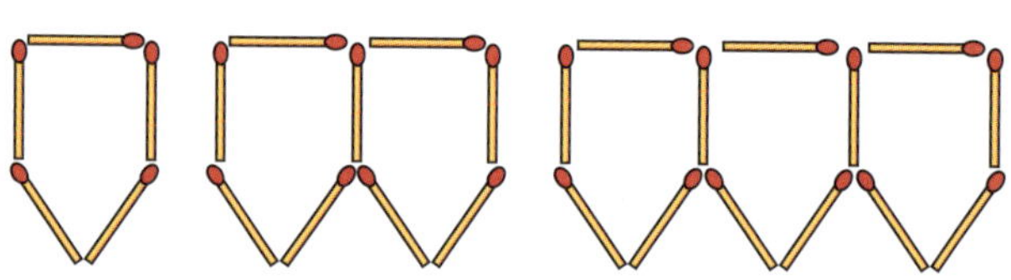

## pentagon

A polygon with five sides.

Regular Pentagon    Irregular Pentagon

See also *polygon*.

## perimeter

The distance around the outside of a shape; the boundary.

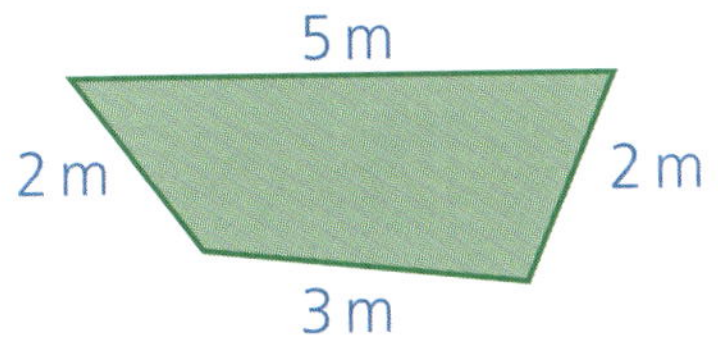


- Perimeter = 2 m + 3 m + 2 m + 5 m
  = 12 m

## perpendicular lines

Lines that meet at right angles.

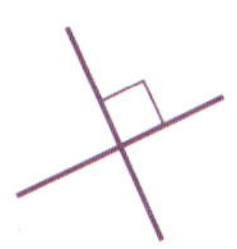

## picture graph

A graph that uses pictures to compare the size of groups.

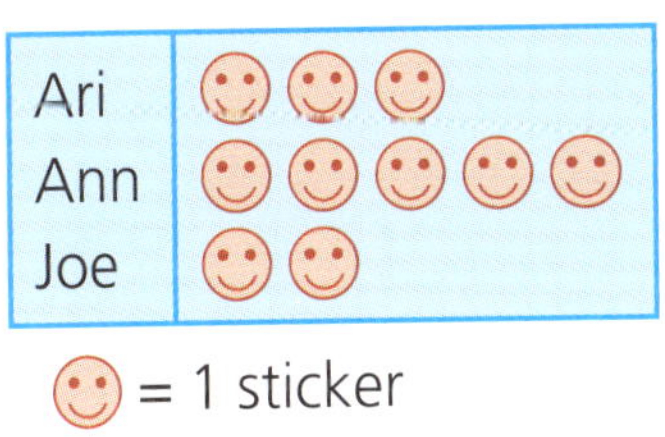


## place value

The column value of a digit.

- 396 =

| Hundreds | Tens | Ones |
|---|---|---|
| 3 | 9 | 6 |

## place-value blocks

1 ones block = 1 

1 tens block = 10    1 hundreds block = 100

- 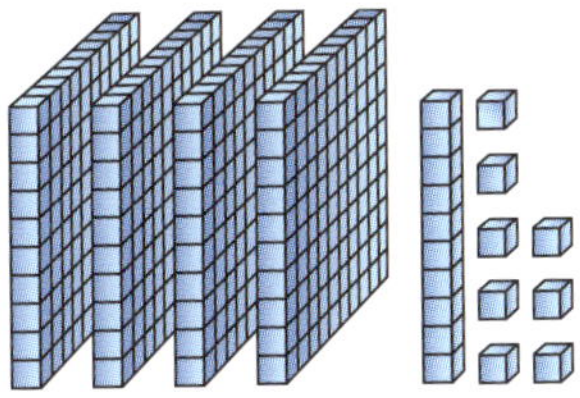 represents 418.

## polygon

A two-dimensional shape with three or more straight sides, such as a triangle, quadrilateral, pentagon etc.

## prism

A three-dimensional object with a uniform cross-section. The ends are identical shapes and all other faces are rectangles. Prisms are named by the shape of their ends.

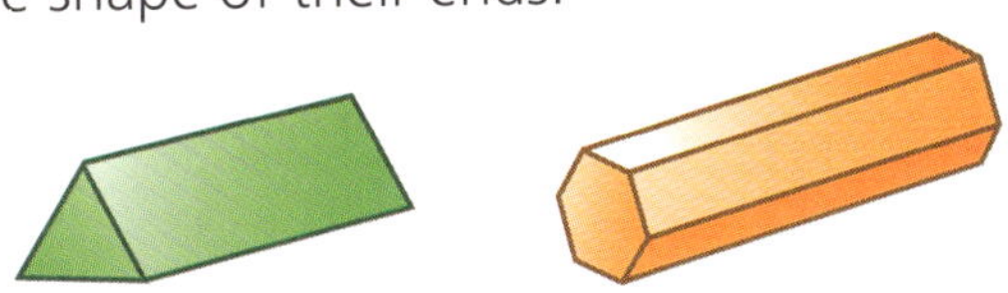

Triangular Prism    Hexagonal Prism

## probability

The probability (or chance) of something happening is its likelihood of happening.

- The probability of rolling an even number on a dice is 50%.

## pyramid

A three-dimensional object that has a polygon for a base and triangles for all other faces. Pyramids are named by the shape of their base.

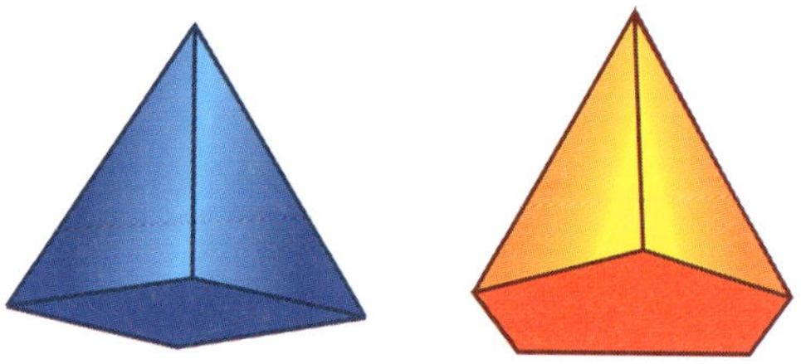

Square Pyramid    Pentagonal Pyramid

## quadrilateral

A two-dimensional shape with four straight sides.

## rectangle

A quadrilateral with four right angles and two pairs of equal and parallel sides.

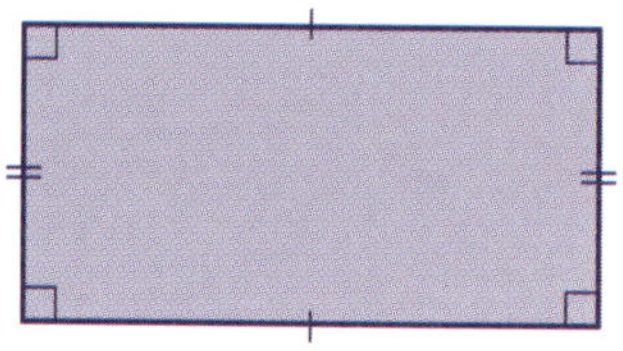

## reflection

See *flip*.

## regular and irregular shapes

Regular shapes have all sides and all angles equal. Irregular shapes do not.

Regular Shape    Irregular Shape

## rhombus

A shape with 4 sides, opposite sides parallel, all sides equal.

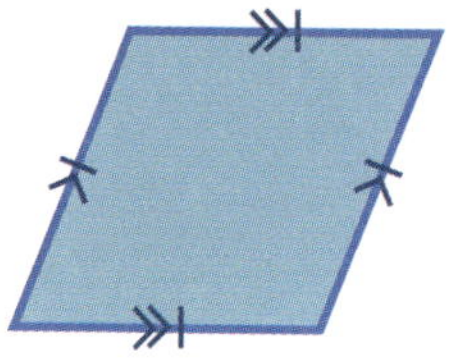

## Roman numerals

A number system devised by the ancient Romans.

Roman numerals use letters for numbers:

| I | V | X | L | C | D | M |
|---|---|---|---|---|---|---|
| 1 | 5 | 10 | 50 | 100 | 500 | 1000 |

- XXVIII = 28

## rotation

See *turn*.

## rounding

Rewriting a number to the nearest 5, 10 or 100.

## skip counting

Counting on, adding the same number each time.

- 5, 10, 15, 20, 25, … is skip counting by 5.

## slide (translation)

To move a shape in any direction without changing its orientation.

- 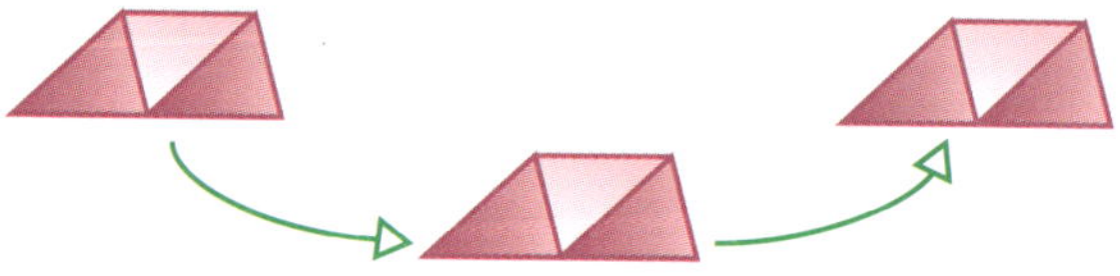

## sphere

A three-dimensional object that is ball-shaped and round. All points on the surface of a sphere are the same distance from its centre.

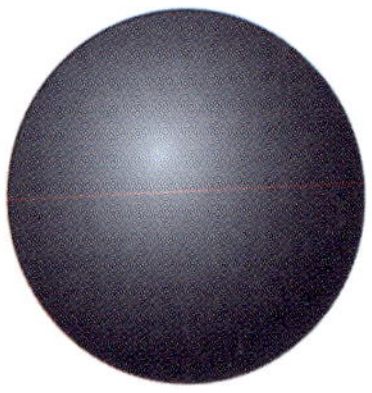

## split strategy

Adding numbers by splitting them into their parts.

- 36 + 52 = 30 + 6 + 50 + 2
  = (30 + 50) + (6 + 2)
  = 80 + 8
  = 88

## square

A quadrilateral that has four equal sides and four right angles.

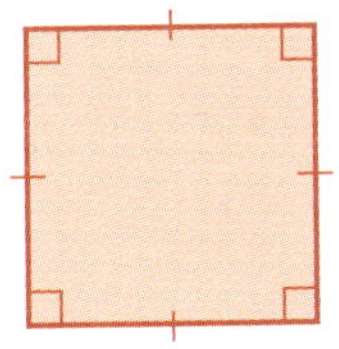

## sum

The answer when you add numbers.

## surface

The outside layer of a three-dimensional object. A surface can be flat or curved.

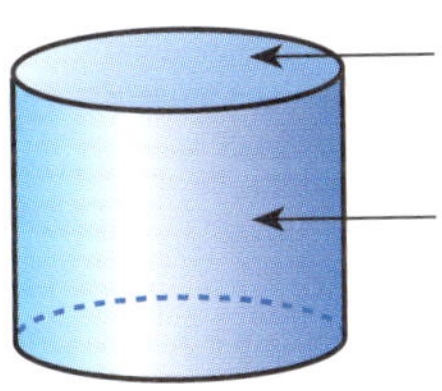


See also *face*.

## survey

A list of questions used to discover information.

## symmetry

If a figure has a line of symmetry, it can be folded so that the two halves exactly overlap.

Each half is a mirror image of the other.

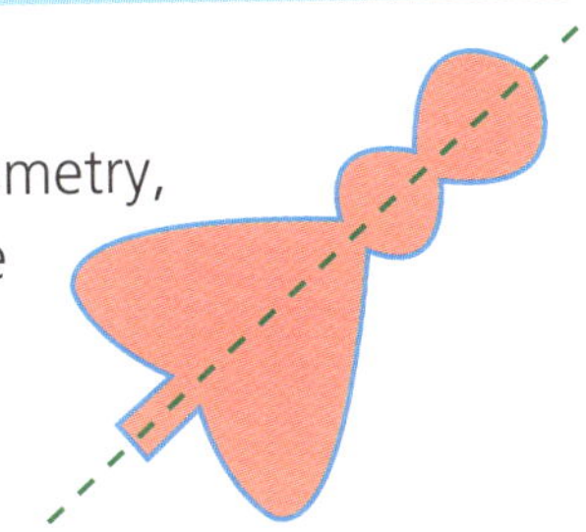


## table

A simple way to display information in rows and columns.

| | |
|---|---|
| Birds | 12 |
| Lions | 2 |
| Monkeys | 6 |

## tally

To keep count by making a mark for each item. To make counting easy, the marks are drawn in groups of five with each fifth mark crossed over the other four marks.

- 卌 卌 卌 ||| = 18

## tangram

A traditional Chinese puzzle. A square is cut into seven pieces that can be rearranged to make different pictures.

- 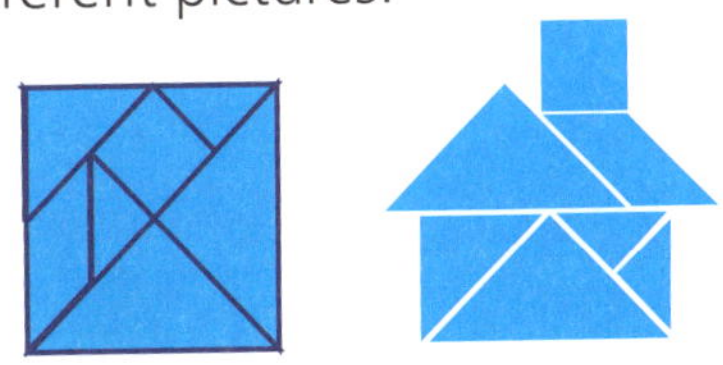

## three-dimensional (3D) objects

Objects are three-dimensional. They have length, width and height.

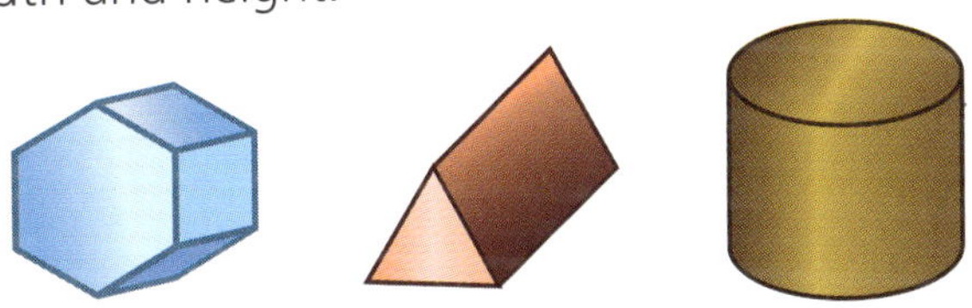

## time

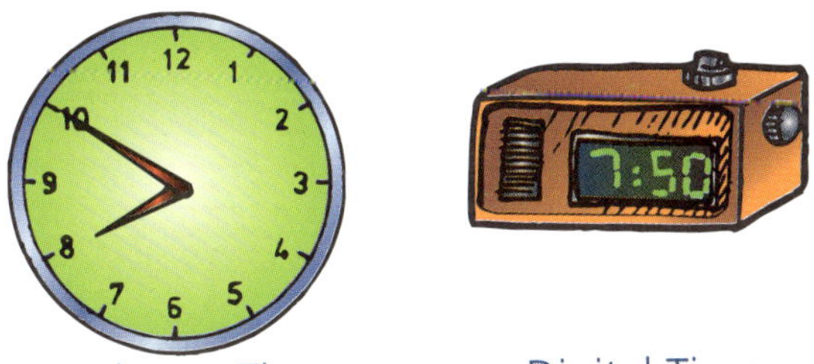


## time line

Shows a sequence of events in time.

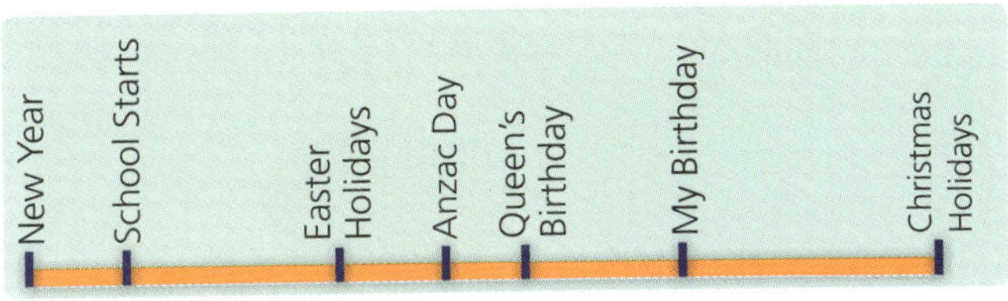

## translation

See *slide*.

## trapezium

A quadrilateral with one pair of parallel sides.

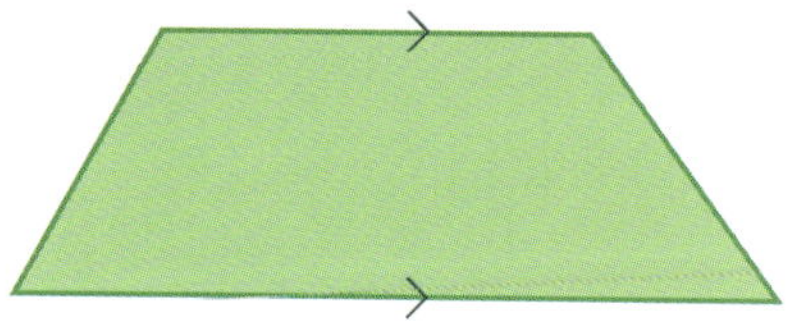

## triangle

A polygon with three sides and three angles.

See also *polygon*.

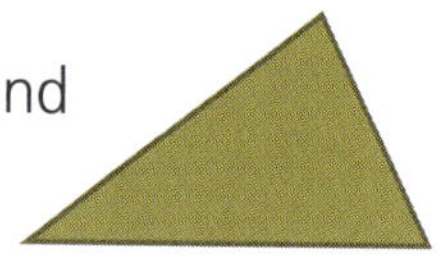

## turn (rotation)

To rotate a shape about a point.

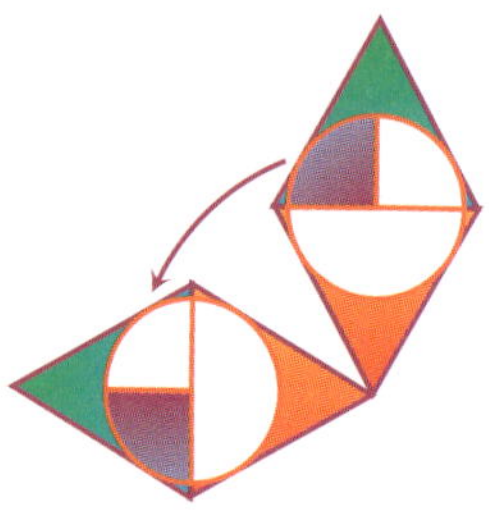

## two-dimensional (2D) shapes

Plane shapes are two-dimensional. They have length and width.

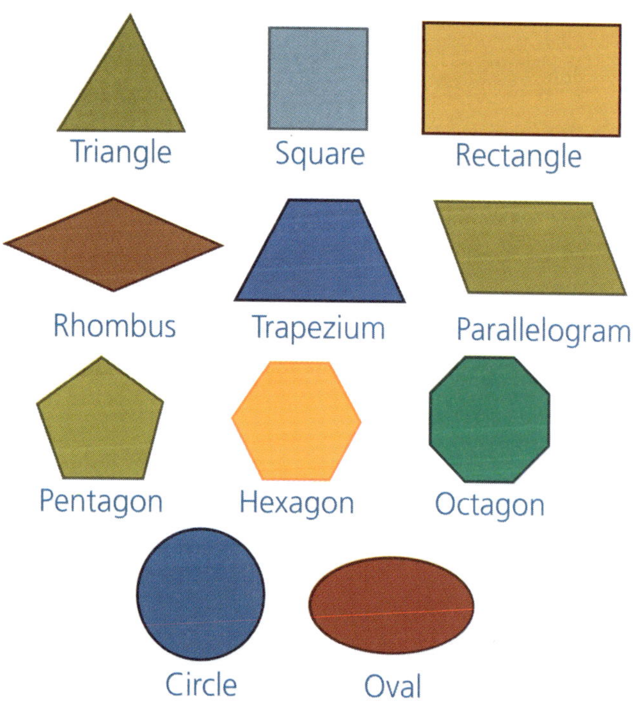


## two-way tables

Tables of data that show categories down the side as well as across the top.

| | Blue | Red | White |
|---|---|---|---|
| Bus | 2 | 5 | 0 |
| Car | 8 | 10 | 21 |
| Truck | 1 | 6 | 2 |

## vertex

A point at which two or more lines meet to form a corner on a plane or object.

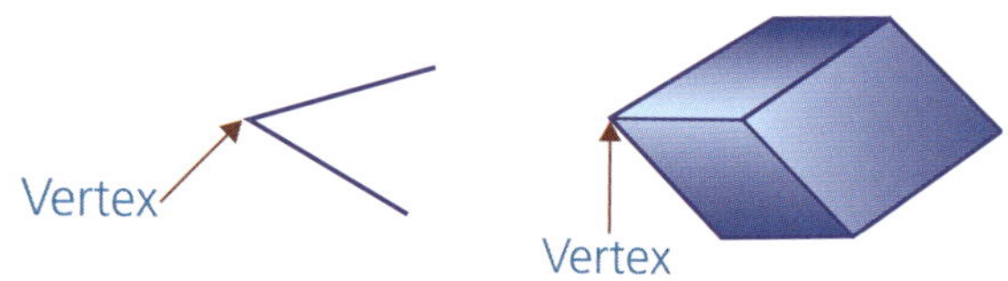


The plural of *vertex* is *vertices*.

## views of objects

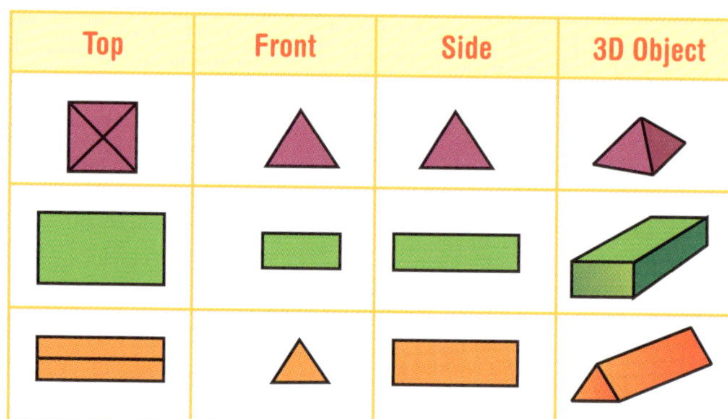

# 1:01 Skip Counting

CONCEPT

Until we know the pattern of numbers, we can count on from the last answer.

**1** Skip count and write the numbers as you go.

| | | | | | | | | | | |
|---|---|---|---|---|---|---|---|---|---|---|
| **a** Each time, count on three more. | 3 | 6 | | | | | | | | |
| **b** Each time, count on four more. | 4 | 8 | | | | | | | | |
| **c** Each time, count on six more. | 6 | 12 | | | | | | | | |
| **d** Each time, count on seven more. | 7 | 14 | | | | | | | | |
| **e** Each time, count on eight more. | 8 | 16 | | | | | | | | |
| **f** Each time, count on nine more. | 9 | 18 | | | | | | | | |
| **g** Each time, count on ten more. | 10 | 20 | | | | | | | | |
| **h** Each time, count on 100 more. | 100 | 200 | | | | | | | | |

# 1:02 Odd and Even Numbers

CONCEPT

An even number of items can be drawn in pairs.

2, 4, 6, 8, 10, 12, ...

An odd number of items can't be drawn in pairs. There is always one left over.

1, 3, 5, 7, 9, 11, 13, ...

1. Under each group, write **odd** or **even** and write the number.

a 

b 

c 

d 

e 

f 

CONCEPT

Odd numbers end in 1, 3, 5, 7 or 9. 87 is an odd number.

Even numbers end in 2, 4, 6, 8 or 0. 34 is an even number.

2. Colour the odd numbers red and the even numbers blue.

83 100 109 111 118 120 125 127 130

3. Why are numbers ending in 1, 3, 5, 7 or 9 odd numbers?



 • *AUSTRALIAN SIGNPOST MATHS 3* • ISBN 978 1 4886 2184 0

# 1:03 Odd and Even Numbers

**1** Use the hundred chart to answer the questions.

Hundred Chart

| | | | | | | | | | |
|---|---|---|---|---|---|---|---|---|---|
| 1 | 2 | 3 | 4 | 5 | 6 | 7 | 8 | 9 | 10 |
| 11 | 12 | 13 | 14 | 15 | 16 | 17 | 18 | 19 | 20 |
| 21 | 22 | 23 | 24 | 25 | 26 | 27 | 28 | 29 | 30 |
| 31 | 32 | 33 | 34 | 35 | 36 | 37 | 38 | 39 | 40 |
| 41 | 42 | 43 | 44 | 45 | 46 | 47 | 48 | 49 | 50 |
| 51 | 52 | 53 | 54 | 55 | 56 | 57 | 58 | 59 | 60 |
| 61 | 62 | 63 | 64 | 65 | 66 | 67 | 68 | 69 | 70 |
| 71 | 72 | 73 | 74 | 75 | 76 | 77 | 78 | 79 | 80 |
| 81 | 82 | 83 | 84 | 85 | 86 | 87 | 88 | 89 | 90 |
| 91 | 92 | 93 | 94 | 95 | 96 | 97 | 98 | 99 | 100 |

**a** Count by 2s. Colour these numbers on the chart. These are all the even numbers up to 100.

**b** What is the name given to the numbers that are not coloured?

**c** What is the largest even number less than 80?

**d** What is the largest even number less than 67?

**e** What is the largest odd number less than 71?

**2** Why are numbers ending in 2, 4, 6, 8 or 0 even numbers?

**3** How many even numbers are between these numbers?

**a** 6 and 16 **b** 47 and 55 **c** 1 and 100

**d** 31 and 72 **e** 28 and 81 **f** 68 and 70

**4** How many odd numbers are between these numbers?

**a** 6 and 16 **b** 47 and 55 **c** 1 and 100

**d** 27 and 54 **e** 42 and 95 **f** 13 and 69

**5** Circle the even numbers. Underline the odd numbers.

38 53 75 14 87 92 66 36 29 41 50 35 74 100

**6** For each number write **even** or **odd**.

**a** 98 **b** 120

**c** 103 **d** 914

**e** 216 **f** 847

**g** 681 **h** 509

**i** 852 **j** 75

**k** 367 **l** 644

  ISBN 978 1 4886 2184 0

# 1:04 Numbers to 1000

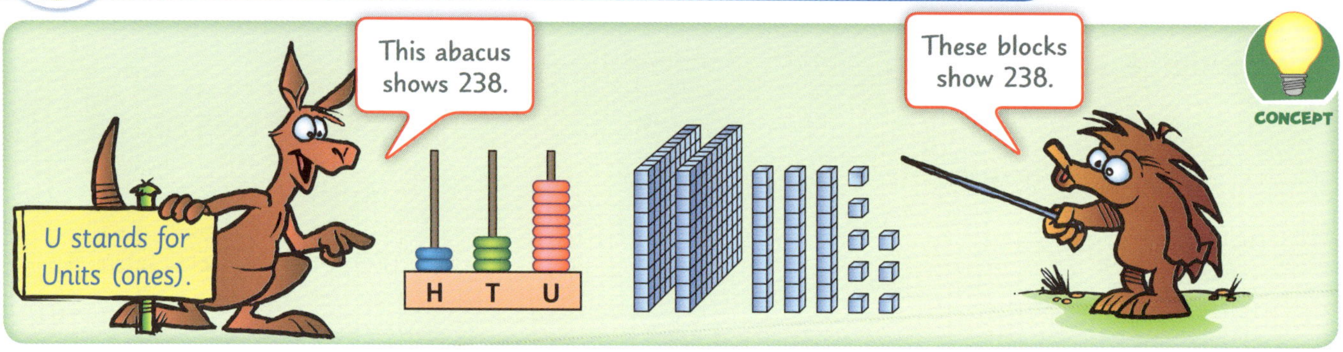


**1** Write the number shown by the place-value blocks or abacus.

**a** 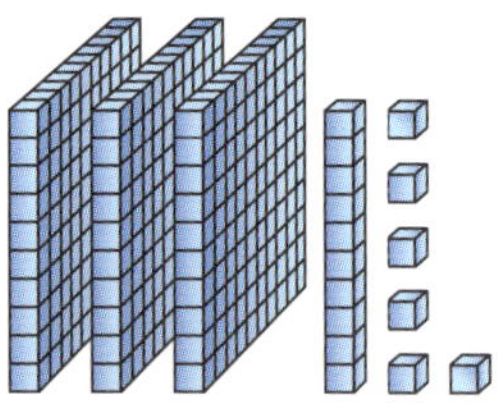

**b** 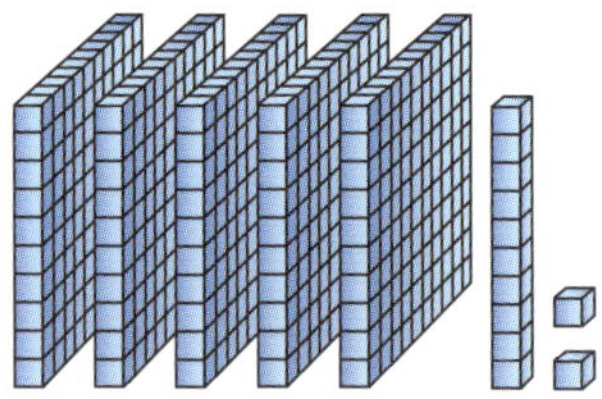

**c** 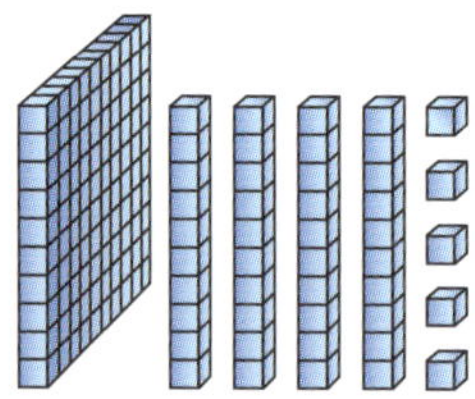

**d** 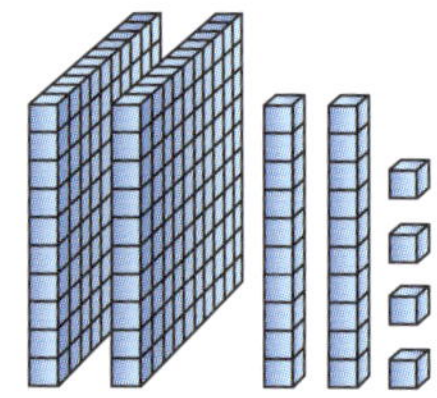

**e** 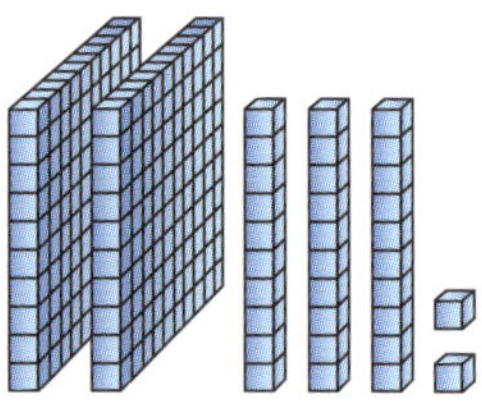

**f** 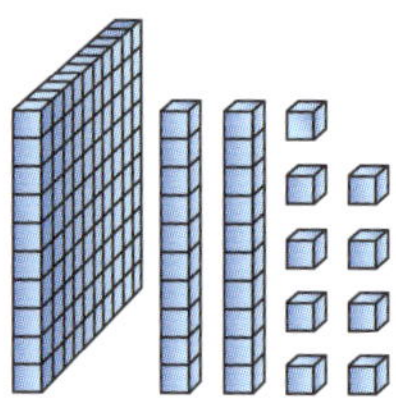

**g** 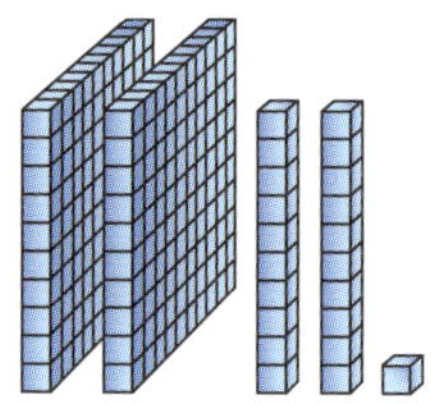

**h** 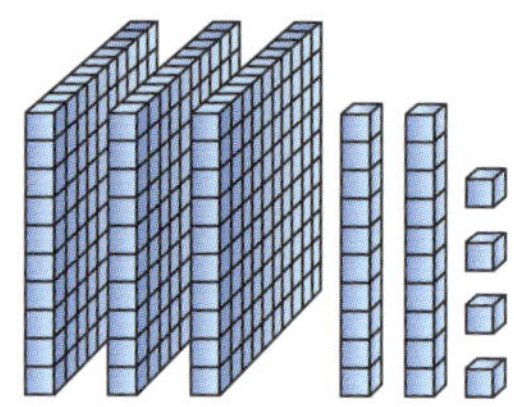

**i** 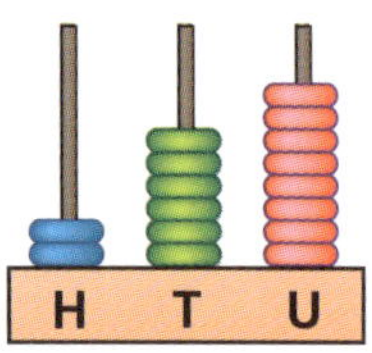


**j** 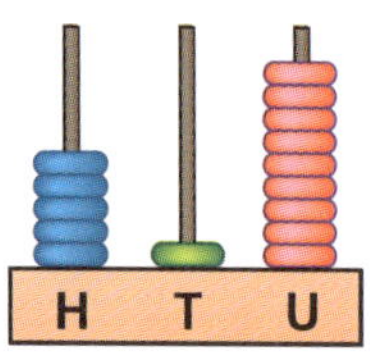


**k** 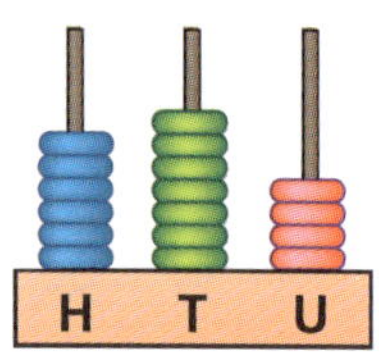


**l** 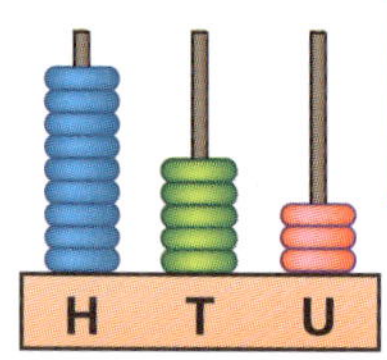

**2** Which number is larger?

**a** 169 or 346 ______ **b** 723 or 481 ______ **c** 962 or 503 ______

**d** 375 or 634 ______ **e** 257 or 572 ______ **f** 491 or 914 ______

**3** Write these in order from smallest to largest.

**a** 137, 653, 446 ______, ______, ______ **b** 974, 237, 491 ______, ______, ______

**c** 819, 106, 567 ______, ______, ______ **d** 683, 749, 250 ______, ______, ______


 • *AUSTRALIAN SIGNPOST MATHS 3* • ISBN 978 1 4886 2184 0

# 1:05 Numbers to 1000

**1** Write the numeral, fill in the numeral expander and write the number in words.

**a**

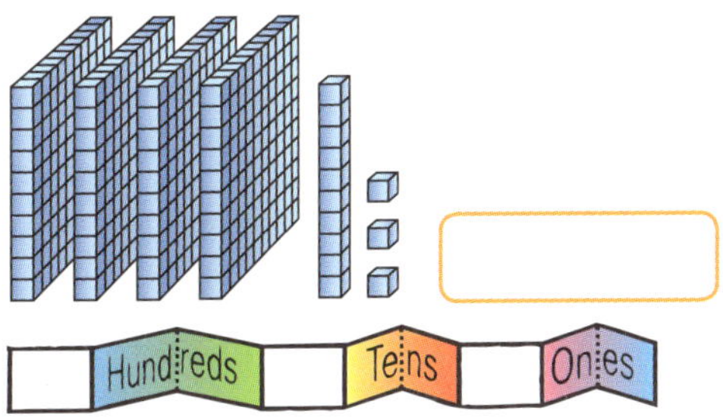


**b**

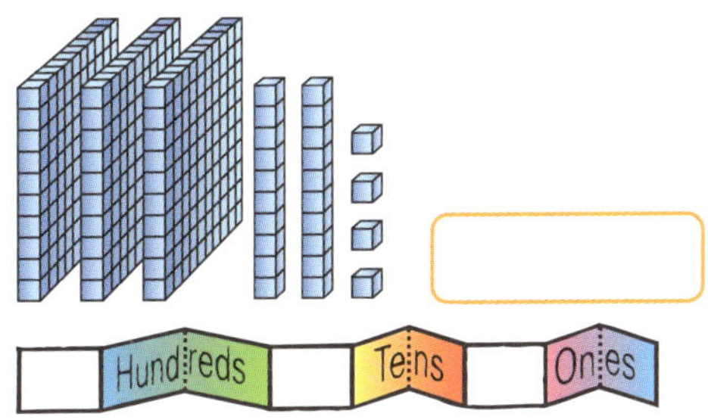


**2** How many digits are in each numeral?

**a** 39 **b** 256 **c** 970 **d** 56 **e** 498

**f** 13 **g** 7 **h** 520 **i** 1 000 **j** 777

**3** Write these numbers as numerals.

**a** two hundred and sixty

**b** one hundred and fifty-two

**c** nine hundred and forty

**d** seven hundred and eighteen

**e** six hundred and seventy-nine

**f** five hundred and thirty-four

**g** eight hundred and sixty-eight

**h** three hundred and six

**4** Write the numbers before and after.

**a** ____, 999, ____ **b** ____, 863, ____ **c** ____, 659, ____

**d** ____, 306, ____ **e** ____, 499, ____ **f** ____, 709, ____

ACTIVITY

- Use place-value blocks to model these numbers.
  - 216
  - 525
  - 848
  - 634
  - 967
  - 388
  - 793
  - 364
  - 190
  - 572
  - 451
  - 1 000

# 1:06 Counting

**Hundred Chart**

| | | | | | | | | | |
|---|---|---|---|---|---|---|---|---|---|
| 1 | 2 | 3 | 4 | 5 | 6 | 7 | 8 | 9 | 10 |
| 11 | 12 | 13 | 14 | 15 | 16 | 17 | 18 | 19 | 20 |
| 21 | 22 | 23 | 24 | 25 | 26 | 27 | 28 | 29 | 30 |
| 31 | 32 | 33 | 34 | 35 | 36 | 37 | 38 | 39 | 40 |
| 41 | 42 | 43 | 44 | 45 | 46 | 47 | 48 | 49 | 50 |
| 51 | 52 | 53 | 54 | 55 | 56 | 57 | 58 | 59 | 60 |
| 61 | 62 | 63 | 64 | 65 | 66 | 67 | 68 | 69 | 70 |
| 71 | 72 | 73 | 74 | 75 | 76 | 77 | 78 | 79 | 80 |
| 81 | 82 | 83 | 84 | 85 | 86 | 87 | 88 | 89 | 90 |
| 91 | 92 | 93 | 94 | 95 | 96 | 97 | 98 | 99 | 100 |

1 Use the hundred chart to answer the questions.

**a** Count by 2s. Colour these numbers yellow.

**b** Starting at 100, count backwards by 10s. Draw a cross on these numbers.

**c** Circle every second even number up to 80. What do you notice?

**d** Count by 8s and tick the first 10 numbers you count. Write them below.

2 What do even numbers end in?

3 When we count by 5s from zero, the numbers end in .

4 When we count by 10s from zero, the numbers end in .

5 Continue each pattern. Check your answers with a calculator.

**a** 223, 233, 243, , , , ,

**b** 815, 810, 805, , , , ,

**c** 126, 124, 122, , , , ,

**d** 1 000, 900, 800, , , , ,

6 Show your answers to Questions 5**a** and 5**b** on the number lines.

**a**

The rule is .

**b**

The rule is .

# 1:07 Counting

**1** **a** Count on from 76 to 100 by 2s.

**b** Count backwards from 1 000 by 100s.

**c** Count on from 645 to 690 by 5s.

**d** Count backwards from 500 to 400 by 10s.

**2** Write the missing numbers.

**a** 865, ____, 845, ____, ____, 815, ____, ____, 785

**b** 625, 620, ____, ____, 605, ____, ____, ____, 585

**c** 412, 410, ____, ____, 404, ____, ____, ____, 396

**3** Write the first 20 even numbers. Circle every second even number and discuss the pattern you see.

**4** Count by 5s and write the first 20 numbers you count. Circle every second number and discuss the pattern.

**5** If you have to count 300 ten-cent coins, what is the best counting strategy to make sure you count them correctly?

**6** Show your answers to Questions 1**a** and 1**b** on the number line.

**a**

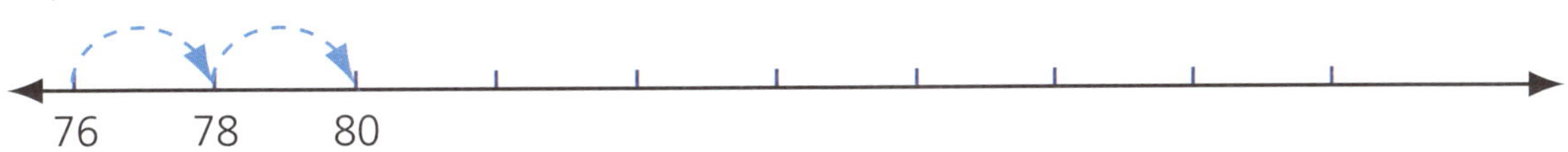


The rule is ____.

**b**

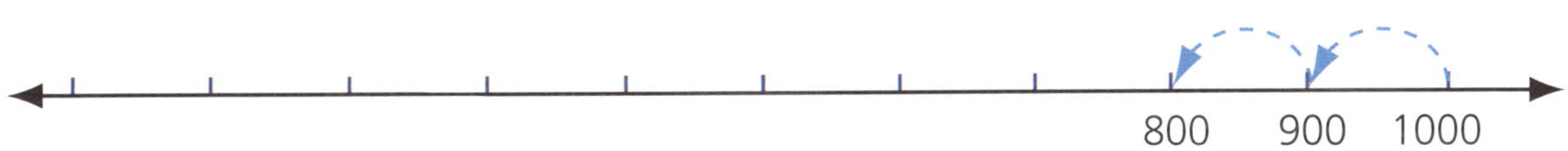


The rule is ____.

**c** Try to do Question 1**c** on your own number line.

# 1:08 Numbers to 1000

CONCEPT

720 is the same as
7 hundreds and 2 tens
or 72 tens or 720 ones.

1. Complete the numeral expanders.

a 479

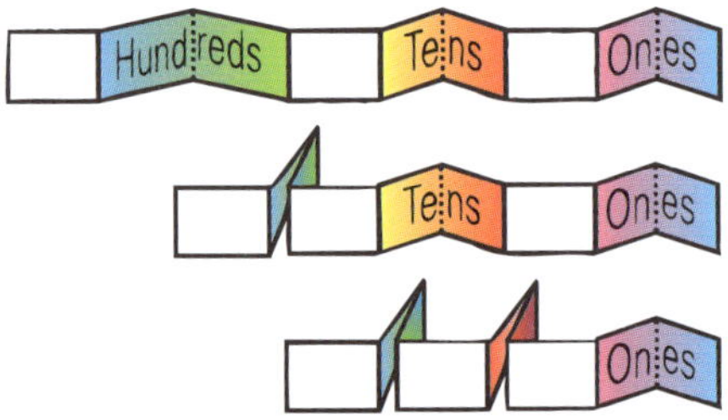


b 568

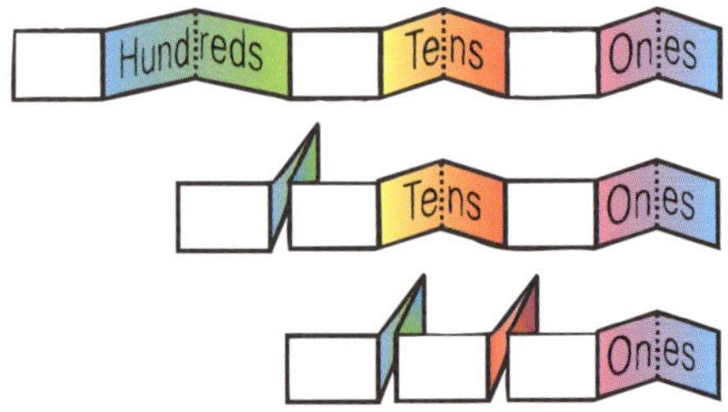


c 231

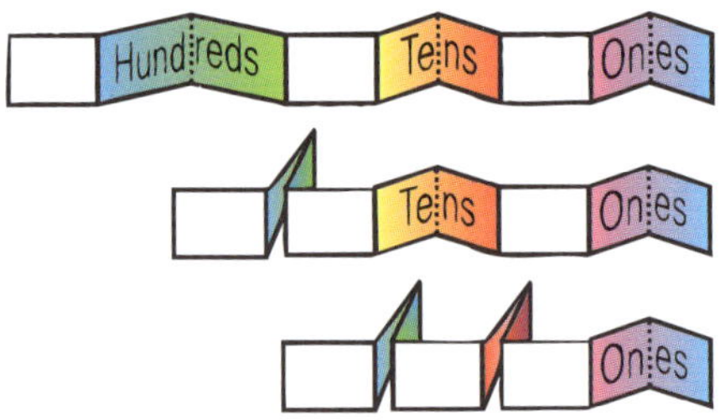


d 307

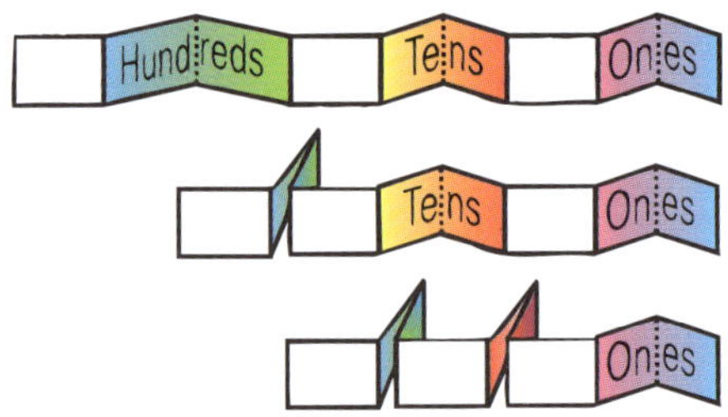


2. Write each number as a numeral.

a six hundred and thirty-two

b eight hundred and seventeen

c four hundred and twenty-nine

d seven hundred and sixty-three

e two hundred and thirty-eight

f five hundred and sixty-two

g nine hundred and forty

h three hundred and fifty-one

3. Write each number in words.

a 156

b 607

c 319

d 841

- Use concrete materials to show the numbers in Question 3. Explain your answer to a partner.

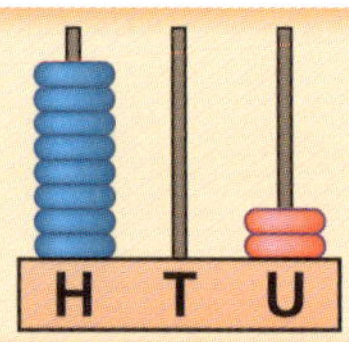


ACTIVITY



© PEARSON AUSTRALIA 2018 • *AUSTRALIAN SIGNPOST MATHS 3* • ISBN 978 1 4886 2184 0

# 1:09 Numbers to 1000

**1** Write the number shown by each abacus.

a
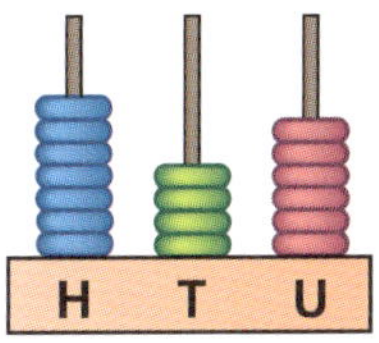


b
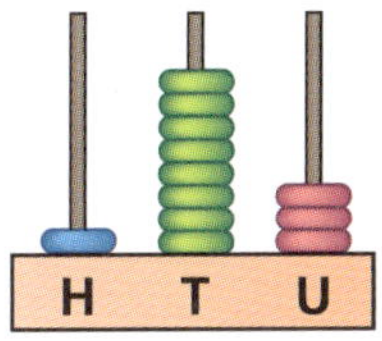


c
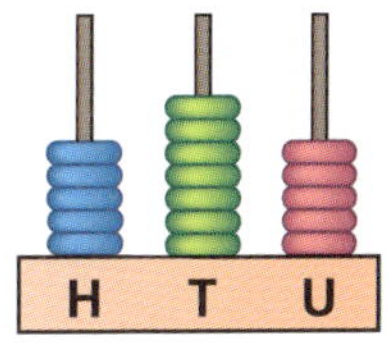


d
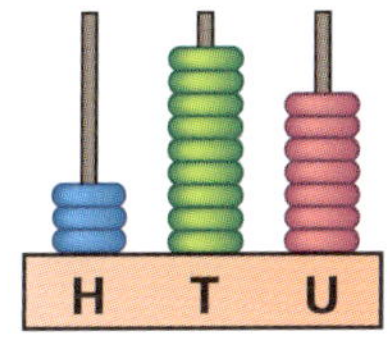


**2** Complete the numeral expanders.

a 375
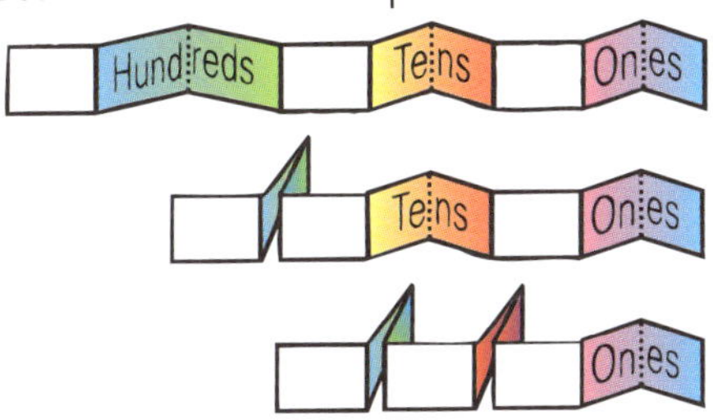


b 198
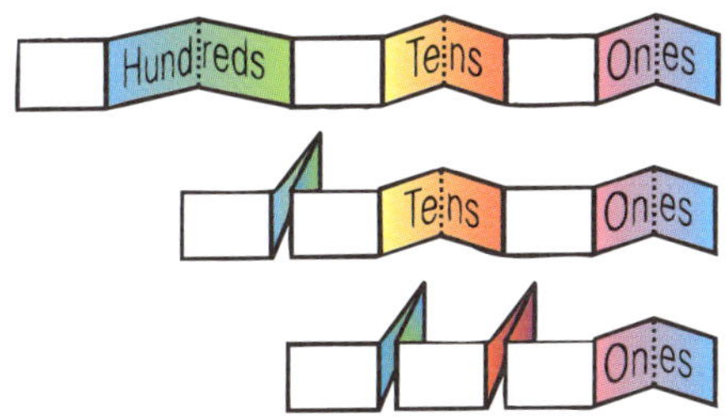


**3** Round each number to the nearest hundred.

a 378 ______ b 842 ______ c 296 ______

d 419 ______ e 675 ______ f 324 ______

g 906 ______ h 547 ______ i 752 ______

**4** Use < or > to show the larger number in each pair.

a 249 ______ 497  b 963 ______ 575  c 237 ______ 999  d 672 ______ 907

e 364 ______ 259  f 743 ______ 816  g 562 ______ 564  h 419 ______ 418

- Use blocks, bundles or other materials to model these numbers.
  - 291
  - 823
  - 457
  - 614
  - 536
  - 749
  - 620
  - 365
  - 918
  - 289
  - 172
  - 1 000

## Higher or Lower

- One player records a secret 3-digit number and says the boundaries for the number, such as "between 200 and 300".
- Other players mark the boundaries on number lines.
- Players take turns to guess the number. After each guess, the holder of the number says whether the secret number is **higher** or **lower** than the guess.
- Players mark this clue for the guess (**higher** or **lower**) on their number lines.
- The game continues until someone guesses the secret number exactly.

 • *AUSTRALIAN SIGNPOST MATHS 3* • ISBN 978 1 4886 2184 0

# 1:10 Fractions of a Whole

CONCEPT

$\frac{3}{4}$ ← Numerator, ← Denominator

The **d**enominator is written **d**own on the bottom.

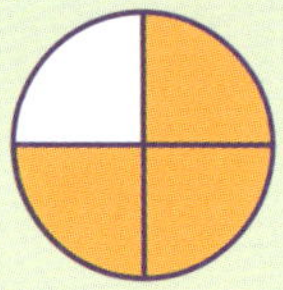

**1** How much of each shape has been coloured?

**a** 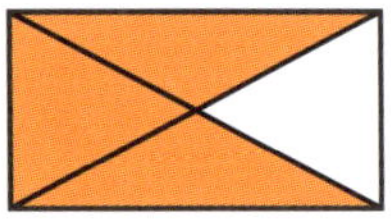 This shows ☐ of ☐ equal parts are coloured.

**b** 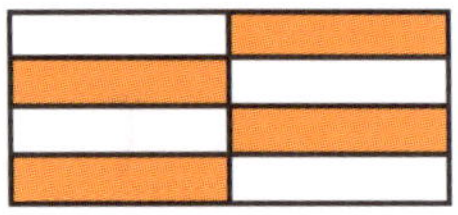 This shows ☐ of ☐ equal parts are coloured.

**c** 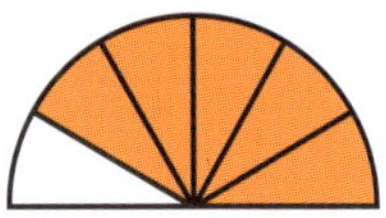 This shows ☐ of ☐ equal parts are coloured.

**2** How much of each shape has been coloured?

**a**  ☐

**b** 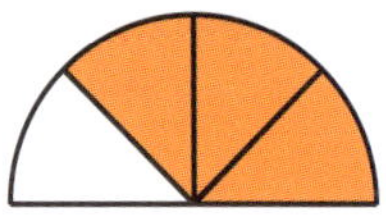 ☐

**c** 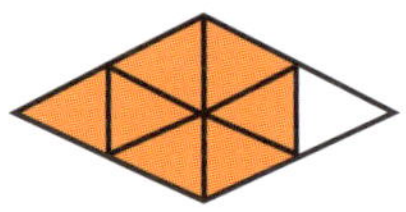 ☐

**d** 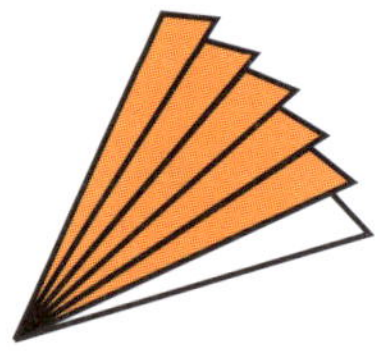 ☐

**e** 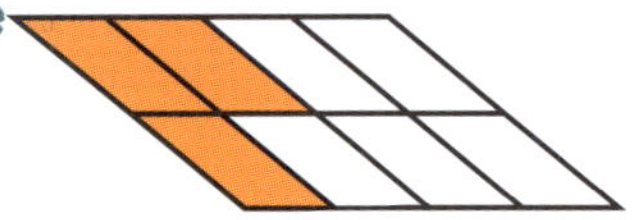 ☐

**f** 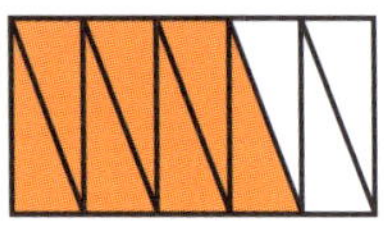 ☐

**g** 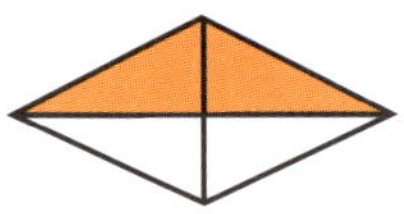 ☐

**h** 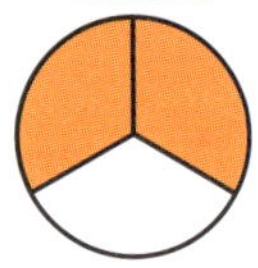 ☐

**3** Colour part of each shape to match the fraction.

**a**  $\frac{1}{3}$

**b**  $\frac{3}{4}$

**c** 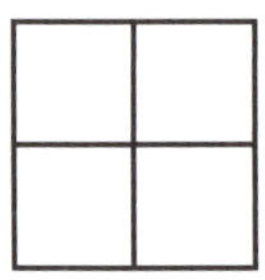 $\frac{1}{4}$

**d** 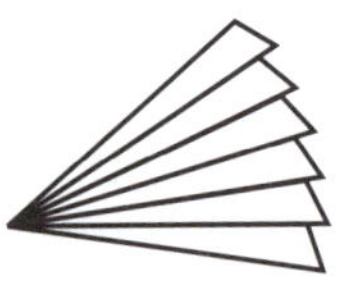 $\frac{3}{6}$

**e**  $\frac{6}{8}$

**f** 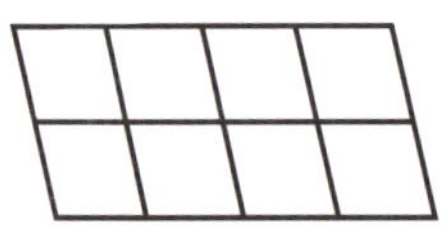 $\frac{5}{8}$

**g**  $\frac{4}{5}$

**h** 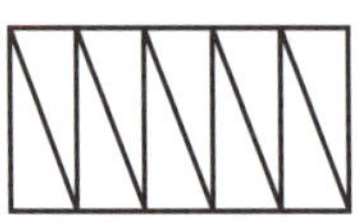 $\frac{7}{10}$

**i** 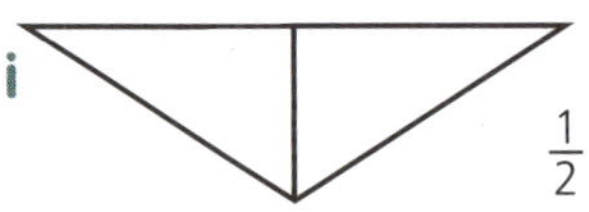 $\frac{1}{2}$

 • *AUSTRALIAN SIGNPOST MATHS 3* • ISBN 978 1 4886 2184 0

# 1:11 Fractions of a Collection

CONCEPT

| Group | Number Coloured | Total in Group | Fraction Coloured |
|---|---|---|---|
|  | 4 | 5 | $\frac{4}{5}$ |
|  | | | |
|  | | | |
|  | | | |

1. Complete the table above.

2. What fraction of each group has been coloured?

a  ☐

b  ☐

c  ☐

d  ☐

e  ☐

f  ☐

g  ☐

h  ☐

3. Colour part of each group to match the fraction.

a  $\frac{4}{6}$

b 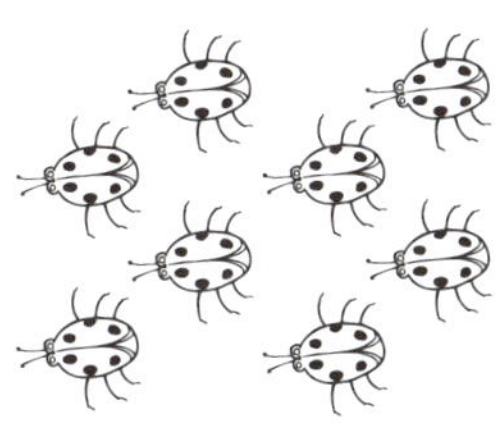 $\frac{7}{8}$

c  $\frac{11}{16}$

d 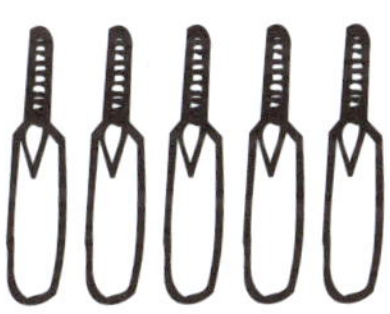 $\frac{3}{5}$

e  $\frac{5}{10}$

f 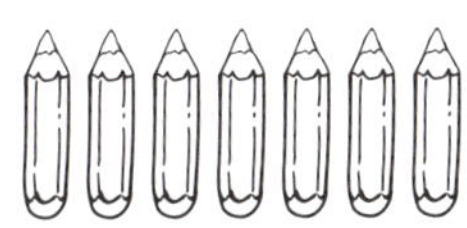 $\frac{3}{7}$

ACTIVITY

- Look through newspapers or magazines and cut out advertisements that show fractions. Paste them onto a sheet of paper to make a fractions collage.

*AUSTRALIAN SIGNPOST MATHS 3* • ISBN 978 1 4886 2184 0

# 1:12 Numbers to 10 000

CONCEPT

1324

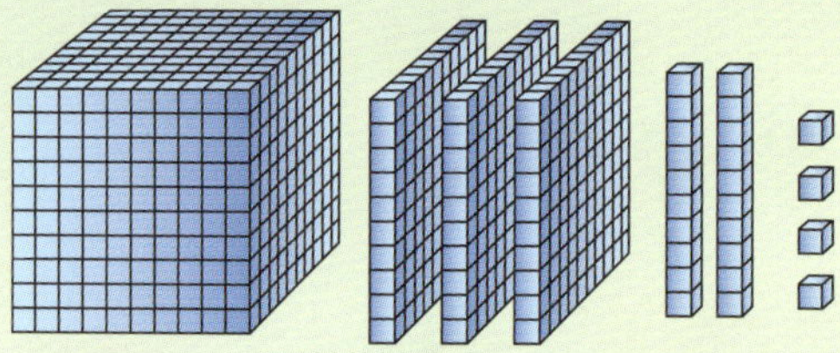

1 Thousands 3 Hundreds 2 Tens 4 Ones

one thousand three hundred and twenty-four

**1** Write the numeral, fill in the numeral expander and write the number in words.

a

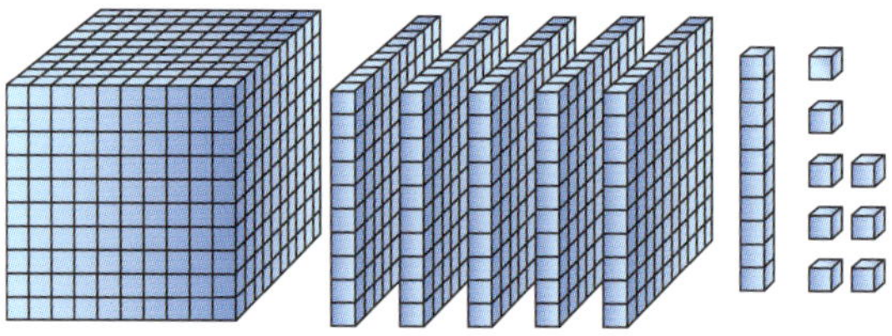

☐ Thousands ☐ Hundreds ☐ Tens ☐ Ones

b

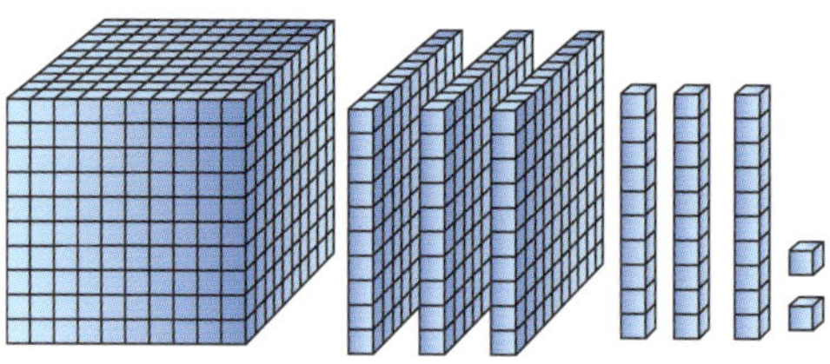

☐ Thousands ☐ Hundreds ☐ Tens ☐ Ones

**2** How many digits are in each numeral?

a 39 ☐ b 1307 ☐ c 9 ☐ d 85 ☐

e 1000 ☐ f 99 ☐ g 2469 ☐ h 1325 ☐

**3** Write the numeral for each number.

a five hundred and forty-five ☐ b twenty-seven ☐

c nine hundred and eight ☐ d one thousand ☐

e six hundred and eleven ☐ f two hundred ☐

**4** Circle the larger number.

a 1365 or 2547 b 3128 or 4263 c 5409 or 4149

d 2157 or 2010 e 4312 or 4196 f 7256 or 10 000

# 1:13 Numbers to 10 000

CONCEPT

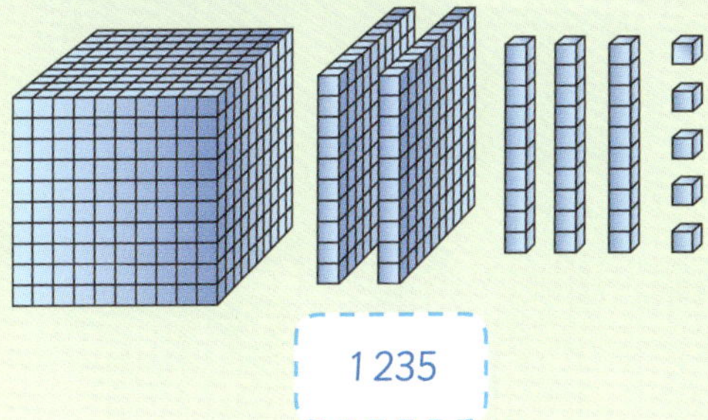


**1** Write the number shown.

a 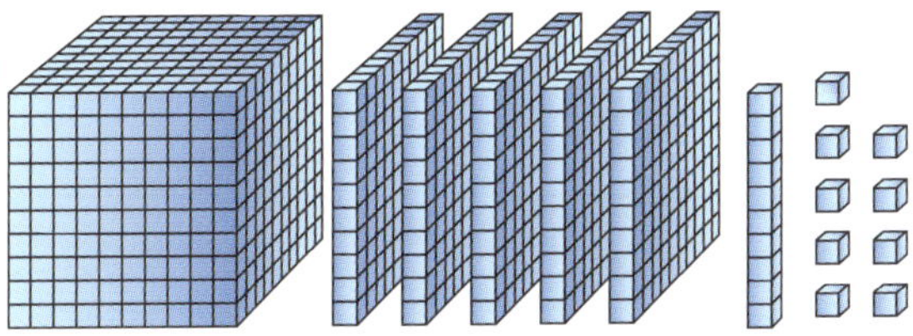

b 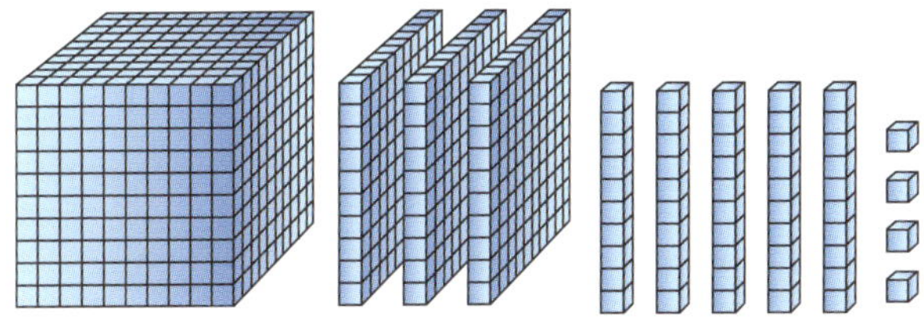

c 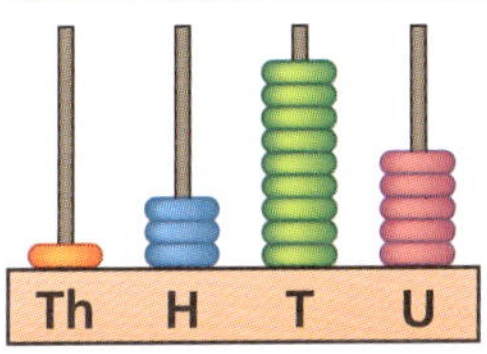


d 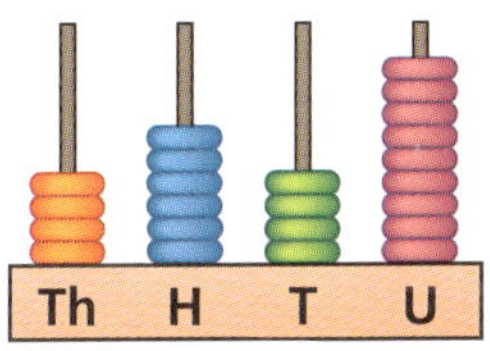


e 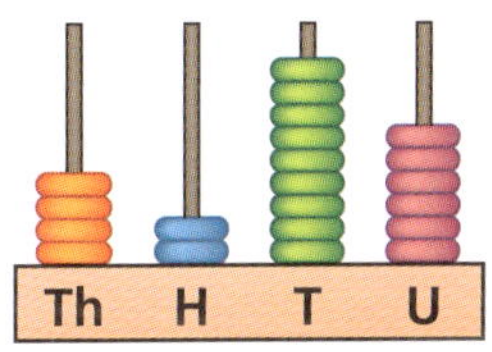


f 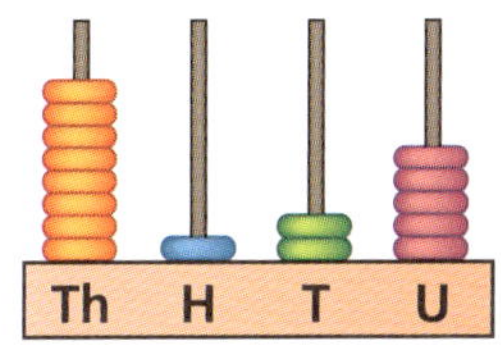


g 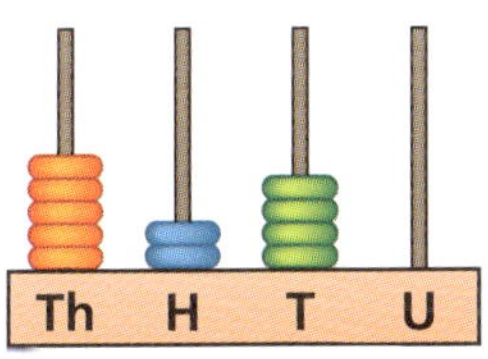


h 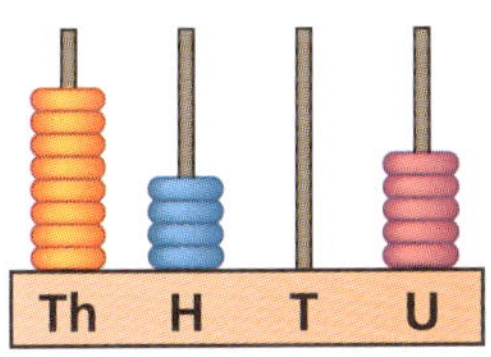


i 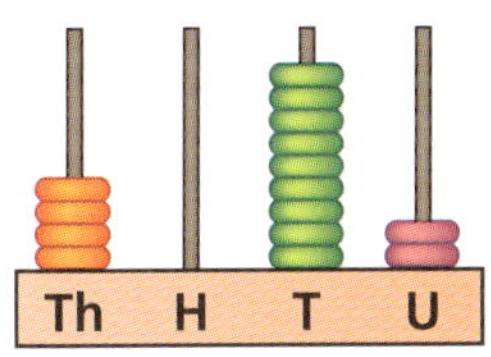

j 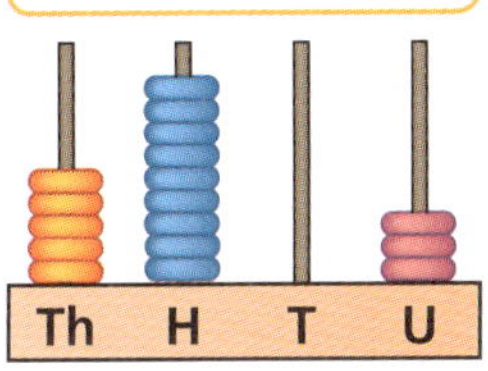


k 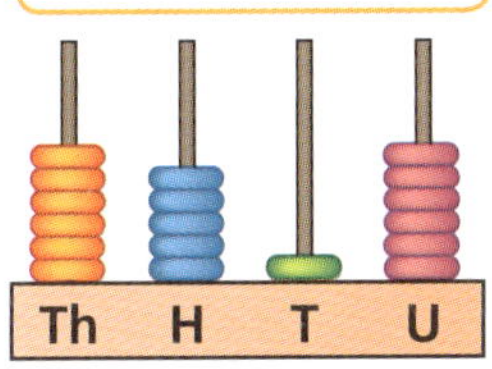


l 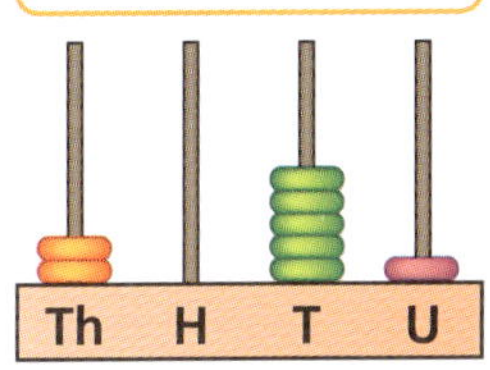


m 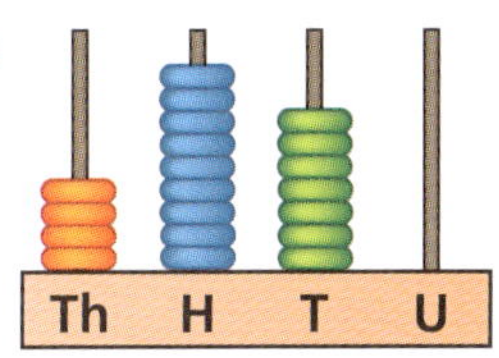


## Number Lotto

FUN SPOT

- Write any digit in each square.
- The teacher calls out 2-digit numbers.
- If the 2-digit number is on your grid, made across or down, cross it off.
- The first player to cross off three numbers is the winner.

| | | |
|---|---|---|
| | | |
| | | |
| | | |

# 1:14 Fractions

CONCEPT

**1** What fraction of each shape is coloured?

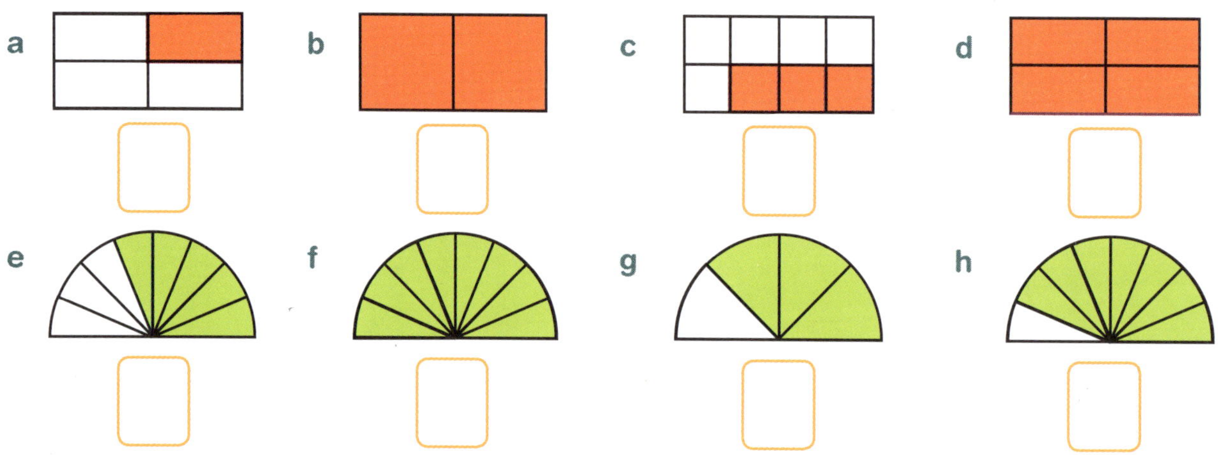


**2** Colour part of each shape to match the fraction.

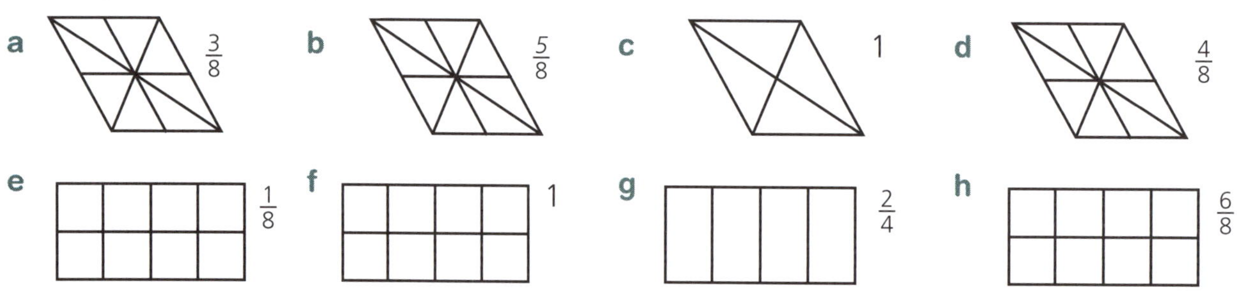


ACTIVITY

- Cut six strips of paper that are each 12 centimetres long.

  Use a ruler to measure, label and colour these fractions on the strips:

  - $\frac{1}{2}$
  - $\frac{1}{3}$
  - $\frac{1}{4}$
  - $\frac{1}{6}$
  - $\frac{1}{12}$

  Compare the fractions on the different strips.

  Discuss and compare the denominators and the lengths.

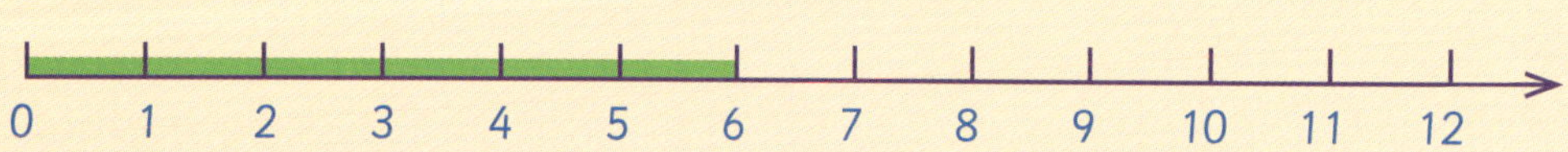

# 1:15 Comparing Fractions

CONCEPT

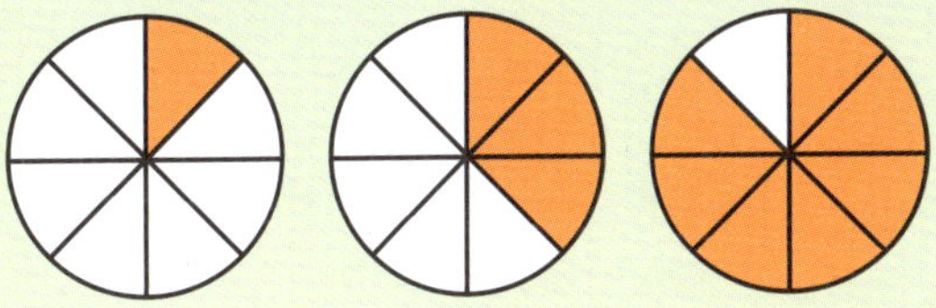

$\frac{1}{8}$ $\frac{3}{8}$ $\frac{7}{8}$

**1** Circle the larger fraction.

a 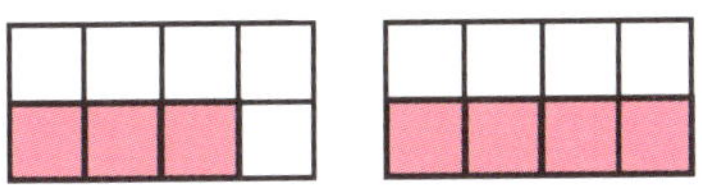

$\frac{3}{8}$ or $\frac{4}{8}$

b 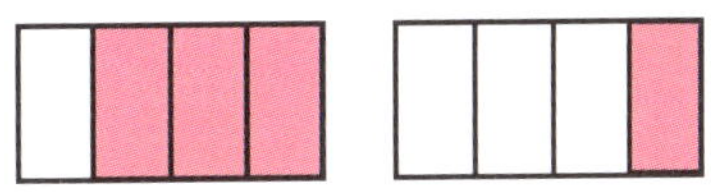

$\frac{3}{4}$ or $\frac{1}{4}$

c 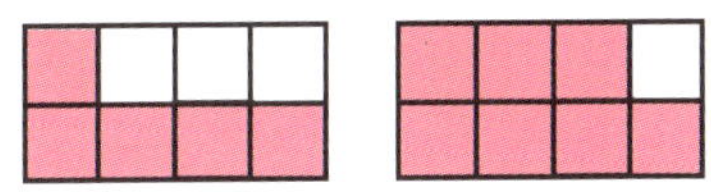

$\frac{5}{8}$ or $\frac{7}{8}$

d 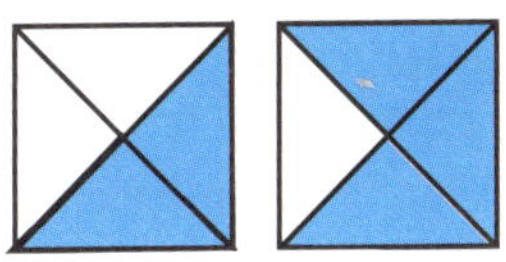

$\frac{2}{4}$ or $\frac{3}{4}$

e 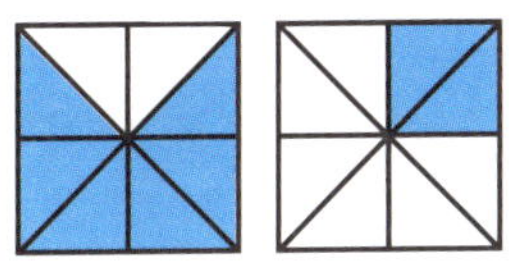

$\frac{6}{8}$ or $\frac{2}{8}$

f 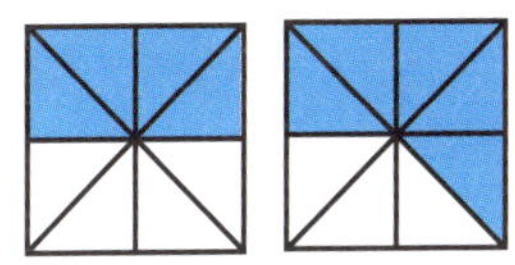

$\frac{4}{8}$ or $\frac{5}{8}$

**2** Colour part of each shape to match the fraction.

a 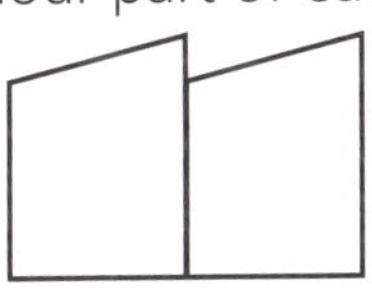 $\frac{1}{2}$

b 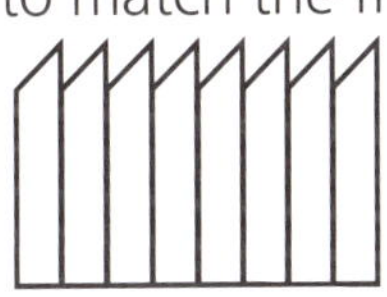 $\frac{5}{8}$

c 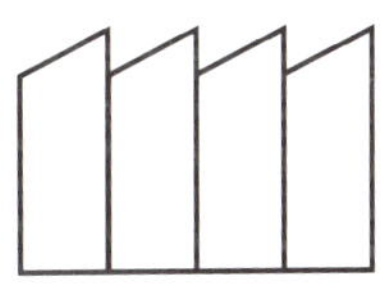 $\frac{3}{4}$

d 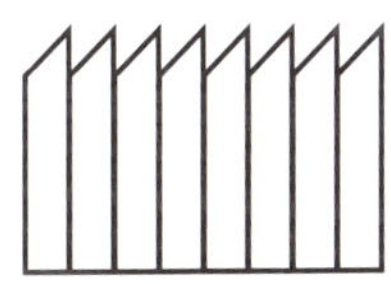 $\frac{2}{8}$

e  $\frac{3}{8}$

f  $\frac{2}{8}$

g  $\frac{6}{8}$

h 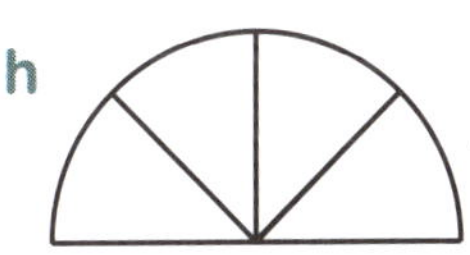 $\frac{1}{4}$

**3** What fraction of each shape is coloured?

a 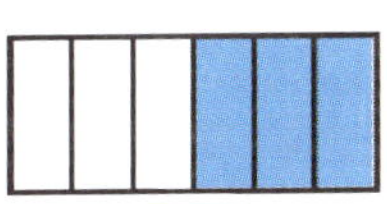

b 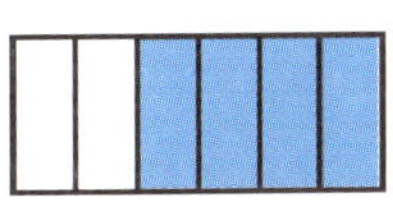

c 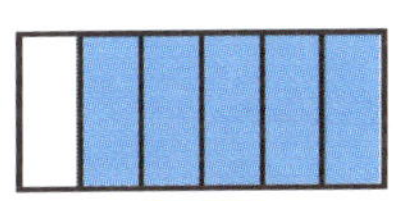

d 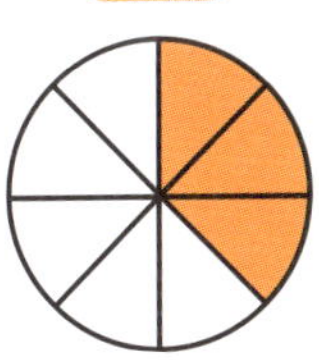

e 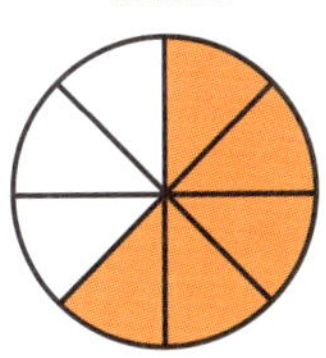

f 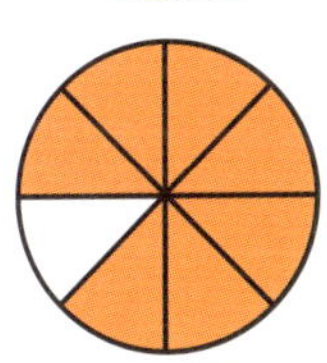

1 out of 6.
That's $\frac{1}{6}$.

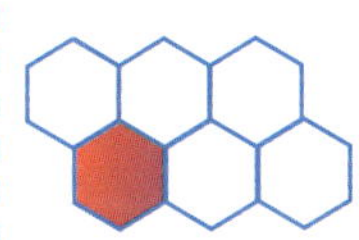

1 out of 8.
That's $\frac{1}{8}$.

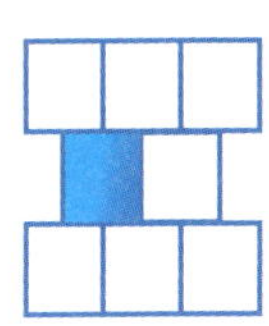

# 1:16 Number Patterns

**1** Complete each number pattern.

**a** 10, 20, 30, ___, ___, ___

**b** 4, 6, 8, ___, ___, ___

**c** 10, 15, 20, ___, ___, ___

**d** 8, 12, 16, ___, ___, ___

**2** Describe each number pattern.

**a** 4, 8, 16, 32

**b** 48, 24, 12, 6

**3** **a**

| | | | | | | | | | |
|---|---|---|---|---|---|---|---|---|---|
| 1 | 2 | 3 | 4 | 5 | 6 | 7 | 8 | 9 | 10 |
| 11 | 12 | 13 | 14 | 15 | 16 | 17 | 18 | 19 | 20 |
| 21 | 22 | 23 | 24 | 25 | 26 | 27 | 28 | 29 | 30 |

Count by fives and circle each number you count. Count by tens and draw an X on each number you count. What do you notice?

**b** Count by twos and circle each number you count. Count by fours and draw an X on each number you count. Count by eights and colour each number you count. Explain what you notice.

| | | | | | | | | | |
|---|---|---|---|---|---|---|---|---|---|
| 1 | 2 | 3 | 4 | 5 | 6 | 7 | 8 | 9 | 10 |
| 11 | 12 | 13 | 14 | 15 | 16 | 17 | 18 | 19 | 20 |
| 21 | 22 | 23 | 24 | 25 | 26 | 27 | 28 | 29 | 30 |
| 31 | 32 | 33 | 34 | 35 | 36 | 37 | 38 | 39 | 40 |

**4** Press these keys on a calculator.

4 + 5 = Keep pressing the = key and record the answers.

ICT

Calculator shows: ___ ___ ___ ___ ___ ___ ___ ___

I noticed that

# 1:17 Numbers to 10 000

CONCEPT

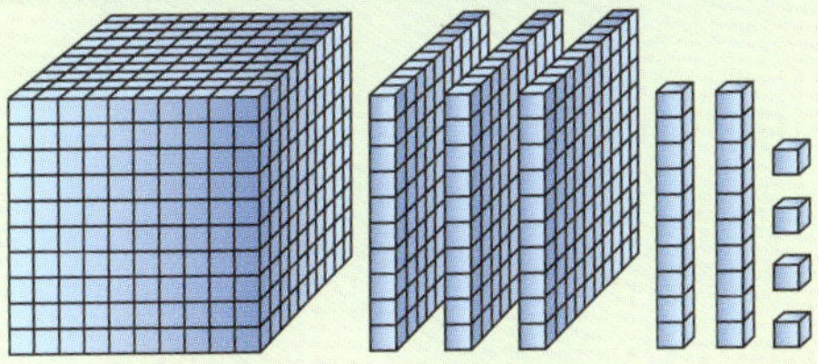

1 324

1 Thousands 3 Hundreds 2 Tens 4 Ones

one thousand three hundred and twenty-four

**1** Write the numeral, fill in the numeral expander and write the number in words.

**a**

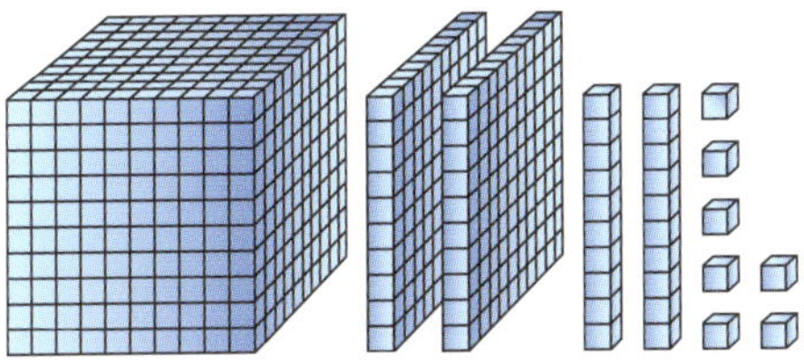

Thousands Hundreds Tens Ones

**b**

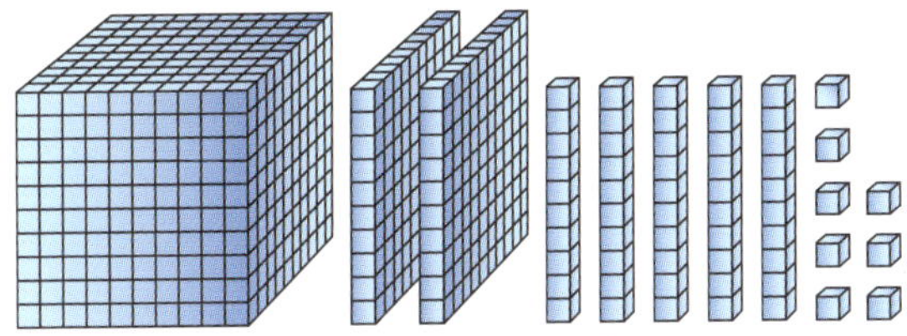

Thousands Hundreds Tens Ones

**c**

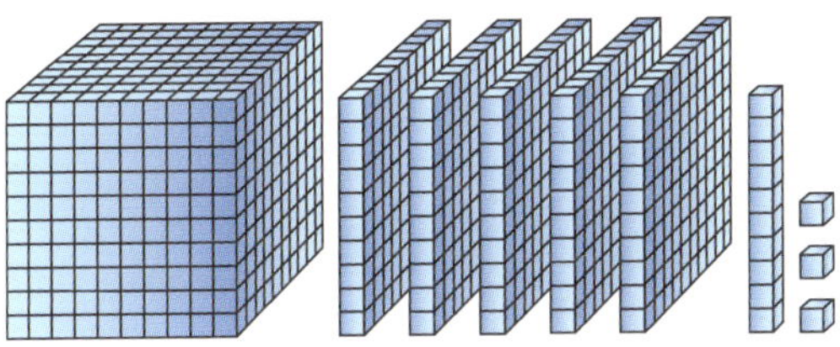

Thousands Hundreds Tens Ones

**d**

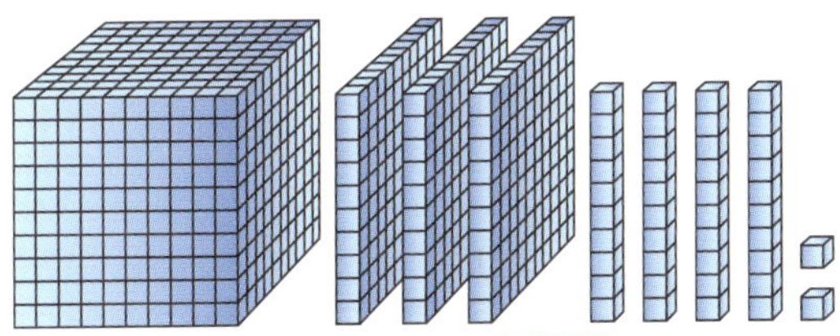

Thousands Hundreds Tens Ones

**2** How many digits are in each numeral?

**a** 786 **b** 1 462 **c** 5 967 **d** 423

**e** 1 075 **f** 36 **g** 4 300 **h** 9 678

**3** Write these numbers as numerals.

**a** four hundred and forty

**b** ten thousand

**c** one thousand and sixty

**d** one thousand and two

**e** one thousand five hundred and forty-three

**f** three thousand seven hundred and eighty-five

**g** six thousand nine hundred and fifty-nine

# 1:18 Ordering Numbers

CONCEPT

**<** means **less than**
(smaller part first)

**>** means **greater than**
(larger part first)

**1** Write these numbers in order from smallest to largest.

**a** 67, 39, 106, 94, 86, 116

**b** 109, 167, 134, 186, 153, 130

**c** 192, 373, 546, 289, 751, 430

**d** 964, 872, 581, 978, 436, 352

**2** Use the digits to write the largest possible 3-digit number.

**a** 2,6,4 **b** 1,9,7 **c** 4,0,6 **d** 3,9,5

**e** 1,8,7 **f** 0,3,5 **g** 8,6,9 **h** 2,7,5

**3** Use the digits to write the smallest possible 3-digit number.

**a** 4,2,3 **b** 5,1,0 **c** 2,6,9 **d** 7,3,8

**e** 6,0,8 **f** 9,4,1 **g** 7,5,0 **h** 6,2,8

**4** Use < or > to link these numbers.

**a** 94 ☐ 86 **b** 134 ☐ 126 **c** 335 ☐ 465

**d** 672 ☐ 756 **e** 549 ☐ 594 **f** 806 ☐ 926

**g** 964 ☐ 999 **h** 483 ☐ 438 **i** 1935 ☐ 1724

63 > 10
Sixty-three is
greater than ten.

41 < 60
Forty-one is
less than sixty.

**5** Write **true** or **false** for each number statement.

**a** 327 < 372 **b** 468 > 459 **c** 975 > 957

**d** 106 > 160 **e** 851 < 815 **f** 407 > 492

**6** Write as many number sentences as you can using the numerals below and > or <.

**a** 3, 4, 8 **b** 4, 7, 9 **c** 1, 4, 6

(Example: 34 > 8, 43 > 8, 8 < 34, 83 > 4)

ACTIVITY

# 1:19 Rounding

CONCEPT

**1** a Is 186 closer to 100 or 200? ☐ b Is 392 closer to 300 or 400? ☐

c Is 529 closer to 500 or 600? ☐ d Is 613 closer to 600 or 700? ☐

e Is 435 closer to 400 or 500? ☐ f Is 264 closer to 200 or 300? ☐

g Is 756 closer to 700 or 800? ☐ h Is 348 closer to 300 or 400? ☐

**2** Round these numbers to the nearest hundred.

| | | | |
|---|---|---|---|
| a 127 ☐ | b 386 ☐ | c 535 ☐ | d 479 ☐ |
| e 216 ☐ | f 794 ☐ | g 890 ☐ | h 607 ☐ |
| i 173 ☐ | j 456 ☐ | k 375 ☐ | l 928 ☐ |
| m 409 ☐ | n 249 ☐ | o 652 ☐ | p 580 ☐ |

**3** a Circle the numbers that round to 200.

264 163 296 220 186
217 237 125 143 205

b Circle the numbers that round to 500.

569 467 416 456 575
483 590 532 439 521

**4** a Is 346 closer to 300 or 400? ☐ b Is 946 closer to 900 or 1 000? ☐

c Is 197 closer to 100 or 200? ☐ d Is 752 closer to 700 or 800? ☐

e Is 650 closer to 600 or 700? ☐ f Is 416 closer to 400 or 500? ☐

**5** Answer **true** or **false** for each statement, rounding to the nearest 100.

a 974 rounds to 900. ☐ b 426 rounds to 500. ☐

c 396 rounds to 400. ☐ d 399 rounds to 400. ☐

## Number Hunt

ACTIVITY

- Search through newspapers, magazines and catalogues to find 4-digit numerals.
- Group together similar examples, such as crowd attendances, prices or years.
- Discuss how these different examples are written and read differently.

ISBN 978 1 4886 2184 0

# 1:20 Fractions

CONCEPT

$\frac{1}{2}$ $\frac{1}{8}$

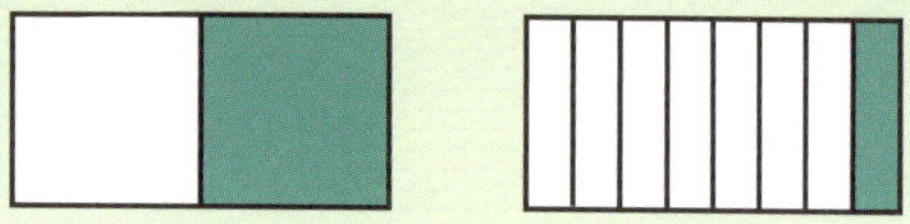

one half one eighth

$\frac{1}{8}$ is smaller than $\frac{1}{2}$ because 1 of 8 equal parts must be smaller than 1 of 2 equal parts.

**1** Colour part of each shape to match the fraction.

**a** 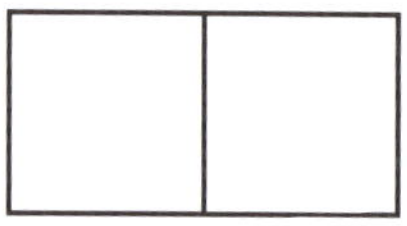 $\frac{1}{2}$ **b** 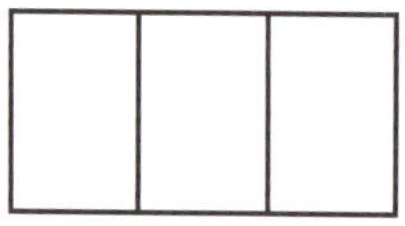 $\frac{1}{3}$ **c** 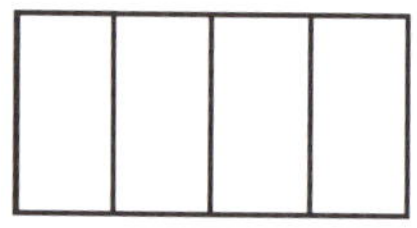 $\frac{1}{4}$

**d** 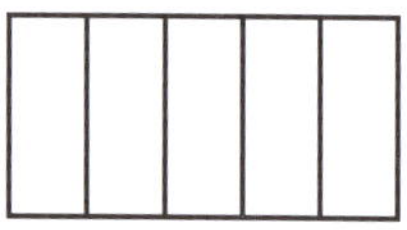 $\frac{1}{5}$ **e** 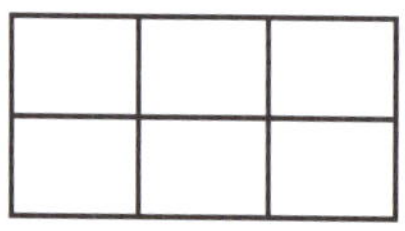 $\frac{1}{6}$ **f** 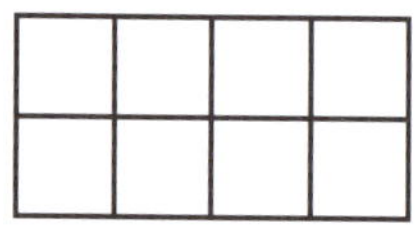 $\frac{1}{8}$

Fractions must be made up of equal parts.

**2** Circle the larger fraction.

**a** $\frac{1}{2}$ or $\frac{1}{8}$ **b** $\frac{1}{4}$ or $\frac{1}{3}$ **c** $\frac{1}{4}$ or $\frac{1}{6}$

**d** $\frac{1}{6}$ or $\frac{1}{3}$ **e** $\frac{1}{5}$ or $\frac{1}{8}$ **f** $\frac{1}{4}$ or $\frac{1}{2}$

**3** Colour part of each group to match the fraction.

**a** 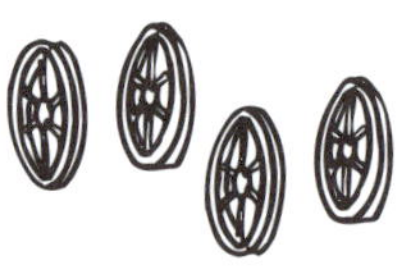 $\frac{1}{4}$ **b**  $\frac{1}{3}$ **c**  $\frac{3}{4}$ **d**  $\frac{5}{6}$

**e**  $\frac{1}{2}$ **f**  $\frac{7}{8}$ **g**  $\frac{2}{3}$ **h** 

FUN SPOT

**4** Circle the diagram that shows $\frac{3}{4}$.

**a** 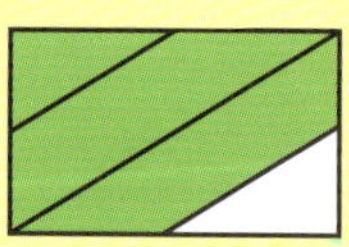  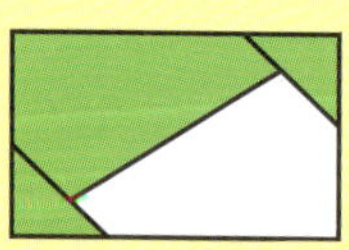 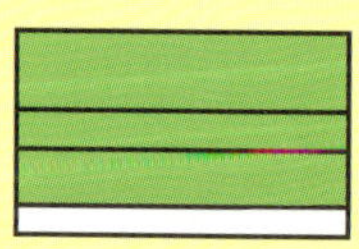

**b**  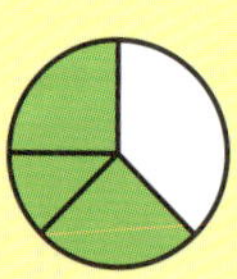  

Don't be tricked!

Explain your answer to a partner.



 • *AUSTRALIAN SIGNPOST MATHS 3* • ISBN 978 1 4886 2184 0

# 1:21 Fractions in Our World

**1** Divide each picture to match the number of parts.

**a** 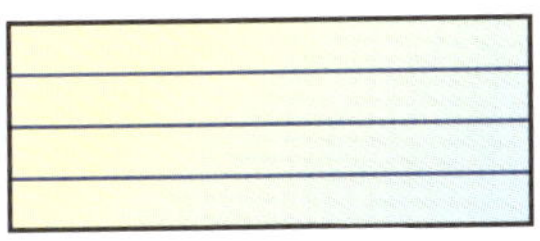
eighths

**b** 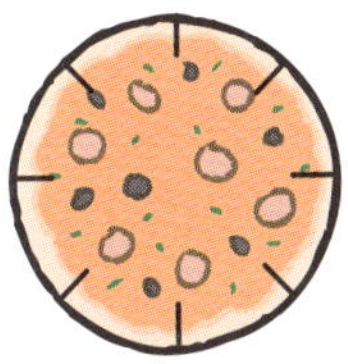
eighths

**c** 
quarters

**d** 
quarters

**2** Colour the time it would take to run the laps. One lap takes five minutes.

**a** 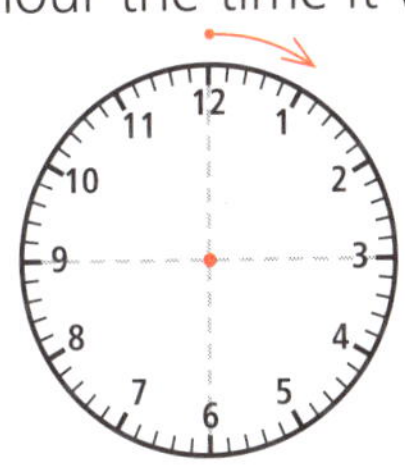
9 laps

**b** 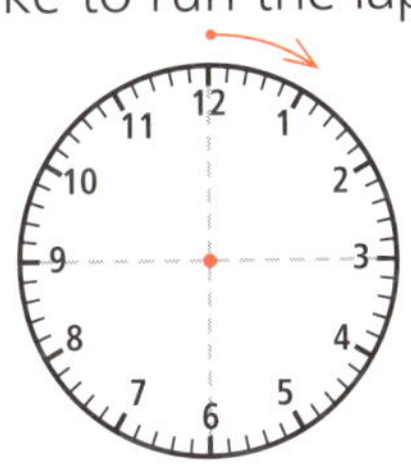
3 laps

**c** 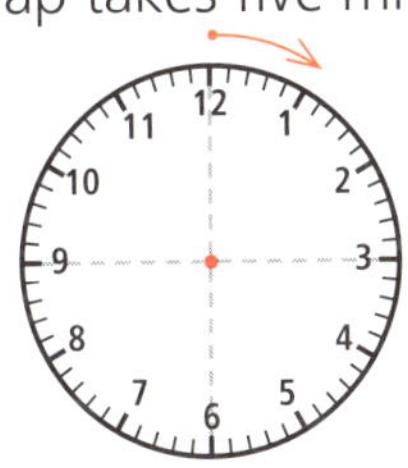
6 laps

**d** 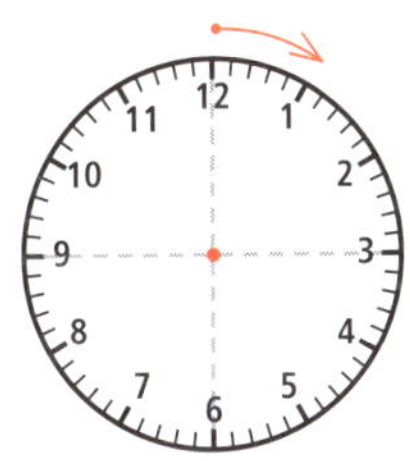
12 laps

**3** Write the missing fractions. Colour part of each line to show one half.

**a** 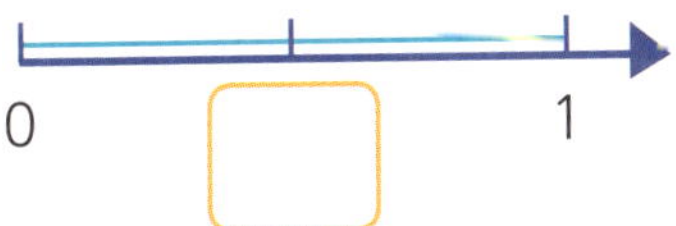


**b** 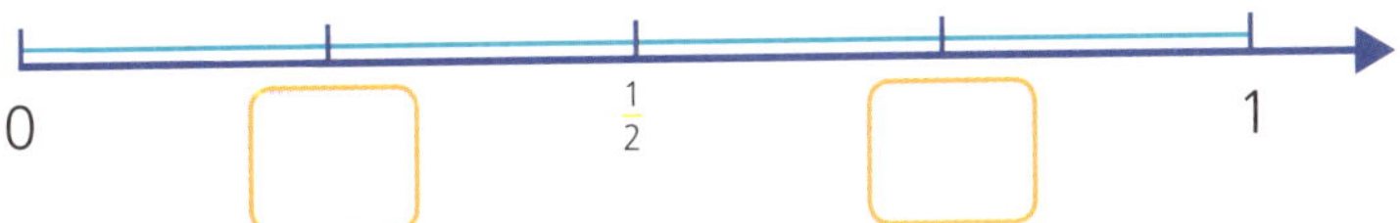


**c** 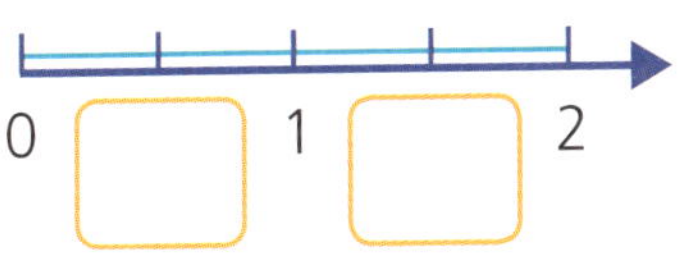


**d** 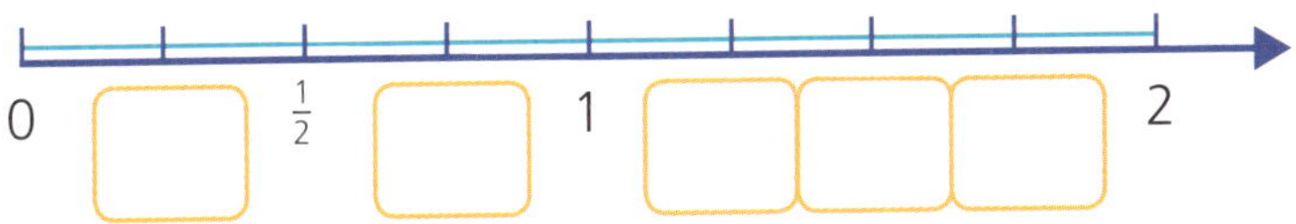


**4** Circle each group to show the fraction.

**a** $\frac{1}{4}$ 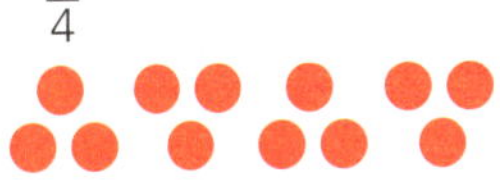

**b** $\frac{1}{8}$ 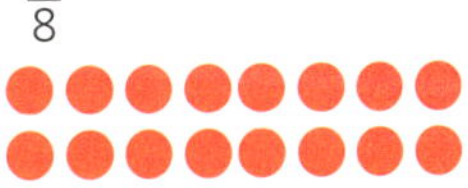

**c** $\frac{1}{3}$ 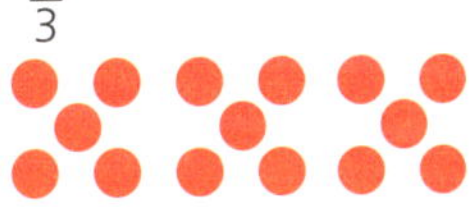

**d** $\frac{1}{2}$ 

ACTIVITY

- With a partner, write a list of situations where we use fractions.

# 1:22 Numbers to 10 000

**1** Write the number shown.

a
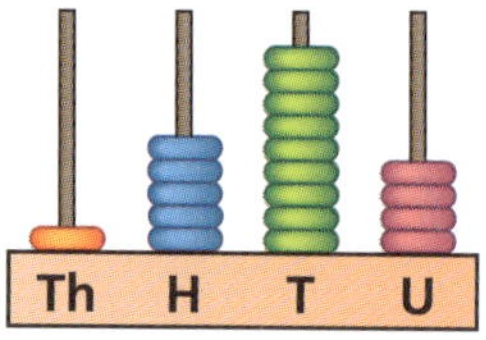


b
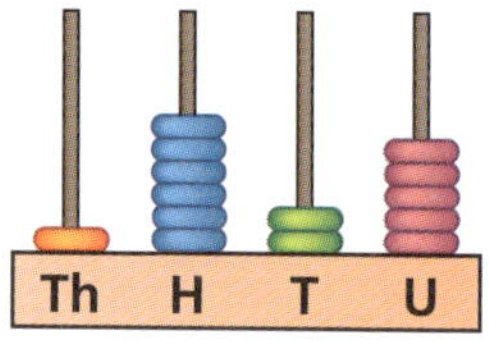

c
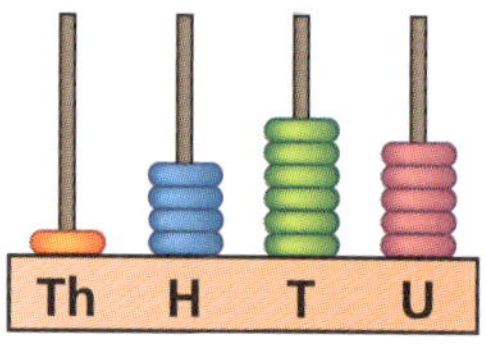


d
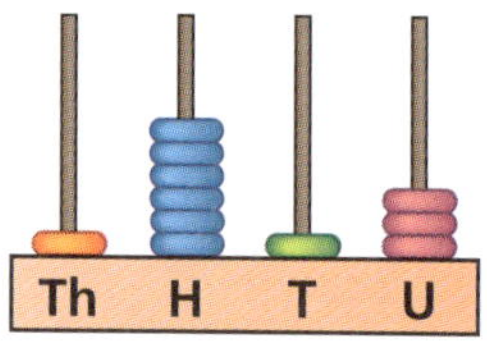


e
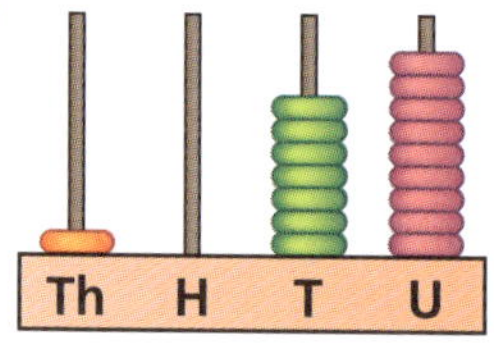


f
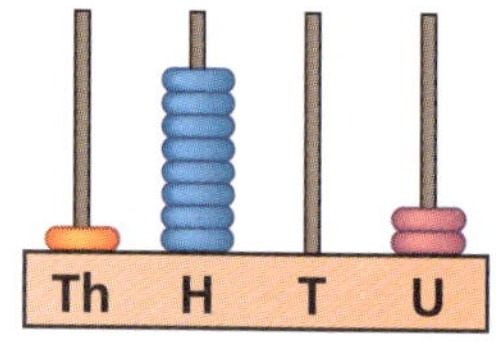


**2** Write the numeral, fill in the numeral expander and write the number in words.

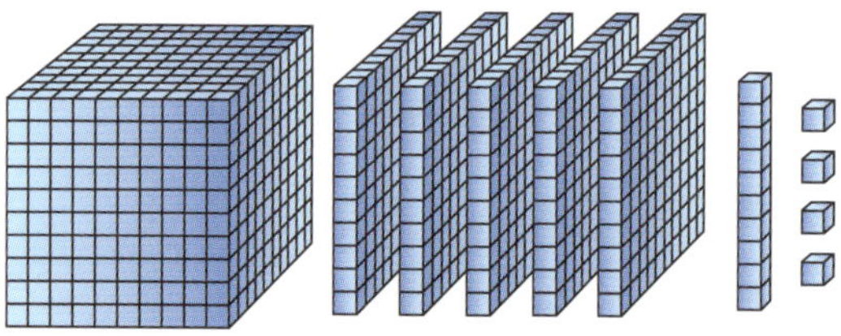

**3** How many digits are in each number?

a 147 b 21 c 1945 d 439
e 64 f 2367 g 4003 h 6

**4** Continue the number patterns.

a 430, 440, 450, ___, ___, ___
b 660, 670, 680, ___, ___, ___
c 819, 829, 839, ___, ___, ___
d 756, 766, 776, ___, ___, ___
e 420, 520, 620, ___, ___, ___
f 339, 439, 539, ___, ___, ___

**5** Write the number that follows.

a 346 b 679 c 999 d 1316
e 4876 f 3207 g 2649 h 3640
i 2199 j 5010 k 9999 l 1001



 ISBN 978 1 4886 2184 0

# 1:23 Place Value to 10 000

CONCEPT

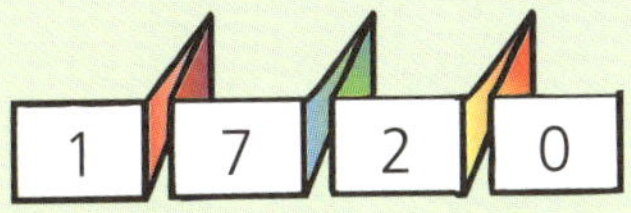


1 720 is the same as
17 hundreds and 2 tens
or 172 tens.

**1** Complete the numeral expanders.

a 1 479

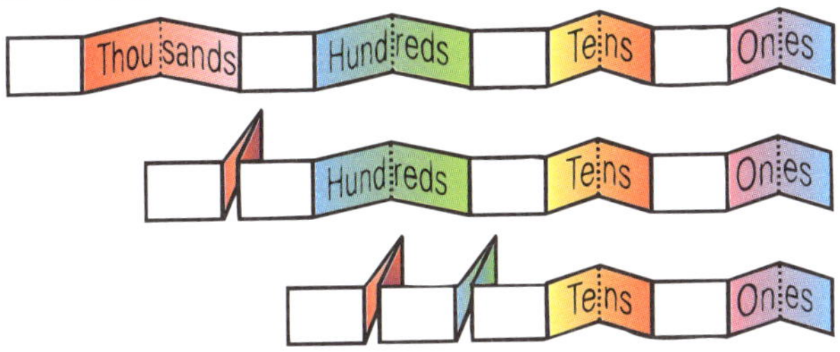


b 3 568

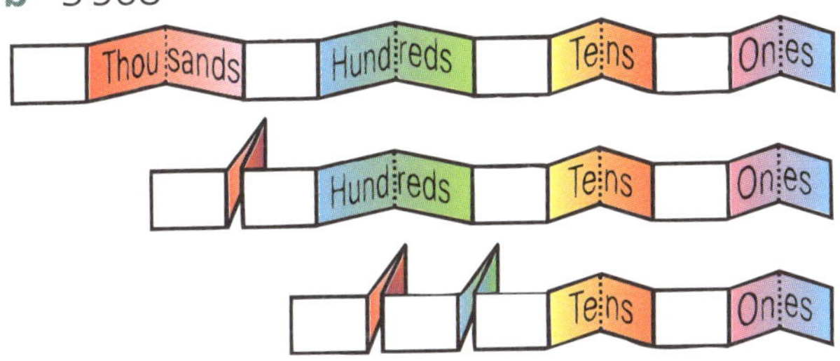


c 4 231

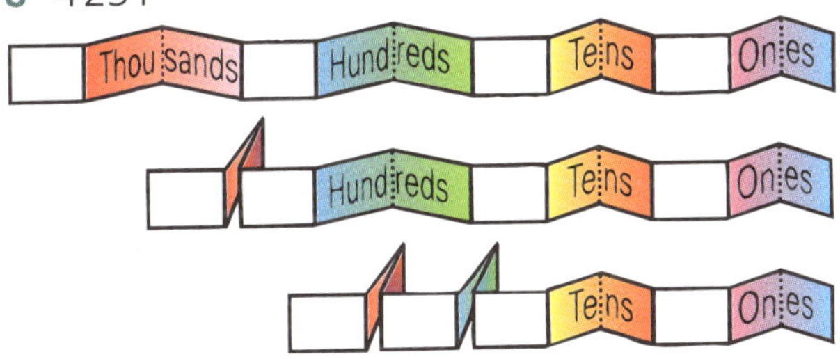


d 2 307

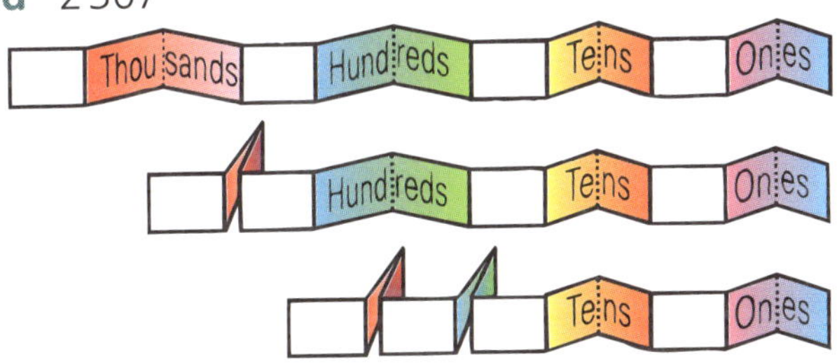


**2** Write the numeral for each number.

a five thousand seven hundred and sixty-three

b eight thousand nine hundred and fifty-six

c six thousand two hundred and seventy-five

d four thousand six hundred and forty-nine

e nine thousand five hundred and eighty-two

**3** Write each number in words.

a 6 824

b 2 319

**4** Round each number to the nearest thousand.

a 4 695 b 3 126 c 9 538

# 1:24 Number Patterns

1 Complete each number pattern.

a 4, 8, 12, ___, ___, ___

b 5, 10, 15, ___, ___, ___

c 3, 6, 9, ___, ___, ___

d 6, 12, 18, ___, ___, ___

2 Write the first five numbers in each number pattern.

a Start at 21 and count by tens. ___, ___, ___, ___, ___

b Start at 40 and count by fives. ___, ___, ___, ___, ___

c Start at 11 and count by odd numbers. ___, ___, ___, ___, ___

d Start at 22 and count by even numbers. ___, ___, ___, ___, ___

3 Count by threes and circle each number you count.
Count by sixes and draw an X on each number you count.
Count by nines and colour each number you count.
Discuss the pattern formed by the circles, Xs and colour.

| | | | | | | | | | |
|---|---|---|---|---|---|---|---|---|---|
| 1 | 2 | 3 | 4 | 5 | 6 | 7 | 8 | 9 | 10 |
| 11 | 12 | 13 | 14 | 15 | 16 | 17 | 18 | 19 | 20 |
| 21 | 22 | 23 | 24 | 25 | 26 | 27 | 28 | 29 | 30 |
| 31 | 32 | 33 | 34 | 35 | 36 | 37 | 38 | 39 | 40 |
| 41 | 42 | 43 | 44 | 45 | 46 | 47 | 48 | 49 | 50 |
| 51 | 52 | 53 | 54 | 55 | 56 | 57 | 58 | 59 | 60 |

4 Describe each number pattern.

a 25, 30, 35, 40

b 79, 69, 59, 49

- Press these keys on a calculator.

2 × × =

- Keep pressing the = key and record the answers.
- Clear the calculator and repeat the exercise, but this time press:

7 + + =

ICT



 • *AUSTRALIAN SIGNPOST MATHS 3* • ISBN 978 1 4886 2184 0

# 1:25 What's the Rule?

CONCEPT

11, 14, 17, ...

The next number would be 20.

70, 60, 50, ...

The next number would be 40.

**1** Write the next number in each pattern.

**a** 1, 3, 5, … ☐  **b** 7, 10, 13, … ☐  **c** 10, 20, 30, … ☐

**d** 20, 16, 12, … ☐  **e** 33, 32, 31, … ☐  **f** 44, 33, 22, … ☐

**g** 10, 14, 18, … ☐  **h** 75, 70, 65, … ☐  **i** 38, 46, 54, … ☐

**2** Write the rule for each pattern in Question 1.

**a** ☐  **b** ☐  **c** ☐

**d** ☐  **e** ☐  **f** ☐

**g** ☐  **h** ☐  **i** ☐

**3** Follow the rule to continue each pattern.

**a** Add 6. 21, ☐, ☐, ☐  **b** Subtract 1. 41, ☐, ☐, ☐

**c** Subtract 2. 74, ☐, ☐, ☐  **d** Add 8. 34, ☐, ☐, ☐

**e** Add 20. 23, ☐, ☐, ☐  **f** Subtract 4. 40, ☐, ☐, ☐

**g** Multiply by 2. 3, ☐, ☐, ☐  **h** Add 15. 5, ☐, ☐, ☐

**i** Subtract 10. 63, ☐, ☐, ☐  **j** Multiply by 2. 1, ☐, ☐, ☐

**4** Write the pattern for the number of sides, then write the rule.

**a**  , …

Pattern: 3, ☐, ☐, ☐, …  Rule: ☐

**b** 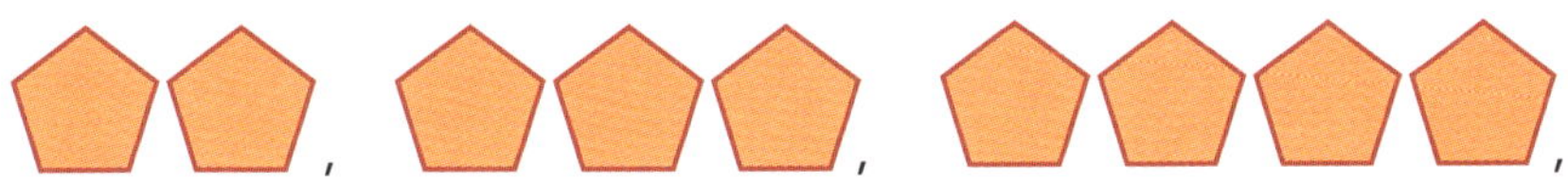 , …

Pattern: 10, ☐, ☐, ☐, …  Rule: ☐

# 1:26 Expanded Notation

CONCEPT

2000 + 100 + 30 + 7

**1** Write the numeral.

a 3000 + 700 + 60 + 9 ☐

b 4000 + 600 + 30 + 7 ☐

c 2000 + 500 + 80 + 1 ☐

d 5000 + 400 + 70 + 6 ☐

e 3000 + 900 + 50 + 8 ☐

f 6000 + 800 + 40 + 3 ☐

**2** Write the number in expanded notation.

a 4247 ☐

b 3169 ☐

c 2675 ☐

d 5192 ☐

**3** Write these numbers in the columns to show place value.

a 4378

b 6249

c 3821

d 5873

e 2634

f 1986

| | Thousands | Hundreds | Tens | Units |
|---|---|---|---|---|
| a | | | | |
| b | | | | |
| c | | | | |
| d | | | | |
| e | | | | |
| f | | | | |

**4** Write the value of each coloured digit.

a 2**9**16 ☐

b **4**037 ☐

c 627**9** ☐

d 56**8**4 ☐

e **3**690 ☐

f 869**5** ☐

g 2 **6**70 ☐

h 58**3**6 ☐

**5** Write the number shown by each abacus.

a
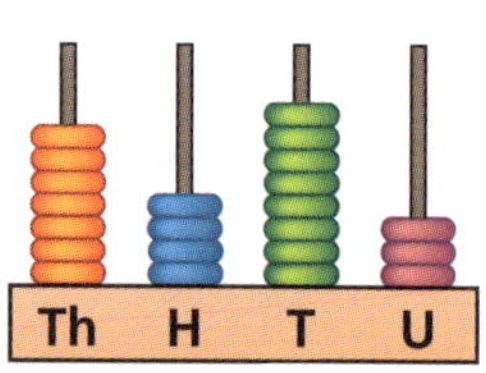

☐

b
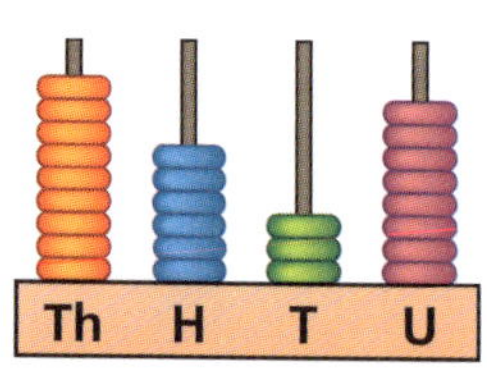

☐

c
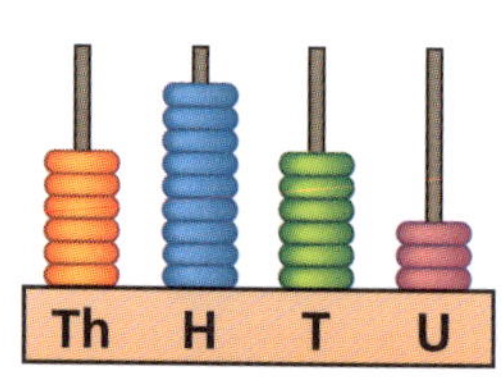

☐

d
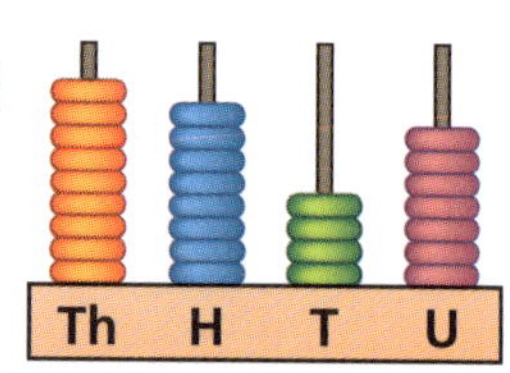

☐



 • *AUSTRALIAN SIGNPOST MATHS 3* • ISBN 978 1 4886 2184 0

# 1:27 Numbers to 10 000

**1** Write the numeral.

a six hundred and forty-two ☐

b nine hundred and seventy-eight ☐

c one thousand five hundred and twenty-three ☐

d five thousand nine hundred and fifty-one ☐

**2** Complete each number pattern.

a 635, 645, 655, ☐, ☐

b 749, 759, 769, ☐, ☐

**3** Circle the smaller number.

a 396 or 936

b 10 000 or 9999

c 1347 or 1374

d 2581 or 2851

e 4135 or 1435

f 6923 or 6723

**4** Circle the largest number in each group.

a 327, 237, 723

b 569, 669, 965

c 631, 613, 603

d 1246, 1264, 1216

e 2375, 2735, 2753

f 4194, 4094, 4049

**5** Write **true** or **false**.

a 634 < 643 ☐

b 906 > 960 ☐

c 1425 > 1597 ☐

d 2178 < 2378 ☐

e 3502 > 3250 ☐

f 4819 < 4981 ☐

g 8010 < 7321 ☐

h 7777 > 5898 ☐

## Cross-number Puzzles

FUN SPOT

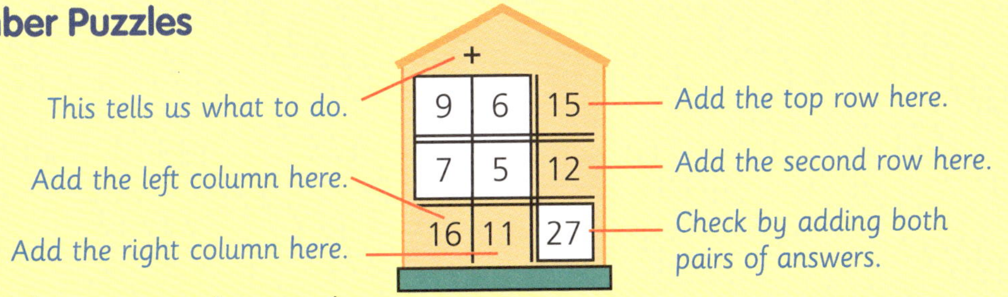


**6** Complete the cross-number puzzles.

a +

| | | |
|---|---|---|
| 6 | 4 | |
| 6 | 7 | |
| | | |

b +

| | | |
|---|---|---|
| 8 | 6 | |
| 4 | 5 | |
| | | |

c +

| | | |
|---|---|---|
| 3 | 9 | |
| 4 | 8 | |
| | | |

d +

| | | |
|---|---|---|
| 9 | 7 | |
| 8 | 5 | |
| | | |

# 1:28 Number Patterns

Describe each number pattern.

**1** a 50 100 150 200 ____________

b 100 200 300 400 ____________

c 47 37 27 17 ____________

d 166 176 186 196 ____________

**2** a 1 $1\frac{1}{2}$ 2 $2\frac{1}{2}$ ____________

b $\frac{1}{4}$ $\frac{2}{4}$ $\frac{3}{4}$ $\frac{4}{4}$ ____________

c $\frac{1}{3}$ $\frac{2}{3}$ $\frac{3}{3}$ $\frac{4}{3}$ ____________

**3** a 2, 4, ___, 8, ___, ___, 14

b 75, 70, 65, ___, 55, ___, 45

c 4, 8, 12, ___, 20, ___, 28

d 120, 130, 140, ___, ___, 170, ___

e 8, ___, 24, 32, 40, ___, 56

f 3, ___, 9, 12, 15, ___, 21

## Magic Circles

ICT

Seki Kowa (1642–1708), a Japanese mathematician, developed a 'magic' circle. In each circle, the sums along each long diameter and around each circumference are all equal.

**4** Look at the first magic circle and complete the others.

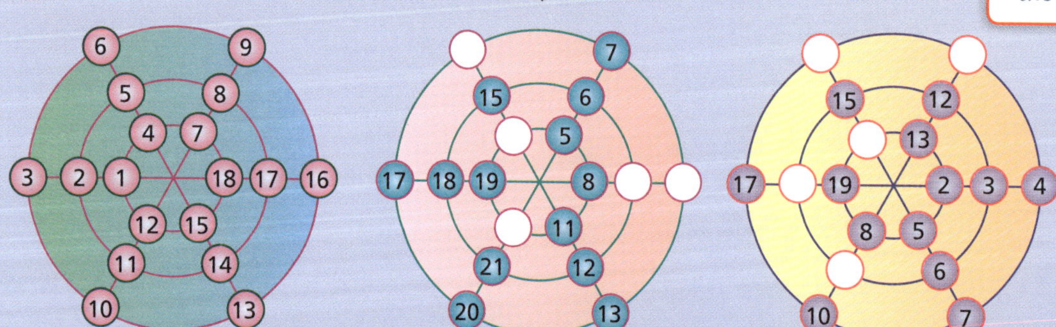


Totals = 57 Totals = 81 Totals = 63

# 1:29 Number Patterns

CONCEPT

How do we make the second number from the first number?

| First Number | 1 | 2 | 3 | 4 | 5 | 6 |
|---|---|---|---|---|---|---|
| Second Number | 16 | 17 | 18 | 19 | 20 | 21 |

**1** Complete each table and write the rule that links the second number to the first.

a

| First Number | 1 | 2 | 3 | 4 | 5 | 6 |
|---|---|---|---|---|---|---|
| Second Number | 8 | 9 | 10 | 11 | | |

Rule: ____________

b

| First Number | 10 | 20 | 30 | 40 | 50 | 60 |
|---|---|---|---|---|---|---|
| Second Number | 4 | 14 | 24 | 34 | | |

Rule: ____________

c

| First Number | 2 | 4 | 6 | 8 | 10 | 12 |
|---|---|---|---|---|---|---|
| Second Number | 1 | 2 | 3 | 4 | | |

Rule: ____________

d

| First Number | 29 | 30 | 31 | 32 | 33 | 34 |
|---|---|---|---|---|---|---|
| Second Number | 31 | 32 | 33 | 34 | | |

Rule: ____________

e

| First Number | 1 | 2 | 3 | 4 | 5 | 6 |
|---|---|---|---|---|---|---|
| Second Number | 2 | 4 | 6 | 8 | | |

Rule: ____________

f

| First Number | 20 | 18 | 16 | 14 | 12 | 10 |
|---|---|---|---|---|---|---|
| Second Number | 10 | 9 | 8 | 7 | | |

Rule: ____________

g

| First Number | 5 | 6 | 7 | 8 | 9 | 10 |
|---|---|---|---|---|---|---|
| Second Number | 25 | 30 | 35 | 40 | | |

Rule: ____________

h

| First Number | 30 | 25 | 20 | 15 | 10 | 5 |
|---|---|---|---|---|---|---|
| Second Number | 6 | 5 | 4 | 3 | | |

Rule: ____________

i

| First Number | 3 | 4 | 5 | 6 | 7 | 8 |
|---|---|---|---|---|---|---|
| Second Number | 30 | 40 | 50 | 60 | | |

Rule: ____________

j

| First Number | 100 | 80 | 60 | 40 | 20 | 10 |
|---|---|---|---|---|---|---|
| Second Number | 10 | 8 | 6 | 4 | | |

Rule: ____________

ACTIVITY

**2** Make your own number patterns using addition and multiplication.

a Addition Rule: ____________ +7

| First Number | 21 | 22 | 23 | 24 | 25 | 26 |
|---|---|---|---|---|---|---|
| Second Number | | | | | | |

b Multiplication Rule: ____________ ×7

| First Number | 1 | 2 | 3 | 4 | 5 | 6 |
|---|---|---|---|---|---|---|
| Second Number | | | | | | |

# 1:30 Numbers to 10 000

CONCEPT

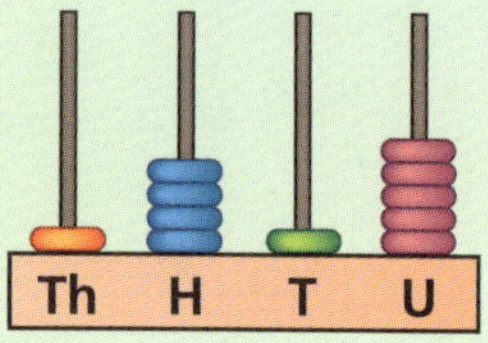


1415

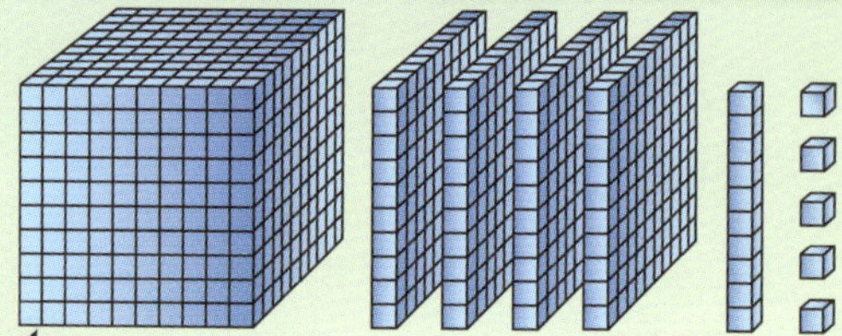

1415

**1** Write the number shown.

a

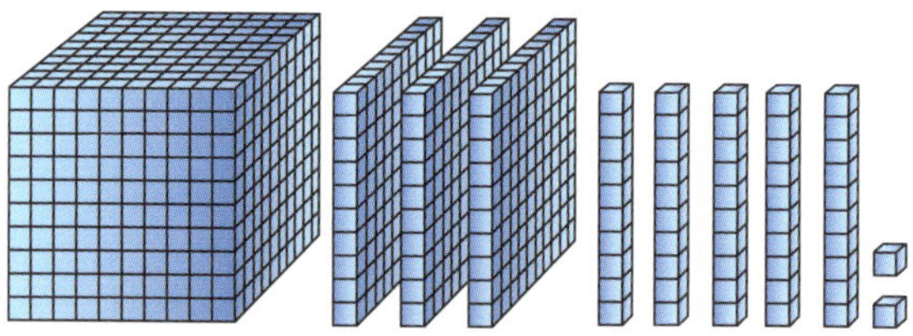

b

c

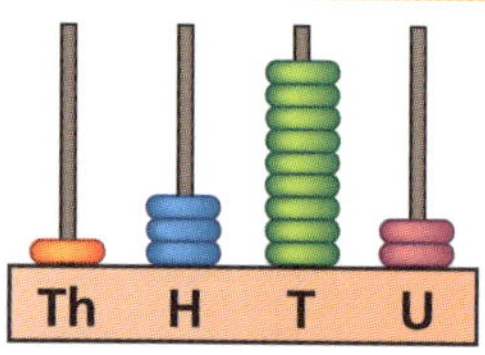


d

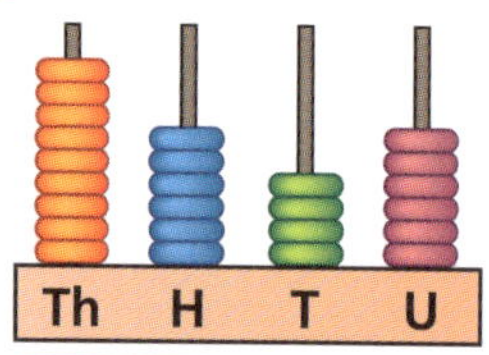


e

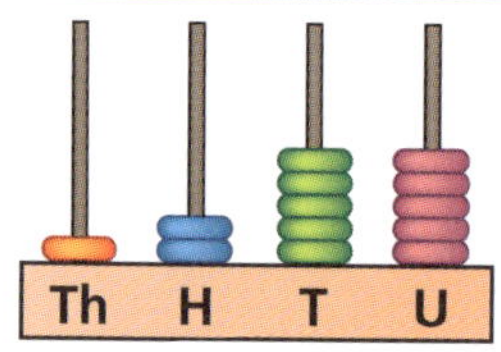


f

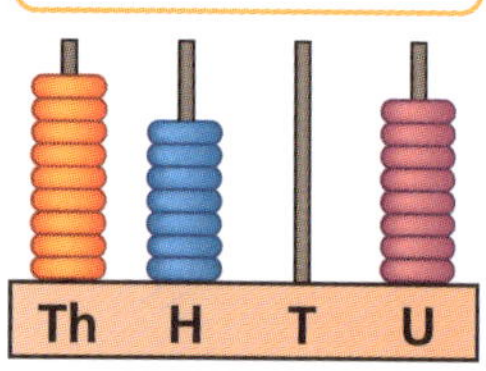


g

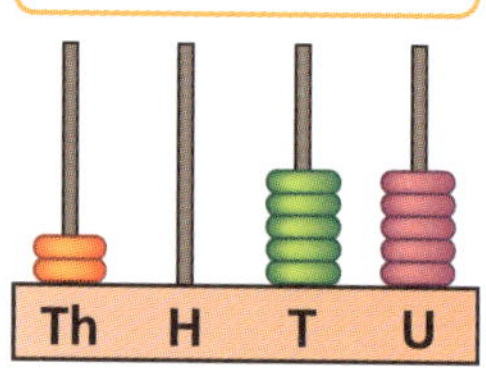


h

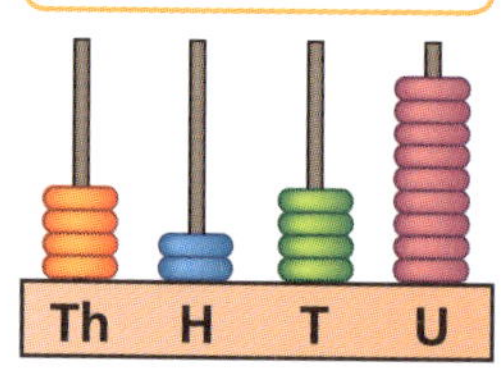


i

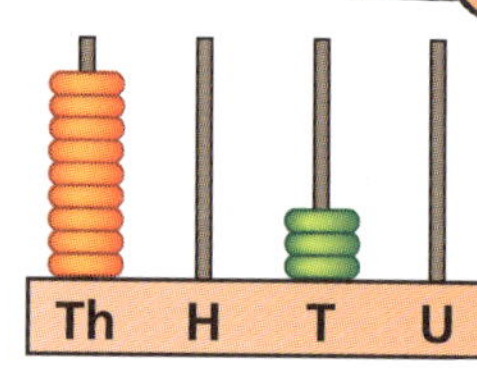


j

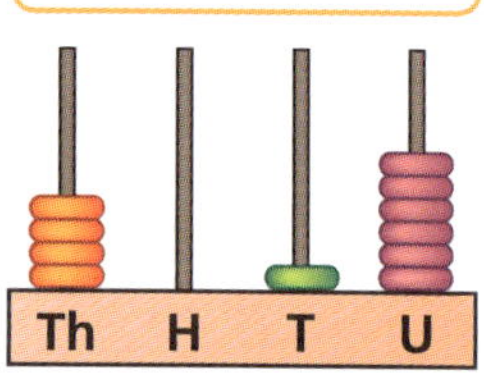


k

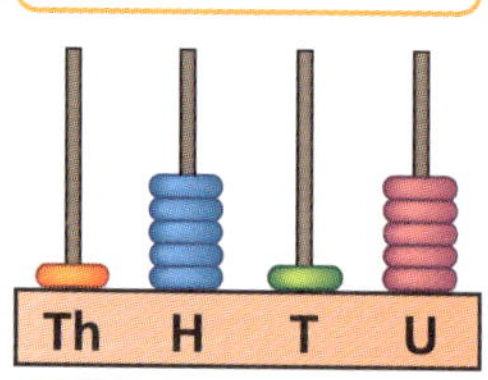


l

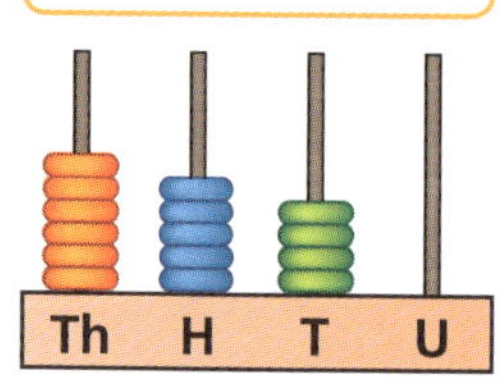


m

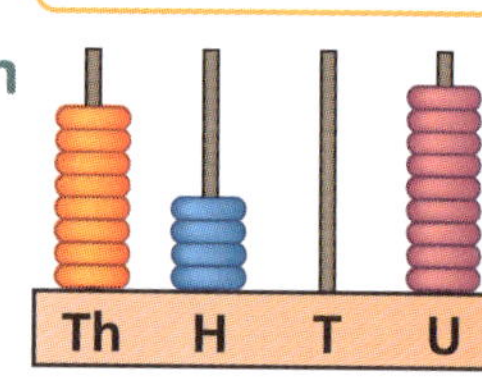


**2** Write the value of the 6 in each number.

a 3659 b 6125 c 4968

## Wipe Out a Digit

ACTIVITY

- Enter a 4-digit number into a calculator.
- Your partner selects any digit to be **wiped out** – that is, changed to zero.
- Try to wipe out that digit by entering only **one** operation into the calculator. Did it work?
- Take turns with your partner. Score 1 point for each successful wipe-out.



 • *AUSTRALIAN SIGNPOST MATHS 3* • ISBN 978 1 4886 2184 0

# 1:31 Expanded Notation

**1** Write the numeral.

a 1000 + 600 + 30 + 9 ☐
b 1000 + 700 + 40 + 1 ☐
c 2000 + 900 + 50 + 6 ☐
d 4000 + 500 + 80 + 3 ☐
e 3000 + 800 + 10 ☐
f 5000 + 300 + 7 ☐

**2** Write the number in expanded notation.

a 2356 ☐
b 1975 ☐
c 6283 ☐
d 4806 ☐

**3** Write these numbers in the columns to show place value.

a 1963
b 2479
c 2307
d 3582
e 5190
f 4063

| | Thousands | Hundreds | Tens | Units |
|---|---|---|---|---|
| a | | | | |
| b | | | | |
| c | | | | |
| d | | | | |
| e | | | | |
| f | | | | |

**4** Write the value of each coloured digit.

a 3**2**16 ☐
b 549**7** ☐
c 65**4**3 ☐
d **1**947 ☐
e 82**6**9 ☐
f **4**920 ☐
g 2**7**58 ☐
h 930**6** ☐

**5** Write the number shown by each abacus.

a
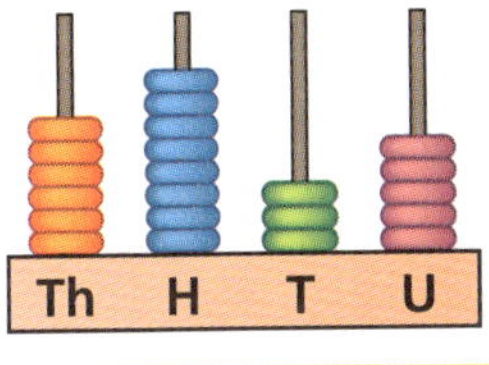


b
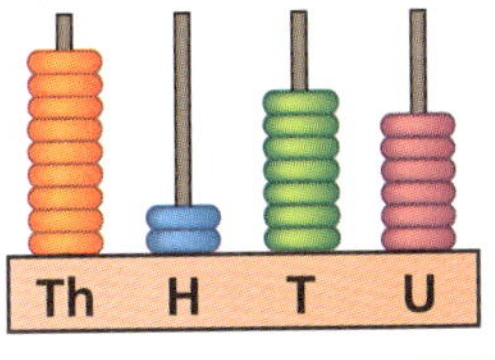


c
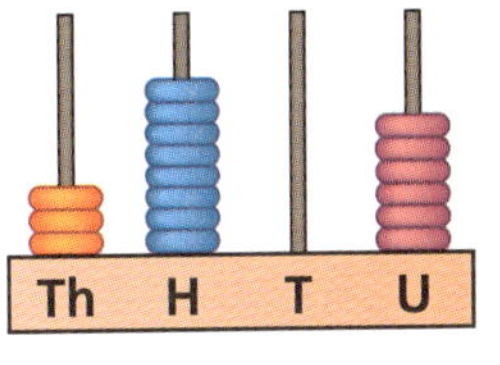


d
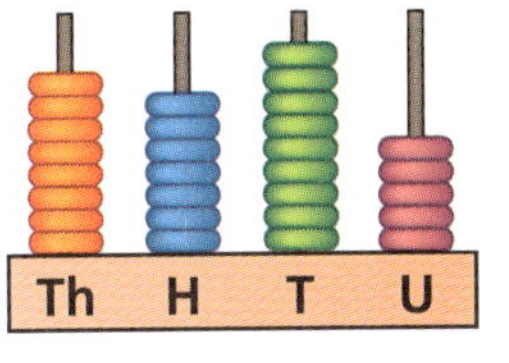


**6** Write the number before:

a 967 ☐
b 1236 ☐
c 2191 ☐
d 4630 ☐
e 5729 ☐
f 6310 ☐
g 3400 ☐
h 7020 ☐

## Dicey Numbers

FUN SPOT

- Take turns to roll three dice.
- Record the 3-digit number made by the dice.
- State the numbers before and after that number.

 • *AUSTRALIAN SIGNPOST MATHS 3* • ISBN 978 1 4886 2184 0

# 1:32 Numbers to 10 000

CONCEPT

3725

| Thousands | Hundreds | Tens | Ones |
|---|---|---|---|
| 3 | 7 | 2 | 5 |

**1** Write the number shown by each abacus.

a 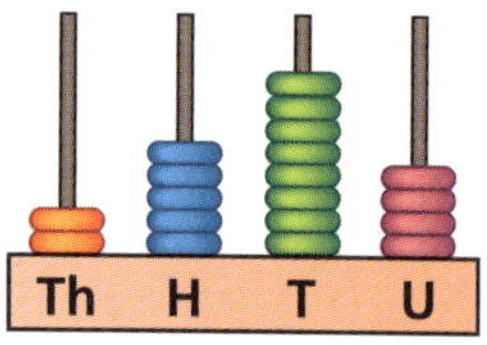


b 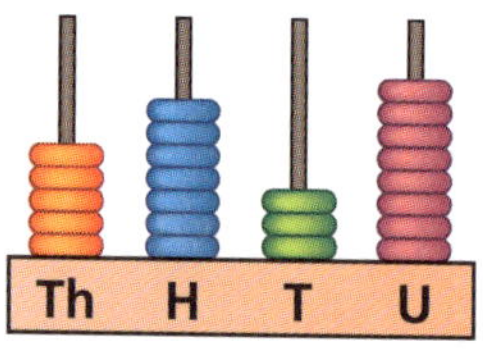


c 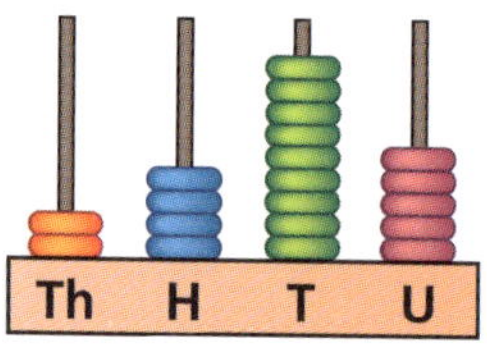


d 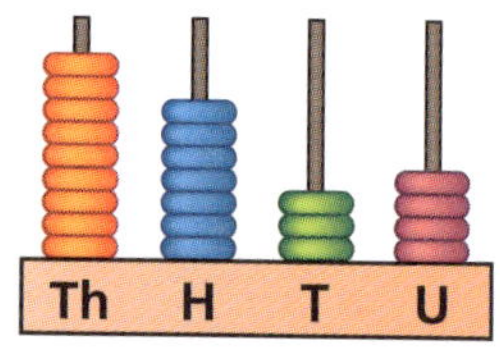


**2** Read the numbers aloud and write them in the place-value chart.

**a** two thousand five hundred and sixty
**b** one thousand two hundred and thirty-four
**c** six thousand four hundred and fifty-two
**d** five thousand six hundred and twenty-nine
**e** seven thousand two hundred and four

| Thousands | Hundreds | Tens | Ones |
|---|---|---|---|
| | | | |
| | | | |
| | | | |
| | | | |
| | | | |

**3** Write the value of each coloured digit.

**a** 3**6**71   **b** 401**2**   **c** 73**9**4
**d** **2**075   **e** 66**5**0   **f** 141**3**

## Digit Game

FUN SPOT

- Each player uses 4 numeral cards to make a number on a place-value chart.
- Players get one numeral card at a time. Before they can see the next card, they must put the card they have onto the place-value chart.
- The player with the largest number at the end wins.

| Thousands | Hundreds | Tens | Ones |
|---|---|---|---|
| 6 | 4 | 3 | 0 |

# 1:33 Place Value to 10 000

CONCEPT

1 Complete the numeral expanders.

a 1764

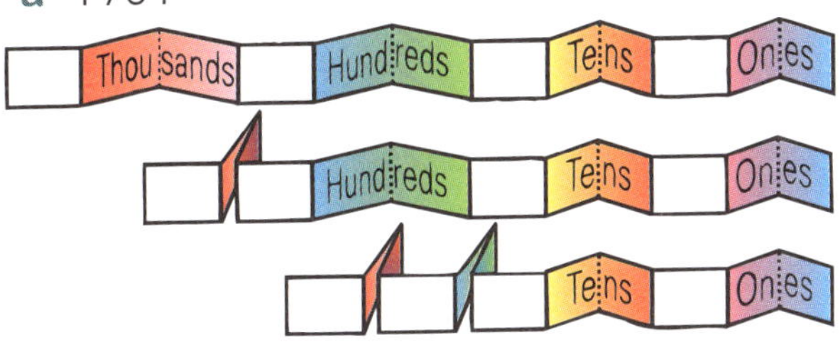


b 3795

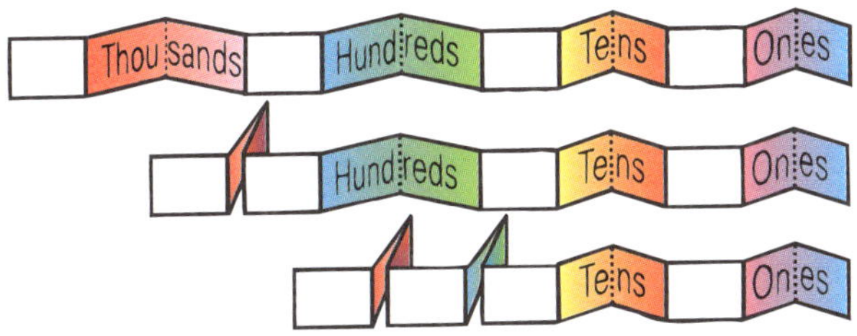


c 4013

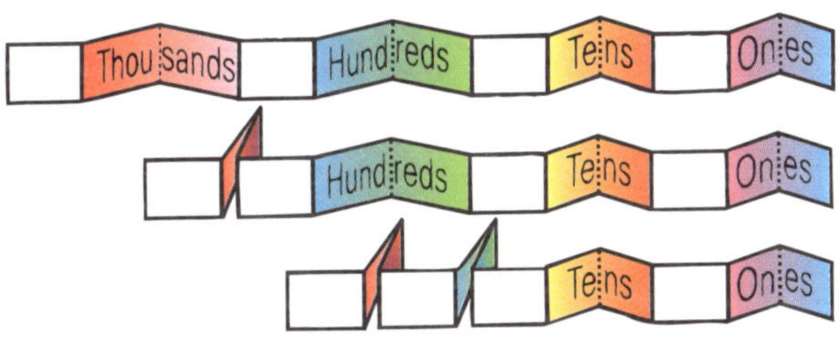


d 9658

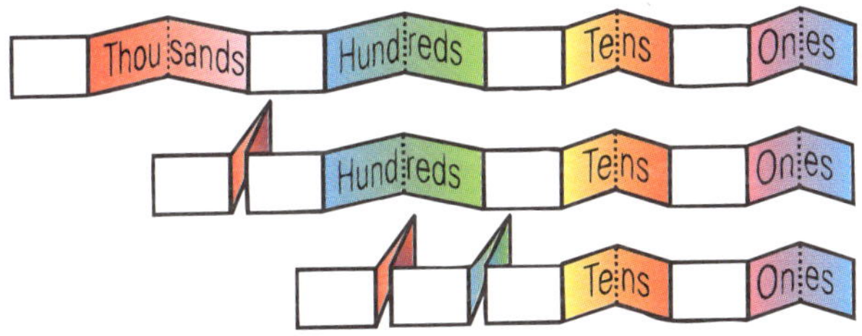


2 How many hundreds could be taken from each number?

a 1576 ☐ b 1253 ☐ c 2301 ☐

d 6192 ☐ e 7471 ☐ f 9368 ☐

3 How many tens could be taken from each number?

a 1742 ☐ b 1624 ☐

c 4906 ☐ d 7264 ☐

4 Write the numeral.

a one thousand two hundred and ten ☐

b five thousand six hundred and forty-one ☐

c six thousand three hundred and one ☐

d two thousand five hundred and sixteen ☐

5 a 3000 = ☐ tens
= ☐ hundreds

b 700 = ☐ tens
= ☐ hundreds

 • *AUSTRALIAN SIGNPOST MATHS 3* • ISBN 978 1 4886 2184 0

# 1:34 Numbers to 10 000

**1** Write the number that matches the clue.

- **a** 5 more than 1965
- **b** 10 more than 2381
- **c** 1 less than 3490
- **d** 1 more than 5369
- **e** 10 less than 1735
- **f** 5 less than 4965
- **g** 1 after 6399
- **h** 1 before 8000
- **i** 100 more than 7306

**2** Use < or > to link these numbers.

- **a** 86 ☐ 92
- **b** 235 ☐ 230
- **c** 409 ☐ 406
- **d** 763 ☐ 769
- **e** 1511 ☐ 1501
- **f** 3637 ☐ 3256
- **g** 5206 ☐ 5260
- **h** 6431 ☐ 6314
- **i** 9725 ☐ 9752
- **j** 4311 ☐ 4131

**3** Write **true** or **false**.

- **a** 2176 < 2376
- **b** 5319 > 5391
- **c** 3094 < 3049
- **d** 4613 > 4361
- **e** 7968 > 7698
- **f** 6158 < 6185

**4** Write the value of each coloured digit.

- **a** 39**1**7
- **b** 2**5**83
- **c** 403**9**
- **d** **6**264
- **e** 17**1**2
- **f** 5**9**65
- **g** 73**5**6
- **h** **7**605
- **i** 26**8**1

**5** Write the number in words.

- **a** 3691
- **b** 5216
- **c** 6509

## Make a Number

ACTIVITY

- One player rolls a 10-sided dice four times and calls out the digit shown each time.
- Players record each digit as it is called out.
- Players use the four digits to make the largest possible number.
- Play the game again, this time making the smallest possible number.

# 1:35 Making Number Patterns

Make these number patterns.

1. Keep adding 9. Notice how the tens and ones digits change.

19 + 9 ☐ + 9 ☐ + 9 ☐ + 9 ☐ + 9 ☐ + 9 ☐ + 9 ☐ + 9 ☐

The pattern …

2. Keep subtracting 8. Notice how the tens and ones digits change.

98 – 8 ☐ – 8 ☐ – 8 ☐ – 8 ☐ – 8 ☐ – 8 ☐ – 8 ☐ – 8 ☐

85 – 8 ☐ – 8 ☐ – 8 ☐ – 8 ☐ – 8 ☐ – 8 ☐ – 8 ☐ – 8 ☐

The pattern …

3. Keep adding 50. Notice how the two digits change.

0 + 50 ☐ + 50 ☐ + 50 ☐ + 50 ☐ + 50 ☐ + 50 ☐ + 50 ☐

13 + 50 ☐ + 50 ☐ + 50 ☐ + 50 ☐ + 50 ☐ + 50 ☐ + 50 ☐

The pattern …

4. Keep adding 25. Notice how the digits change.

0 + 25 ☐ + 25 ☐ + 25 ☐ + 25 ☐ + 25 ☐ + 25 ☐ + 25 ☐

22 + 25 ☐ + 25 ☐ + 25 ☐ + 25 ☐ + 25 ☐ + 25 ☐ + 25 ☐

The pattern …

ACTIVITY

5. I had 100 bottles. 37 bottles were smashed, but I was given another 34 bottles. How many bottles do I have now? ☐

Explain how you would solve this problem.

 • *AUSTRALIAN SIGNPOST MATHS 3* • ISBN 978 1 4886 2184 0

# 1:36 Rounding

1

| | Number | Nearest Ten |
|---|---|---|
| a | 6 | |
| b | 23 | |
| c | 57 | |
| d | 78 | |

| | Number | Nearest Hundred |
|---|---|---|
| e | 107 | |
| f | 283 | |
| g | 440 | |
| h | 563 | |

2 Estimate the answer by rounding each number to the nearest ten.

a 33 + 62 ☐ b 29 + 51 ☐ c 61 + 32 ☐ d 17 + 82 ☐

e 48 + 27 ☐ f 60 + 13 ☐ g 47 + 38 ☐ h 81 + 22 ☐

i 58 + 31 ☐ j 74 + 19 ☐ k 93 – 66 ☐ l 86 – 29 ☐

m 39 – 28 ☐ n 48 – 18 ☐ o 97 – 68 ☐ p 75 – 37 ☐

3 Use a calculator to find the exact answers in Question 2.

4

| | Number sentence | Joshua's estimate | Actual answer | Difference |
|---|---|---|---|---|
| a | 76 + 39 | 120 | | |
| b | 52 + 71 | 120 | | |
| c | 38 + 33 | 70 | | |
| d | 66 + 28 | 100 | | |

5

| | Number sentence | Professor's estimate | Actual answer | Difference |
|---|---|---|---|---|
| a | 56 – 18 | 40 | | |
| b | 59 – 22 | 40 | | |
| c | 82 – 46 | 30 | | |
| d | 78 – 27 | 50 | | |

6

| | Number sentence | Amber's estimate | Actual answer | Difference |
|---|---|---|---|---|
| a | 163 + 196 | 400 | | |
| b | 237 + 348 | 500 | | |
| c | 631 – 420 | 200 | | |
| d | 788 – 159 | 600 | | |

# 2:01 Australian Money

1 Write the value of each coin.

2 Match each banknote to the correct label.

$20
twenty dollars

$50
fifty dollars

$100
one hundred dollars

$10
ten dollars

$5
five dollars

3 Write the value of the banknotes in order from smallest to largest.

# 2:02 Addition and Subtraction Facts

**1** Complete the additions and subtractions. Discuss any patterns you can see.

**a**
4 + 1 =
1 + 4 =
5 – 4 =
5 – 1 =

**b**
4 + 2 =
2 + 4 =
6 – 4 =
6 – 2 =

**c**
1 + 9 =
9 + 1 =
10 – 1 =
10 – 9 =

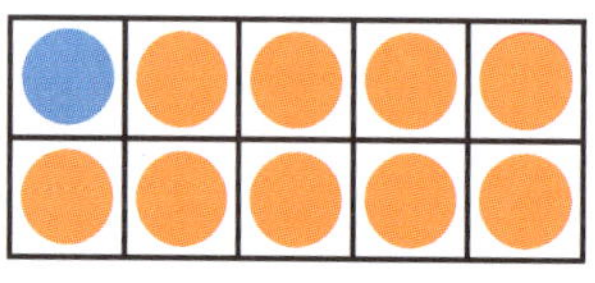

**d**
2 + 8 =
8 + 2 =
10 – 2 =
10 – 8 =

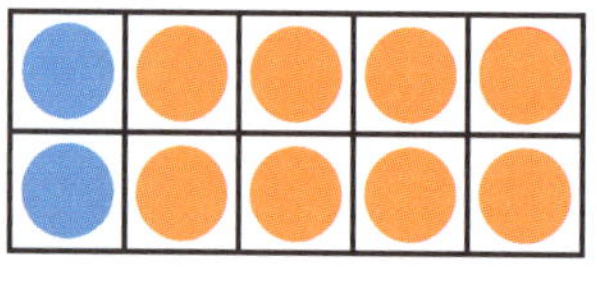

**2** Complete this addition grid. Discuss any patterns you can see.
Colour matching pairs, such as 6 + 7 = 7 + 6 = 13.

| + | 0 | 1 | 2 | 3 | 4 | 5 | 6 | 7 | 8 | 9 | 10 |
|---|---|---|---|---|---|---|---|---|---|---|---|
| 0 | | | | | | | | | | | |
| 1 | | | | | | | | | | | |
| 2 | | | | | | 7 | | | | | |
| 3 | | | | | | | | | | | |
| 4 | | | | | | | | | | | |
| 5 | | | 7 | | | | | | | | |
| 6 | | | | | | | | 13 | | | |
| 7 | | | | | | | 13 | | | | |
| 8 | | | | | | | | | | | |
| 9 | | | | | | | | | | | |
| 10 | | | | | | | | | | | |

**3** Use the number line to complete the number facts. Look for patterns.

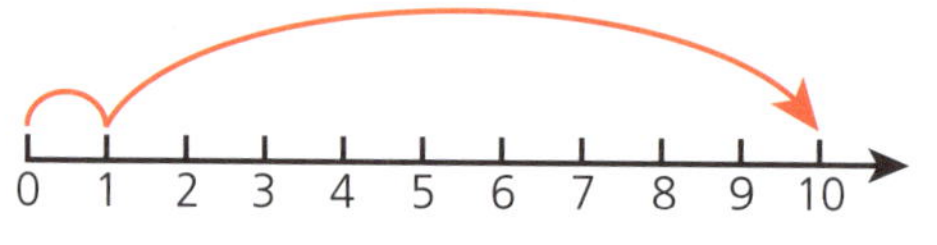


**a**
1 + 9 =
2 + 8 =
3 + 7 =
4 + 6 =
5 + 5 =
6 + 4 =
7 + 3 =
8 + 2 =
9 + 1 =

**b**
10 – 9 =
10 – 8 =
10 – 7 =
10 – 6 =
10 – 5 =
10 – 4 =
10 – 3 =
10 – 2 =
10 – 1 =

**INVESTIGATION**

- Create your own organised lists of addition and subtraction facts up to 10 + 10.
- Discuss any patterns you can find.

# 2:03 Strategies

## Addition and Subtraction Strategies

CONCEPT

- Counting on

9 + 3

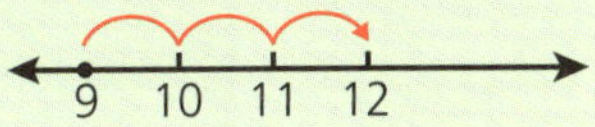


9 – 7

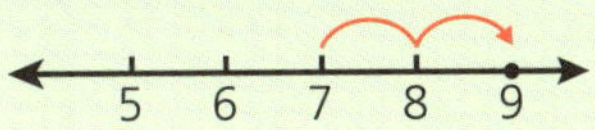


- Counting back

6 – 4

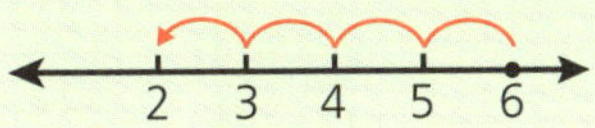


- Bridging to 10

8 + 6 = 8 + 2 + 4 = 10 + 4

- Near doubles

6 + 7 = 6 + 6 + 1

- Patterns

2 + 3 = 5 so 5 – 3 = 2

**1** Use the best strategy to complete each number fact.

| | | | |
|---|---|---|---|
| **a** 5 + 8 = ☐ | **b** 9 + 5 = ☐ | **c** 7 + 8 = ☐ | **d** 9 – 4 = ☐ |
| **e** 6 – 4 = ☐ | **f** 10 – 5 = ☐ | **g** 9 – 8 = ☐ | **h** 3 + 7 = ☐ |
| **i** 3 + 4 = ☐ | **j** 10 – 8 = ☐ | **k** 8 + 9 = ☐ | **l** 7 + 6 = ☐ |
| **m** 10 – 6 = ☐ | **n** 7 + 5 = ☐ | | |

**2** **a** 9 + 4 = ☐ so 13 – 9 = ☐ **b** 9 + 7 = ☐ so 16 – 7 = ☐

**c** 6 + 2 = ☐ so 8 – 6 = ☐

**3** Write **true** or **false**.

**a** 7 + 3 = 3 + 7 ☐ **b** 10 + 7 = 7 – 10 ☐

## Playing Card Addition

FUN SPOT

- You will need a partner and a pack of cards. Remove the picture cards.
- Deal half of the pack to each player. All the cards are face down.
- Each player turns over their top card.
- The first player to correctly call out the total of the two cards keeps both cards. Number cards are worth their number and Aces are worth 1.
- Keep turning over cards until the pack is finished. The player with the most cards is the winner!

© PEARSON AUSTRALIA 2018 • *AUSTRALIAN SIGNPOST MATHS 3* • ISBN 978 1 4886 2184 0

# 2:04 Number Facts, × 2

**1** Use the number models to complete the multiplications.

a $1 \times 2 =$ ☐

b $2 \times 2 =$ ☐

c $3 \times 2 =$ ☐

d $4 \times 2 =$ ☐

e $5 \times 2 =$ ☐

f $6 \times 2 =$ ☐

g $7 \times 2 =$ ☐

h $8 \times 2 =$ ☐

i $9 \times 2 =$ ☐

j $10 \times 2 =$ ☐

k $1 \times 4 =$ ☐

l $2 \times 4 =$ ☐

m $3 \times 4 =$ ☐

n $4 \times 4 =$ ☐

o $5 \times 4 =$ ☐

**2** a $8 \times 2 =$ ☐, $3 \times 2 =$ ☐, $8 \times 2$ plus $3 \times 2 = 11 \times 2 =$ ☐

b $7 \times 2 =$ ☐, $5 \times 2 =$ ☐, $7 \times 2$ plus $5 \times 2 = 12 \times 2 =$ ☐

c $4 \times 2 =$ ☐, $9 \times 2 =$ ☐, $4 \times 2$ plus $9 \times 2 = 13 \times 2 =$ ☐

d $10 \times 2 =$ ☐, $6 \times 2 =$ ☐, $10 \times 2$ minus $6 \times 2 = 4 \times 2 =$ ☐

**3** a $6 \times 2$ plus $2 = 7 \times 2 =$ ☐

b $7 \times 2$ plus $2 = 8 \times 2 =$ ☐

c $10 \times 2$ plus $2 = 11 \times 2 =$ ☐

d $8 \times 2$ plus $7 \times 2 = 15 \times 2 =$ ☐

**4** a

| | 3 | 10 | 2 | 1 | 5 | 4 | 7 | 9 | 8 | 6 |
|---|---|---|---|---|---|---|---|---|---|---|
| × 2 | 6 | 20 | | | | | | | | |

b

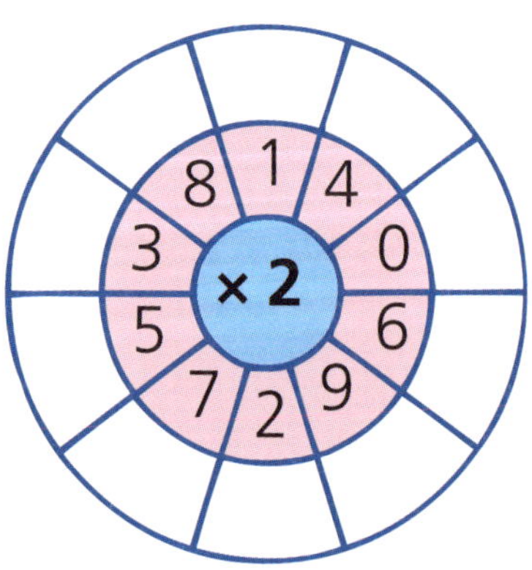


c

| | 6 | 4 | | | | | | | | |
|---|---|---|---|---|---|---|---|---|---|---|
| × 2 | 12 | 8 | 16 | 4 | 10 | 20 | 2 | 14 | 18 | 6 |

# 2:05 Addition by Looking for 10s

**1**

| a | b | c | d | e | f |
|---|---|---|---|---|---|
| 3 | 3 | 1 | 4 | 6 | 7 |
| 4 | 5 | 3 | 2 | 9 | 3 |
| + 6 | + 5 | + 7 | + 8 | + 1 | + 5 |

**2**

a 6 cows, 5 sheep, 4 pigs.
How many altogether? [ = ]

b 5 cars, 9 trucks, 5 bikes.
What is the total? [ = ]

c 3 pencils, 7 pens, 4 crayons.
How many altogether? [ = ]

d 2 rulers, 8 books, 6 pencils.
What is the total? [ = ]

**3**

| a | b | c | d | e |
|---|---|---|---|---|
| 2 | 4 | 7 | 5 | 9 |
| 6 | 1 | 9 | 7 | 7 |
| 8 | 6 | 3 | 5 | 1 |
| + 3 | + 2 | + 1 | + 2 | + 3 |

**4**

| a | b | c | d | e |
|---|---|---|---|---|
| 3 | 2 | 5 | 8 | 9 |
| 2 | 7 | 4 | 3 | 0 |
| 6 | 8 | 1 | 5 | 7 |
| + 3 | + 3 | + 3 | + 7 | + 5 |

**5**

| a | b | c | d | e |
|---|---|---|---|---|
| 6 | 9 | 4 | 6 | 6 |
| 3 | 5 | 9 | 4 | 3 |
| 2 | 6 | 3 | 1 | 5 |
| 8 | 4 | 8 | 8 | 5 |
| 4 | 1 | 2 | 1 | 7 |
| + 4 | + 3 | + 7 | + 4 | + 4 |

## Three For Three

- Roll 3 dice and add the numbers. Colour the answer on the grid.
- How many dice rolls do you need to colour 3 squares in a line?
- Play with a partner. The first player to colour 3 squares in a line is the winner.

| | | |
|---|---|---|
| 15 | 10 | 13 |
| 9 | 16 | 11 |
| 12 | 8 | 14 |

| | | |
|---|---|---|
| 15 | 10 | 13 |
| 9 | 16 | 11 |
| 12 | 8 | 14 |

# 2:06 Patterns in Adding and Subtracting

INVESTIGATION

1 When we add 9, what happens to the tens and ones digits?

- The tens digit: ______
- The ones digit: ______

So, to add 9 we can add ____ and then subtract ____.

Example: 37 + 9 = 37 + 10 – 1 = 47 – 1 = 46

| |
|---|
| 9 |
| 9 + 9 = 18 |
| 18 + 9 = 27 |
| 27 + 9 = 36 |
| 36 + 9 = 45 |
| 45 + 9 = 54 |

2 When we subtract 9, what happens to the tens and ones digits?

- The tens digit: ______
- The ones digit: ______

So, to subtract 9 we can subtract ____ and then add ____.

Example: 42 – 9 = 42 – 10 + 1 = 32 + 1 = 33

3 Race a partner to complete these. Use the strategy above. Use a calculator to check your answers.

a 7 + 9 = ____ b 16 + 9 = ____ c 28 + 9 = ____ d 42 + 9 = ____

e 17 – 9 = ____ f 34 – 9 = ____ g 99, 90, 81, ____, ____, ____, ____

4 Discuss and explore strategies for adding or subtracting 8. Race a partner to complete these.

a 6 + 8 = ____ b 19 + 8 = ____ c 53 + 8 = ____ d 75 + 8 = ____

e 23 – 8 = ____ f 67 – 8 = ____ g 32, 40, 48, ____, ____, ____, ____

CONCEPT

- When we add 2 to an even number, we jump to the next even number.
- When we subtract 2 from an even number, we jump to the previous even number.

5 Race a partner to complete these. Use the strategy above.

a 24 + 2 = ____ b 68 + 2 = ____ c 82 + 2 = ____ d 32 + 2 = ____

e 66 – 2 = ____ f 40 – 2 = ____ g 54 – 2 = ____ h 86 – 2 = ____

INVESTIGATION

Explore other patterns you can use to add and subtract other 1-digit numbers. Discuss them with a partner.



 • *AUSTRALIAN SIGNPOST MATHS 3* • ISBN 978 1 4886 2184 0

# 2:07 Patterns Involving Subtraction

**1** Use the hundred chart to find these numbers.

**a** 5 less than 32 ☐  **b** 5 less than 42 ☐

**c** 8 less than 65 ☐  **d** 8 less than 55 ☐

**e** 6 less than 82 ☐  **f** 6 less than 92 ☐

| | | | | | | | | | |
|---|---|---|---|---|---|---|---|---|---|
| 1 | 2 | 3 | 4 | 5 | 6 | 7 | 8 | 9 | 10 |
| 11 | 12 | 13 | 14 | 15 | 16 | 17 | 18 | 19 | 20 |
| 21 | 22 | 23 | 24 | 25 | 26 | 27 | 28 | 29 | 30 |
| 31 | 32 | 33 | 34 | 35 | 36 | 37 | 38 | 39 | 40 |
| 41 | 42 | 43 | 44 | 45 | 46 | 47 | 48 | 49 | 50 |
| 51 | 52 | 53 | 54 | 55 | 56 | 57 | 58 | 59 | 60 |
| 61 | 62 | 63 | 64 | 65 | 66 | 67 | 68 | 69 | 70 |
| 71 | 72 | 73 | 74 | 75 | 76 | 77 | 78 | 79 | 80 |
| 81 | 82 | 83 | 84 | 85 | 86 | 87 | 88 | 89 | 90 |
| 91 | 92 | 93 | 94 | 95 | 96 | 97 | 98 | 99 | 100 |

**2** **a** Colour the numbers on the chart that are 8 less than a number ending with 0.

**b** Circle the numbers on the chart that are 7 less than a number ending with 5.

**3** **a**

| | 13 | 23 | 33 | 43 | 53 | 63 | 73 | 83 | 93 | 103 | 113 |
|---|---|---|---|---|---|---|---|---|---|---|---|
| – 4 | | | | | | | | | | | |
| – 6 | | | | | | | | | | | |

**b**

| | 12 | 22 | 32 | 42 | 52 | 62 | 72 | 82 | 92 | 102 | 112 |
|---|---|---|---|---|---|---|---|---|---|---|---|
| – 5 | | | | | | | | | | | |
| – 9 | | | | | | | | | | | |

**4** Write a number sentence to find each answer.

**a** Heather has 23 pins. How many more does she need to have 28?

☐ ☐ ☐ = ☐ (with – in the second box)

**b** Jack needs 55 cents. He has 40 cents. How much more does he need?

☐ – ☐ = ☐

**c** 61 nails were lost. Luke found 56. How many are still missing?

☐ – ☐ = ☐

**5** Complete the patterns.

**a**
11 – 4 = 7
21 – 4 = 17
31 – 4 = ☐
☐ – 4 = 37
51 – ☐ = 47

**b**
14 – 8 = 6
24 – 8 = 16
34 – 8 = ☐
☐ – 8 = 36
54 – ☐ = 46

**c**
15 – 6 = 9
25 – 6 = 19
35 – 6 = ☐
45 – ☐ = 39
☐ – 6 = 49

**6** Write as many number sentences as you can that link these numbers.

INVESTIGATION

**a** 3, 4 and 7  **b** 4, 7 and 11  **c** 7, 8 and 15  **d** 6, 14 and 20

# 2:08 Number Facts, × 5, × 10

**1** Use the number models to complete the multiplications.

a 1 × 5 = ☐
b 2 × 5 = ☐
c 3 × 5 = ☐
d 4 × 5 = ☐
e 5 × 5 = ☐
f 6 × 5 = ☐
g 7 × 5 = ☐
h 8 × 5 = ☐
i 9 × 5 = ☐
j 10 × 5 = ☐

k 1 × 10 = ☐
l 2 × 10 = ☐
m 3 × 10 = ☐
n 4 × 10 = ☐
o 5 × 10 = ☐

**2** Use the place-value blocks to complete the multiplications.

a
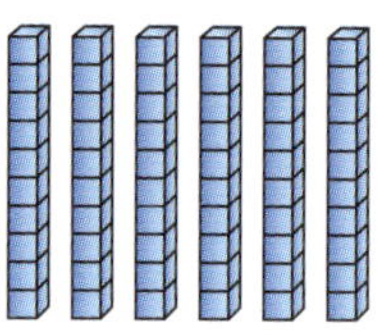
6 groups of 10

$$\begin{array}{r} 10 \\ \times \quad 6 \\ \hline \square \end{array}$$

b
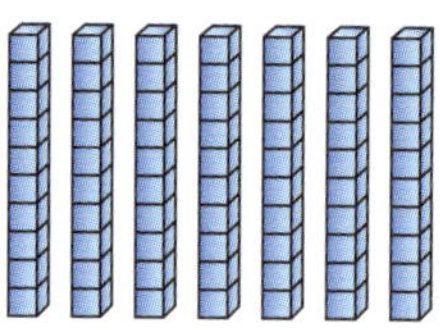
7 groups of 10

$$\begin{array}{r} 10 \\ \times \quad 7 \\ \hline \square \end{array}$$

c
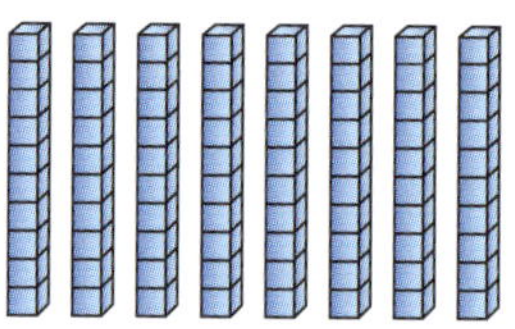
8 groups of 10

$$\begin{array}{r} 10 \\ \times \quad 8 \\ \hline \square \end{array}$$

d
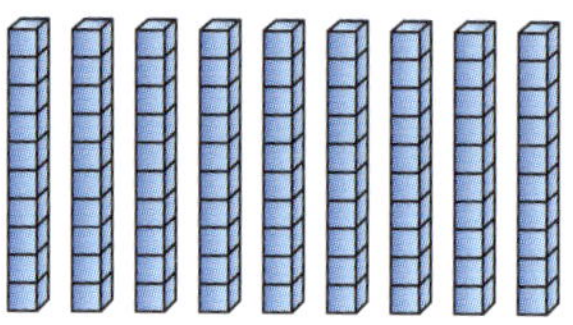
9 groups of 10

$$\begin{array}{r} 10 \\ \times \quad 9 \\ \hline \square \end{array}$$

e
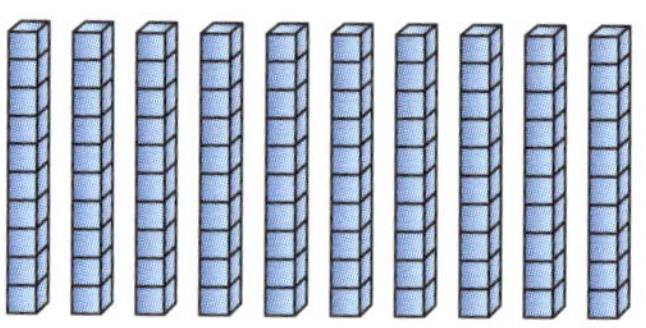
10 groups of 10

$$\begin{array}{r} 10 \\ \times \quad 10 \\ \hline \square \end{array}$$

**3** Try to complete these tables from memory.

a

| | 3 | 10 | 2 | 1 | 5 | 4 | 7 | 9 | 8 | 6 |
|---|---|---|---|---|---|---|---|---|---|---|
| × 10 | | | | | | | | | | |

b

| | 3 | 10 | 2 | 1 | 5 | 4 | 7 | 9 | 8 | 6 |
|---|---|---|---|---|---|---|---|---|---|---|
| × 5 | | | | | | | | | | |

c Is 3 × 5 half of 3 × 10? ☐

d Is 5 × 5 half of 10 × 5? ☐

e Is 9 × 5 half of 9 × 10? ☐



 • *AUSTRALIAN SIGNPOST MATHS 3* • ISBN 978 1 4886 2184 0

# Number Facts, ×1, ×0

**1** Use the pictures to complete the multiplications.

**a**

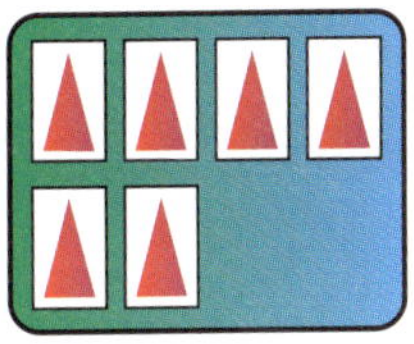

6 groups of 1

$$\begin{array}{r} 1 \\ \times\ 6 \\ \hline \end{array}$$

**b**

7 groups of 1

$$\begin{array}{r} 1 \\ \times\ 7 \\ \hline \end{array}$$

**c**

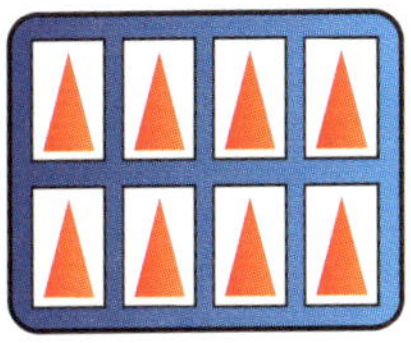

8 groups of 1

$$\begin{array}{r} 1 \\ \times\ 8 \\ \hline \end{array}$$

**d**

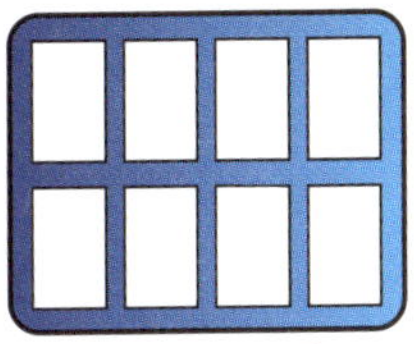

8 groups of 0

$$\begin{array}{r} 0 \\ \times\ 8 \\ \hline \end{array}$$

**e**

7 groups of 0

$$\begin{array}{r} 0 \\ \times\ 7 \\ \hline \end{array}$$

**f**

6 groups of 0

$$\begin{array}{r} 0 \\ \times\ 6 \\ \hline \end{array}$$

- Any number multiplied by 1 always gives that number.
- Any number multiplied by 0 always gives zero.

CONCEPT

**2**
**a** 2 × 1 = ☐ **b** 10 × 2 = ☐ **c** 7 × 2 = ☐ **d** 2 × 5 = ☐
**e** 2 × 2 = ☐ **f** 4 × 10 = ☐ **g** 6 × 5 = ☐ **h** 6 × 2 = ☐
**i** 8 × 5 = ☐ **j** 7 × 10 = ☐ **k** 3 × 2 = ☐ **l** 4 × 5 = ☐
**m** 9 × 2 = ☐ **n** 10 × 5 = ☐ **o** 9 × 1 = ☐ **p** 8 × 0 = ☐

**3**
**a** 5 × 2 = ☐ **b** 9 × 10 = ☐ **c** 7 × 5 = ☐ **d** 7 × 1 = ☐
**e** 5 × 5 = ☐ **f** 6 × 10 = ☐ **g** 4 × 2 = ☐ **h** 9 × 5 = ☐
**i** 8 × 2 = ☐ **j** 8 × 10 = ☐ **k** 3 × 5 = ☐ **l** 4 × 0 = ☐

**4** Try to complete these tables from memory.

**a**

| | 3 | 10 | 2 | 1 | 5 | 4 | 7 | 9 | 8 | 6 |
|---|---|---|---|---|---|---|---|---|---|---|
| ×1 | | | | | | | | | | |

**b**

| | 3 | 10 | 2 | 1 | 5 | 4 | 7 | 9 | 8 | 6 |
|---|---|---|---|---|---|---|---|---|---|---|
| ×0 | | | | | | | | | | |

# 2:10 Mental Strategies

CONCEPT

Different types of questions need different strategies.

| | | |
|---|---|---|
| **A** | Using patterns | 5 + 4 = 9 so 500 + 400 = 900 |
| **B** | Changing the order | 16 + 8 + 4 = 16 + 4 + 8 = 20 + 8 = 28 |
| **C** | Bridging to 10s | 45 + 8 = 45 + 5 + 3 = 50 + 3 = 53 |
| **D** | Compensation | 63 + 19 = 63 + 20 – 1 = 83 – 1 = 82 |
| **E** | Split | 23 + 35 = 20 + 30 + 3 + 5 = 50 + 8 = 58 |
| **F** | Jump | 58 + 23 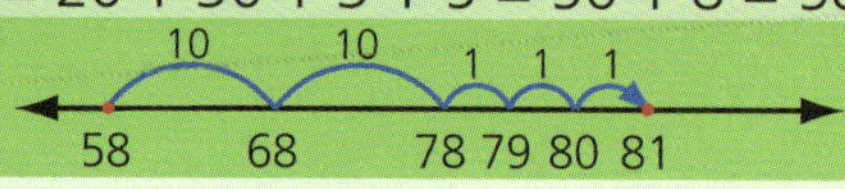 58 + 23 = 81 |



**1** Using patterns

**a** 500 + 200 = ☐ **b** 200 + 700 = ☐ **c** 400 + 300 = ☐ **d** 200 + 400 = ☐

**2** Changing the order

**a** 16 + 7 + 4 = ☐ **b** 27 + 4 + 3 = ☐ **c** 65 + 8 + 5 = ☐ **d** 28 + 4 + 2 = ☐

**3** Bridging to 10s

**a** 38 + 8 = ☐ **b** 29 + 4 = ☐ **c** 37 + 6 = ☐ **d** 54 + 7 = ☐

**4** Compensation

**a** 37 + 19 = ☐ **b** 63 + 29 = ☐ **c** 48 + 19 = ☐ **d** 25 + 39 = ☐

**5** Split

**a** 32 + 41 = ☐ **b** 55 + 34 = ☐ **c** 23 + 16 = ☐

**d** 14 + 34 = ☐ **e** 36 + 21 = ☐ **f** 47 + 33 = ☐

**6** Jump

**a** 67 + 22 = ☐

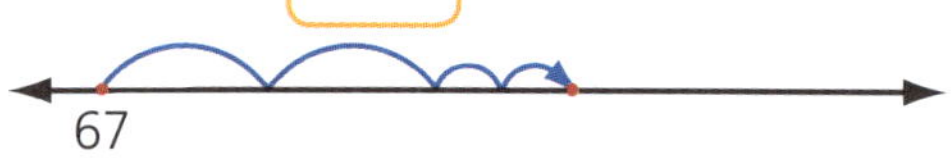



**b** 53 + 25 = ☐

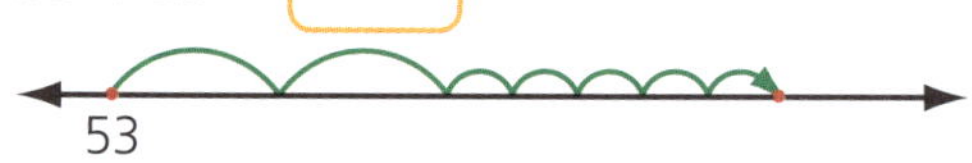



**c** 27 + 34 = ☐

**d** 44 + 18 = ☐

INVESTIGATION

**7** Complete the additions. Write the strategy you used (from **A** to **F** above).

**a** 51 + 37 = ☐ ☐ **b** 36 + 3 + 4 = ☐ ☐ **c** 44 + 9 = ☐ ☐

**d** 38 + 7 = ☐ ☐ **e** 300 + 500 = ☐ ☐ **f** 39 + 27 = ☐ ☐

Discuss how and why you chose each strategy.

# 2:11 Subtraction from 2-Digit Numbers

CONCEPT

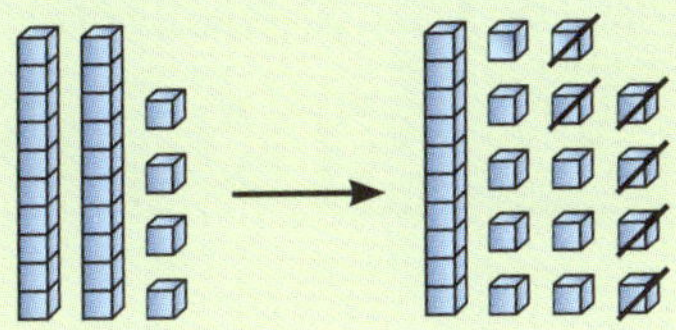

24 − 6 = 18

**1** Use place-value blocks to answer each question.

a 25 − 8 = ☐ b 32 − 7 = ☐ c 22 − 5 = ☐

d 34 − 6 = ☐ e 43 − 9 = ☐ f 28 − 4 = ☐

g 39 − 7 = ☐ h 24 − 5 = ☐ i 32 − 9 = ☐

j $\begin{array}{r} 34 \\ -\ \ 8 \\ \hline \end{array}$ k $\begin{array}{r} 21 \\ -\ \ 6 \\ \hline \end{array}$ l $\begin{array}{r} 31 \\ -\ \ 4 \\ \hline \end{array}$ m $\begin{array}{r} 40 \\ -\ \ 9 \\ \hline \end{array}$

CONCEPT

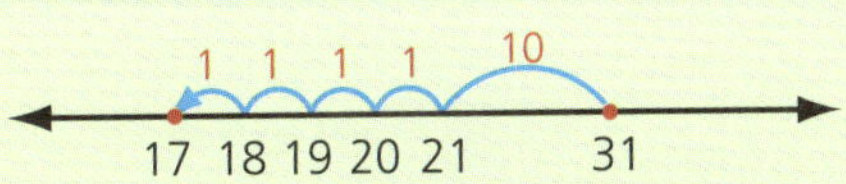


31 − 14 = 17

**2** Use the jump strategy to answer each question.

a 41 − 12 = ☐

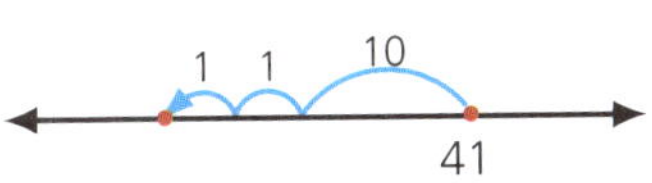


b 34 − 6 = ☐

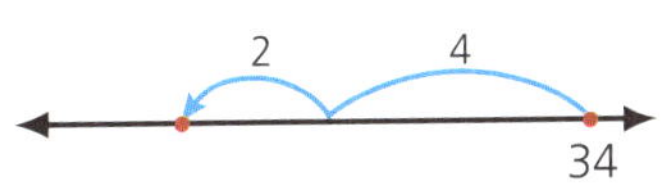


c 23 − 14 = ☐

d 33 − 14 = ☐ e 55 − 7 = ☐ f 42 − 23 = ☐

g 81 − 4 = ☐ h 73 − 15 = ☐ i 64 − 27 = ☐

 • *AUSTRALIAN SIGNPOST MATHS 3* • ISBN 978 1 4886 2184 0

# 2:12 Multiplication Facts

**1** Use the dice to find the answers.

a $6 \times 2 =$ ☐ b $7 \times 5 =$ ☐ c $7 \times 2 =$ ☐ d $8 \times 10 =$ ☐

e $7 \times 10 =$ ☐ f $8 \times 2 =$ ☐ g $5 \times 5 =$ ☐ h $10 \times 5 =$ ☐

i $9 \times 2 =$ ☐ j $6 \times 5 =$ ☐ k $6 \times 10 =$ ☐ l $9 \times 10 =$ ☐

m $8 \times 5 =$ ☐ n $11 \times 2 =$ ☐ o $9 \times 5 =$ ☐ p $11 \times 5 =$ ☐

**2** Complete each pair of multiplication facts.

a $2 \times 1 =$ ☐
$1 \times 2 =$ ☐

b $2 \times 5 =$ ☐
$5 \times 2 =$ ☐

c $2 \times 10 =$ ☐
$10 \times 2 =$ ☐

d $1 \times 5 =$ ☐
$5 \times 1 =$ ☐

e $5 \times 0 =$ ☐
$0 \times 5 =$ ☐

f $5 \times 10 =$ ☐
$10 \times 5 =$ ☐

**3** Measure the time you take to complete each set of multiplication facts. Write the time.

| a | b | c | d |
|---|---|---|---|
| $9 \times 0 =$ ☐ | $8 \times 0 =$ ☐ | $6 \times 0 =$ ☐ | $7 \times 0 =$ ☐ |
| $7 \times 1 =$ ☐ | $6 \times 1 =$ ☐ | $9 \times 1 =$ ☐ | $8 \times 1 =$ ☐ |
| $6 \times 5 =$ ☐ | $9 \times 5 =$ ☐ | $8 \times 5 =$ ☐ | $7 \times 5 =$ ☐ |
| $10 \times 5 =$ ☐ | $10 \times 10 =$ ☐ | $10 \times 2 =$ ☐ | $10 \times 1 =$ ☐ |
| $3 \times 2 =$ ☐ | $2 \times 2 =$ ☐ | $5 \times 2 =$ ☐ | $4 \times 2 =$ ☐ |
| $5 \times 5 =$ ☐ | $4 \times 5 =$ ☐ | $3 \times 5 =$ ☐ | $2 \times 5 =$ ☐ |
| $4 \times 10 =$ ☐ | $5 \times 10 =$ ☐ | $2 \times 10 =$ ☐ | $3 \times 10 =$ ☐ |
| $6 \times 2 =$ ☐ | $9 \times 2 =$ ☐ | $7 \times 2 =$ ☐ | $8 \times 2 =$ ☐ |
| $8 \times 5 =$ ☐ | $7 \times 5 =$ ☐ | $9 \times 5 =$ ☐ | $6 \times 5 =$ ☐ |
| $7 \times 10 =$ ☐ | $8 \times 10 =$ ☐ | $6 \times 10 =$ ☐ | $9 \times 10 =$ ☐ |
| ☐ seconds | ☐ seconds | ☐ seconds | ☐ seconds |

# 2:13 Number Facts, × 3

**1** Use the number models to complete the multiplications. Practise counting by 3s.

a $1 \times 3 =$ ☐
b $2 \times 3 =$ ☐
c $3 \times 3 =$ ☐
d $4 \times 3 =$ ☐
e $5 \times 3 =$ ☐
f $6 \times 3 =$ ☐
g $7 \times 3 =$ ☐
h $8 \times 3 =$ ☐
i $9 \times 3 =$ ☐
j $10 \times 3 =$ ☐

**2** Write the answers. Remember that $3 \times 1 = 1 \times 3$.

a $3 \times 0 =$ ☐ b $0 \times 3 =$ ☐ c $3 \times 10 =$ ☐ d $10 \times 3 =$ ☐
e $3 \times 2 =$ ☐ f $2 \times 3 =$ ☐ g $3 \times 1 =$ ☐ h $1 \times 3 =$ ☐
i $3 \times 5 =$ ☐ j $5 \times 3 =$ ☐ k $8 \times 3 =$ ☐ l $4 \times 3 =$ ☐
m $9 \times 3 =$ ☐ n $6 \times 3 =$ ☐ o $3 \times 3 =$ ☐ p $7 \times 3 =$ ☐

**3**
a $5 \times 1 =$ ☐ b $1 \times 5 =$ ☐ c $5 \times 2 =$ ☐ d $2 \times 5 =$ ☐
e $5 \times 0 =$ ☐ f $0 \times 5 =$ ☐ g $5 \times 10 =$ ☐ h $10 \times 5 =$ ☐
i $2 \times 10 =$ ☐ j $10 \times 2 =$ ☐ k $5 \times 5 =$ ☐ l $10 \times 10 =$ ☐

**4**
a 3 pens in a set
8 sets
How many pens? ×
☐

b 3 dollars per person
9 people
How many dollars? ×
☐

**5** Try to complete these tables from memory.

a

| | 4 | 7 | 9 | 5 | 8 | 6 |
|---|---|---|---|---|---|---|
| × 3 | | | | | | |

b

| | 4 | 7 | 9 | 5 | 8 | 6 |
|---|---|---|---|---|---|---|
| × 5 | | | | | | |

c

| | 4 | 7 | 9 | 5 | 8 | 6 |
|---|---|---|---|---|---|---|
| × 2 | | | | | | |

d

| | 4 | 7 | 9 | 5 | 8 | 6 |
|---|---|---|---|---|---|---|
| × 10 | | | | | | |

 • *AUSTRALIAN SIGNPOST MATHS 3* • ISBN 978 1 4886 2184 0

# 2:14 Addition to 99, No Trading

CONCEPT

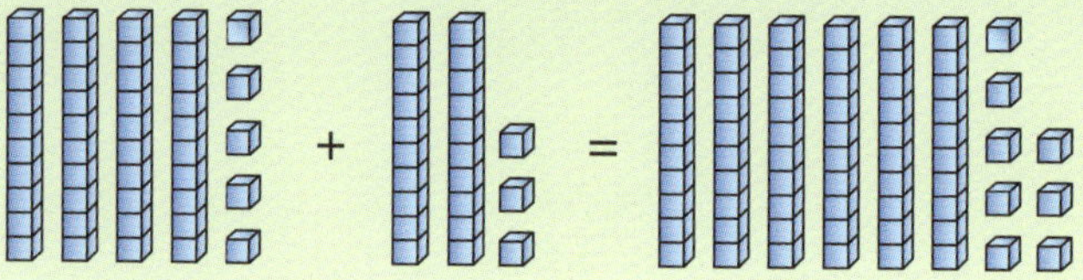

45 + 23 = 68

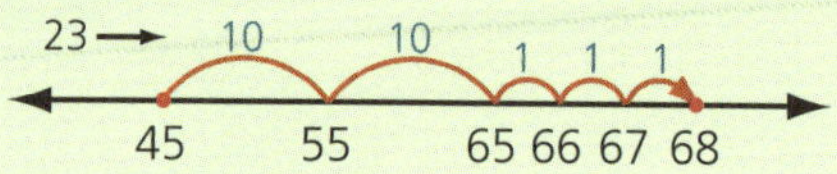


- Start with the larger number.
- Jump forward by tens, then by ones.

**1** Complete the additions.

a
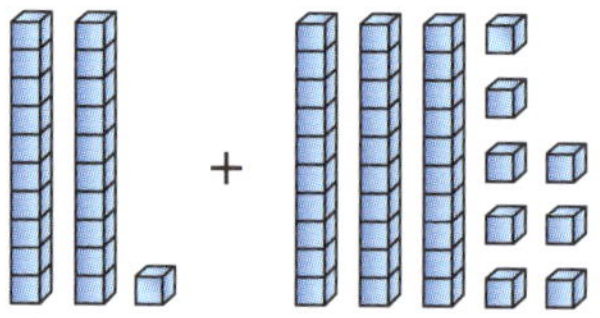
☐ + ☐ = ☐

b
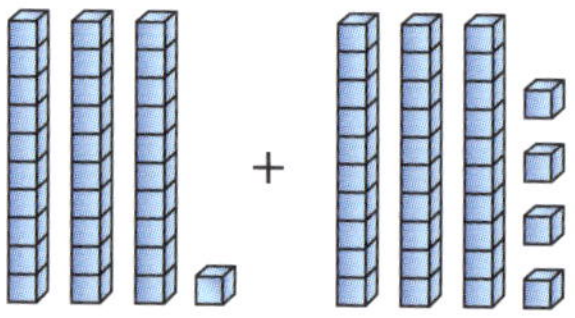
☐ + ☐ = ☐

c
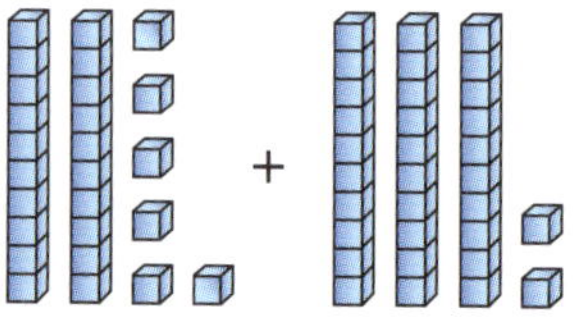
☐ + ☐ = ☐

d
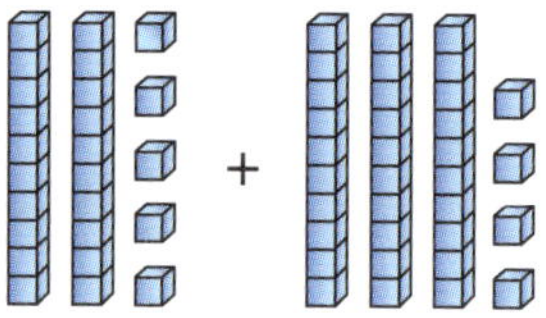
☐ + ☐ = ☐

e
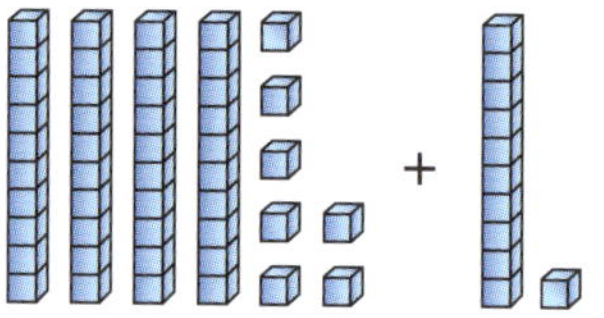
☐ + ☐ = ☐

f
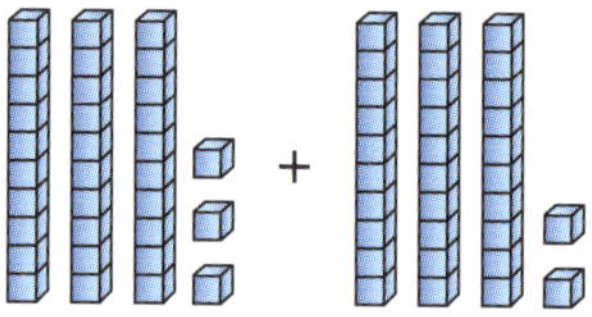
☐ + ☐ = ☐

**2** Use place-value blocks or the jump strategy to complete the additions.

a 64 + 25 = ☐  b 32 + 46 = ☐  c 29 + 50 = ☐

d 43 + 36 = ☐  e 56 + 41 = ☐  f 52 + 34 = ☐

g 71 + 25 = ☐  h 35 + 30 = ☐  i 27 + 61 = ☐

**3** a
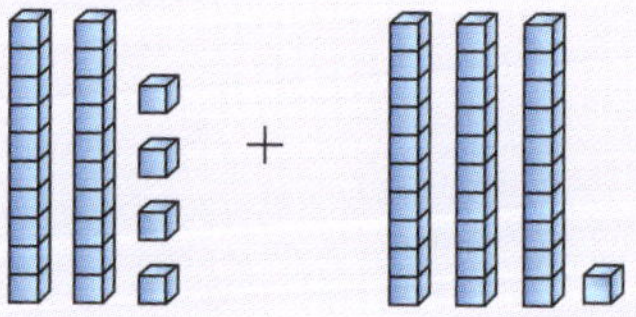
☐ + ☐ = ☐

b
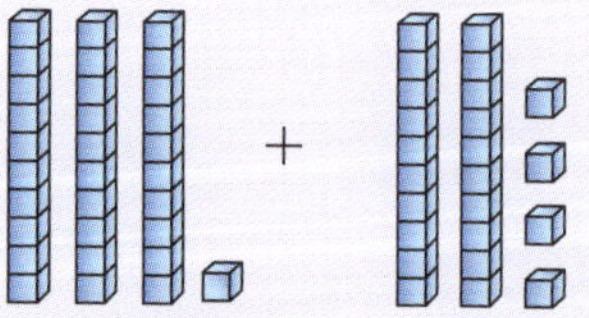
☐ + ☐ = ☐

Does 31 − 24 equal 24 − 31? ☐

INVESTIGATION

c Explain why 24 + 31 = 31 + 24.

- Use a calculator to check all your answers.



 • *AUSTRALIAN SIGNPOST MATHS 3* • ISBN 978 1 4886 2184 0

# 2:15 Addition to 99, No Trading

CONCEPT

| I show | I think | I write |
|---|---|---|
| 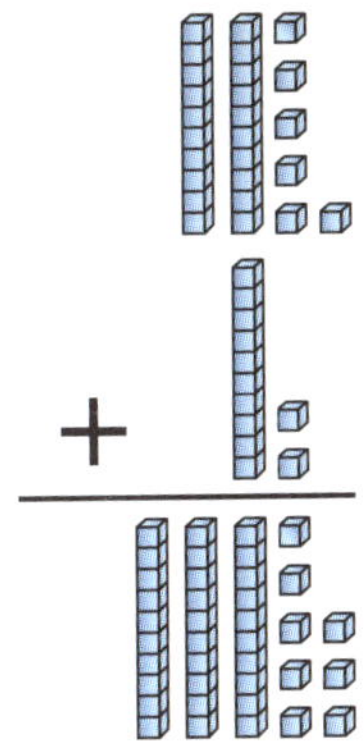 | 2 tens and 6 ones<br>+<br>1 tens and 2 ones<br>=<br>3 tens and 8 ones | Tens \| Ones<br>2 \| 6<br>+ 1 \| 2<br>3 \| 8 |

*The split strategy:*

26 + 12 = 20 + 6 + 10 + 2

= (20 + 10) + (6 + 2)

= 30 + 8 or 38

**1** Model these problems and write them as number sentences.

**a** 5 eggs
12 eggs
How many eggs altogether?

☐ + ☐ = ☐

**b** 13 horses
32 horses
How many horses altogether?

☐ + ☐ = ☐

**2** Use the split strategy or place-value blocks to answer these.

**a**

| Tens | Ones |
|---|---|
| 1 | 6 |
| + 1 | 1 |
| | |

**b**

| Tens | Ones |
|---|---|
| 2 | 0 |
| + 2 | 3 |
| | |

**c**

| Tens | Ones |
|---|---|
| 1 | 4 |
| + 2 | 3 |
| | |

**d**

| Tens | Ones |
|---|---|
| 3 | 1 |
| + 2 | 2 |
| | |

**e**

| Tens | Ones |
|---|---|
| 3 | 8 |
| + 1 | 0 |
| | |

**f**

| Tens | Ones |
|---|---|
| 1 | 7 |
| + 2 | 2 |
| | |

**g**

| Tens | Ones |
|---|---|
| 2 | 3 |
| + 2 | 5 |
| | |

**h**

| Tens | Ones |
|---|---|
| 1 | 4 |
| + 1 | 5 |
| | |

**3** **a** 3 tens and 5 ones
+ 1 ten and 2 ones

**b** 2 tens and 4 ones
+ 2 tens and 5 ones

**c** 1 ten and 3 ones
+ 2 tens and 6 ones

 • *AUSTRALIAN SIGNPOST MATHS 3* • ISBN 978 1 4886 2184 0

# 2:16 Multiplication

CONCEPT

Every week, for 3 weeks, Charlotte was given 5 stars for neat work. How many stars was she given altogether?

$$\begin{array}{r} 5 \\ \times\ 3 \\ \hline \square \end{array}$$

Model these problems using counters or blocks.

**1** **a** 8 groups of 2 pens ☐ × ☐ = ☐ **b** 3 rows of 10 stars ☐ × ☐ = ☐ **c** 2 rows of 4 soldiers ☐ × ☐ = ☐

**2** **a** 5 apples in a row
3 rows
How many apples?

$$\begin{array}{r} 5 \\ \times\ 3 \\ \hline \square \end{array}$$

**b** 10 pencils in a set
9 sets
How many pencils?

$$\begin{array}{r} \times\ \phantom{0} \\ \hline \square \end{array}$$

**3**

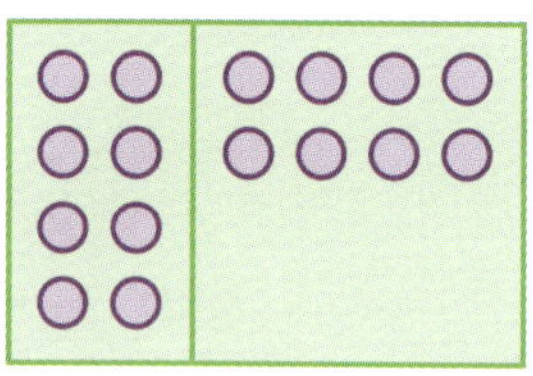

**a** In this picture, is 4 rows of 2 equal to 2 rows of 4? ☐
**b** 2 × 4 = ☐ **c** 4 × 2 = ☐
**d** Is 2 × 4 = 4 × 2 true? ☐

**4**

**a** $\begin{array}{r} 4 \\ \times\ 1 \\ \hline \end{array}$ **b** $\begin{array}{r} 1 \\ \times\ 4 \\ \hline \end{array}$ **c** $\begin{array}{r} 10 \\ \times\ 5 \\ \hline \end{array}$ **d** $\begin{array}{r} 5 \\ \times\ 10 \\ \hline \end{array}$ **e** $\begin{array}{r} 10 \\ \times\ 7 \\ \hline \end{array}$

**f** $\begin{array}{r} 2 \\ \times\ 10 \\ \hline \end{array}$ **g** $\begin{array}{r} 10 \\ \times\ 2 \\ \hline \end{array}$ **h** $\begin{array}{r} 5 \\ \times\ 2 \\ \hline \end{array}$ **i** $\begin{array}{r} 2 \\ \times\ 5 \\ \hline \end{array}$ **j** $\begin{array}{r} 10 \\ \times\ 4 \\ \hline \end{array}$

ACTIVITY

| | | | | | | | | | |
|---|---|---|---|---|---|---|---|---|---|
| 1 | 2 | 3 | 4 | 5 | 6 | 7 | 8 | 9 | 10 |
| 11 | 12 | 13 | 14 | 15 | 16 | 17 | 18 | 19 | 20 |
| 21 | 22 | 23 | 24 | 25 | 26 | 27 | 28 | 29 | 30 |
| 31 | 32 | 33 | 34 | 35 | 36 | 37 | 38 | 39 | 40 |
| 41 | 42 | 43 | 44 | 45 | 46 | 47 | 48 | 49 | 50 |

**5** **a** Colour in the table of twos.
**b** Circle the table of fives.
**c** Tick the table of tens.
**d** Discuss the patterns you have made.



 • *AUSTRALIAN SIGNPOST MATHS 3* • ISBN 978 1 4886 2184 0

# 2:17 Number Facts, Multiplication

**1** Model these problems and record the answers.

a 7 rows of 5 hats ☐ ☐ ☐ ☐ ☐

b 4 groups of 10 pens ☐ ☐ ☐ ☐ ☐

c 5 groups of 5 eggs ☐ ☐ ☐ ☐ ☐

d 5 girls in a team
8 teams ×
How many girls? ☐

e 5 boys in a row
6 rows ×
How many boys? ☐

**2**

| | | | |
|---|---|---|---|
| a $9 \times 2 =$ ☐ | b $4 \times 5 =$ ☐ | c $8 \times 10 =$ ☐ | d $7 \times 2 =$ ☐ |
| e $4 \times 1 =$ ☐ | f $6 \times 2 =$ ☐ | g $3 \times 10 =$ ☐ | h $5 \times 2 =$ ☐ |
| i $3 \times 5 =$ ☐ | j $3 \times 1 =$ ☐ | k $7 \times 10 =$ ☐ | l $8 \times 0 =$ ☐ |

**3**

| | | | | |
|---|---|---|---|---|
| a $5 \times 2$ | b $5 \times 7$ | c $5 \times 9$ | d $0 \times 6$ | e $2 \times 3$ |
| f $9 \times 3$ | g $10 \times 7$ | h $10 \times 4$ | i $2 \times 10$ | j $1 \times 9$ |

**4** On this hundred chart:
- colour the table of fives yellow
- colour the table of tens blue
- circle the table of twos.

| | | | | | | | | | |
|---|---|---|---|---|---|---|---|---|---|
| 1 | 2 | 3 | 4 | 5 | 6 | 7 | 8 | 9 | 10 |
| 11 | 12 | 13 | 14 | 15 | 16 | 17 | 18 | 19 | 20 |
| 21 | 22 | 23 | 24 | 25 | 26 | 27 | 28 | 29 | 30 |
| 31 | 32 | 33 | 34 | 35 | 36 | 37 | 38 | 39 | 40 |
| 41 | 42 | 43 | 44 | 45 | 46 | 47 | 48 | 49 | 50 |
| 51 | 52 | 53 | 54 | 55 | 56 | 57 | 58 | 59 | 60 |
| 61 | 62 | 63 | 64 | 65 | 66 | 67 | 68 | 69 | 70 |
| 71 | 72 | 73 | 74 | 75 | 76 | 77 | 78 | 79 | 80 |
| 81 | 82 | 83 | 84 | 85 | 86 | 87 | 88 | 89 | 90 |
| 91 | 92 | 93 | 94 | 95 | 96 | 97 | 98 | 99 | 100 |

**5** This multiplication grid shows the table of 2s.

ACTIVITY

a Write in the table of 5s.

b Write in the table of 10s, 1s and 0s.

| × | 0 | 1 | 2 | 3 | 4 | 5 | 6 | 7 | 8 | 9 | 10 |
|---|---|---|---|---|---|---|---|---|---|---|---|
| 0 | | | 0 | | | | | | | | |
| 1 | | | 2 | | | | | | | | |
| 2 | 0 | 2 | 4 | 6 | 8 | 10 | 12 | 14 | 16 | 18 | 20 |
| 3 | | | 6 | | | | | | | | |
| 4 | | | 8 | | | | | | | | |
| 5 | | | 10 | | | | | | | | |
| 6 | | | 12 | | | | | | | | |
| 7 | | | 14 | | | | | | | | |
| 8 | | | 16 | | | | | | | | |
| 9 | | | 18 | | | | | | | | |
| 10 | | | 20 | | | | | | | | |

  ISBN 978 1 4886 2184 0

# 2:18 Multiplication Facts

**1** Complete each pair of multiplication facts.

**a** 2 × 3 = ☐
3 × 2 = ☐

**b** 3 × 5 = ☐
5 × 3 = ☐

**c** 3 × 10 = ☐
10 × 3 = ☐

**d** 1 × 3 = ☐
3 × 1 = ☐

**e** 3 × 0 = ☐
0 × 3 = ☐

**f** 3 × 6 = ☐
6 × 3 = ☐

**2** Measure the time you take to complete each set of multiplication facts. Write the time.

| a | b | c | d |
|---|---|---|---|
| 9 × 0 = ☐ | 3 × 0 = ☐ | 6 × 0 = ☐ | 7 × 0 = ☐ |
| 7 × 1 = ☐ | 6 × 1 = ☐ | 9 × 1 = ☐ | 3 × 1 = ☐ |
| 6 × 3 = ☐ | 3 × 3 = ☐ | 8 × 3 = ☐ | 7 × 3 = ☐ |
| 10 × 5 = ☐ | 10 × 10 = ☐ | 6 × 3 = ☐ | 10 × 1 = ☐ |
| 3 × 3 = ☐ | 2 × 3 = ☐ | 5 × 3 = ☐ | 4 × 3 = ☐ |
| 5 × 5 = ☐ | 4 × 5 = ☐ | 3 × 5 = ☐ | 2 × 5 = ☐ |
| 4 × 10 = ☐ | 5 × 10 = ☐ | 2 × 10 = ☐ | 3 × 10 = ☐ |
| 4 × 3 = ☐ | 9 × 3 = ☐ | 7 × 3 = ☐ | 8 × 3 = ☐ |
| 8 × 3 = ☐ | 7 × 3 = ☐ | 9 × 3 = ☐ | 6 × 3 = ☐ |
| 7 × 10 = ☐ | 8 × 10 = ☐ | 6 × 10 = ☐ | 9 × 10 = ☐ |
| ☐ seconds | ☐ seconds | ☐ seconds | ☐ seconds |

**3** Use the dice to find the answers.

| | | | |
|---|---|---|---|
| **a** 2 × 3 = ☐ | **b** 7 × 3 = ☐ | **c** 1 × 3 = ☐ | **d** 4 × 3 = ☐ |
| **e** 3 × 5 = ☐ | **f** 8 × 3 = ☐ | **g** 5 × 5 = ☐ | **h** 10 × 3 = ☐ |
| **i** 3 × 9 = ☐ | **j** 6 × 5 = ☐ | **k** 3 × 3 = ☐ | **l** 6 × 3 = ☐ |
| **m** 8 × 5 = ☐ | **n** 5 × 3 = ☐ | **o** 9 × 5 = ☐ | **p** 11 × 3 = ☐ |

# 2:19 Relating × and ÷

CONCEPT

5 groups of 3 make ☐.   5 × 3 = 15

How many groups of 3 in 15? ☐   15 ÷ 3 = 5

**1** Complete each question and write the related number sentences.

**a**

5 groups of 2 make ☐.

How many 2s in 10? ☐

5 × 2 = ☐   10 ÷ 2 = ☐

**b**

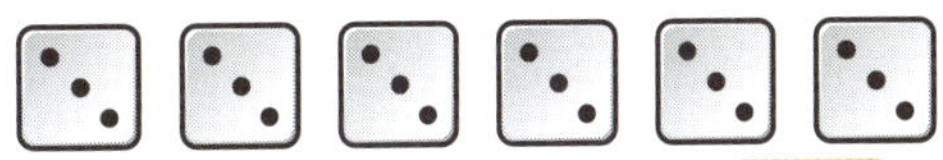

6 groups of 3 make ☐.

How many 3s in 18? ☐

6 × 3 = ☐   18 ÷ 3 = ☐

**c**

5 × 5 = ☐

25 ÷ 5 = ☐

5 × ☐ = 25

**d**

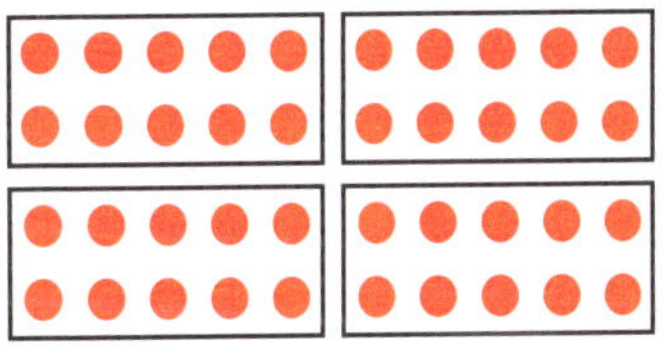

4 × 10 = ☐

40 ÷ 10 = ☐

☐ × 10 = 40

**e**

7 × 5 = ☐

35 ÷ 5 = ☐

☐ × 5 = 35

**f**

8 × 2 = ☐

16 ÷ 2 = ☐

☐ × 2 = 16

**2**
- **a** 9 × 3 = 27 so 27 ÷ 3 = ☐
- **b** 6 × 3 = 18 so 18 ÷ 3 = ☐
- **c** 4 × 3 = 12 so 12 ÷ 3 = ☐
- **d** 7 × 3 = 21 so 21 ÷ 3 = ☐
- **e** 5 × 3 = 15 so 15 ÷ 3 = ☐
- **f** 8 × 3 = 24 so 24 ÷ 3 = ☐
- **g** 10 × 3 = 30 so 30 ÷ 3 = ☐

**3**
- **a** 6 × 5 = 30 so 30 ÷ 5 = ☐
- **b** 7 × 2 = 14 so 14 ÷ 2 ☐
- **c** 9 × 5 = 45 so 45 ÷ 5 = ☐
- **d** 7 × 10 = 70 so 70 ÷ 10 = ☐
- **e** 7 × 5 = 35 so 35 ÷ 5 = ☐
- **f** 10 × 5 = 50 so 50 ÷ 5 = ☐
- **g** 8 × 5 = 40 so 40 ÷ 5 = ☐

 • *AUSTRALIAN SIGNPOST MATHS 3* • ISBN 978 1 4886 2184 0

# 2:20 Money

A

B

C
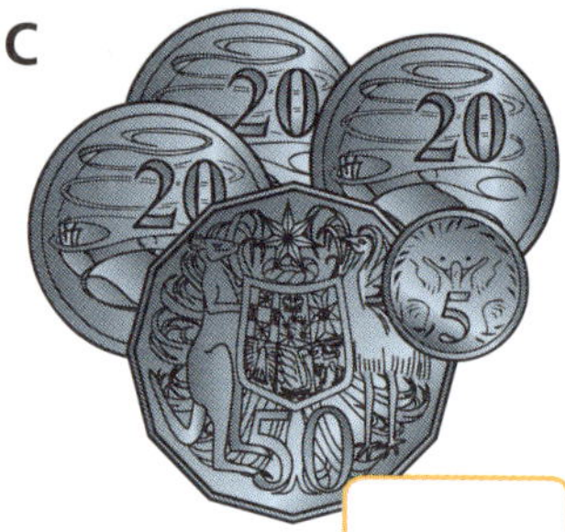

D

E

F

G

H

1. Which group of coins is worth:
   a the same as A?
   b the same as F?
   c less than D?
   d more than B?

2. Write the total value of the coins in:
   a A and D
   b E and F
   c C and G

3. Which group of banknotes is worth:
   a the same as D?
   b the same as H?
   c less than D?
   d more than E?

A

B

C

D

E

F

G

H

# 2:21 Shopping

To make an amount of money, start with the higher value coins.

**Steps**:
Start with $1 and count on.
- Use 50c to make $1.50.
- Use 10c to make $1.60.
- Use 5c to make $1.65.

CONCEPT

1 Colour the coins you could use to buy each item.

a

90c

b

85c

c

$1.00

d

$1.60

e

$1.95

f

$1.20

FUN SPOT

2 Choose coins to make each amount of money.

| | | | | | | | |
|---|---|---|---|---|---|---|---|
| a $3.80 | | | | | | | |
| b $2.95 | | | | | | | |
| c $4.35 | | | | | | | |

# 2:22 Relating Addition and Subtraction

CONCEPT

Jenny needed to find the difference in length between these two ribbons.

33 – 15 = ☐ can also be thought of as 15 + ☐ = 33.

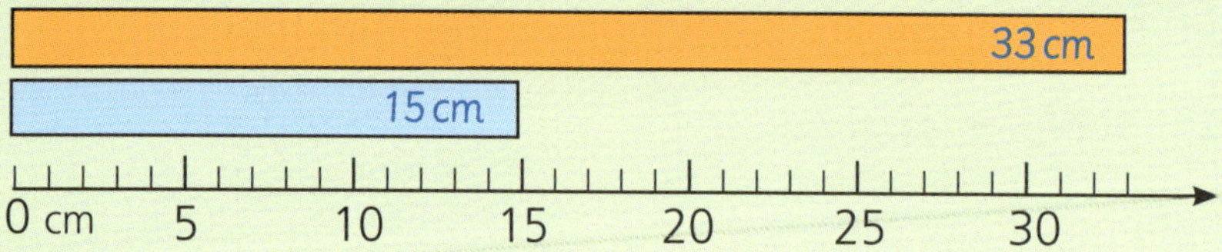

The length difference between the two ribbons is 18 cm.

**1** Find the length differences between ribbons.

**a**

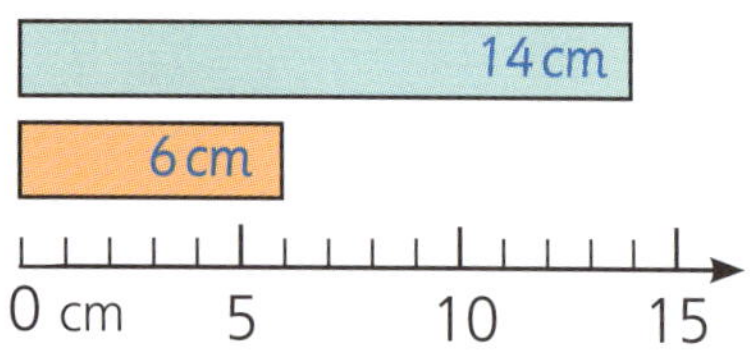


14 – 6 = ☐ so 6 + ☐ = 14

Difference = ☐ cm

**b**

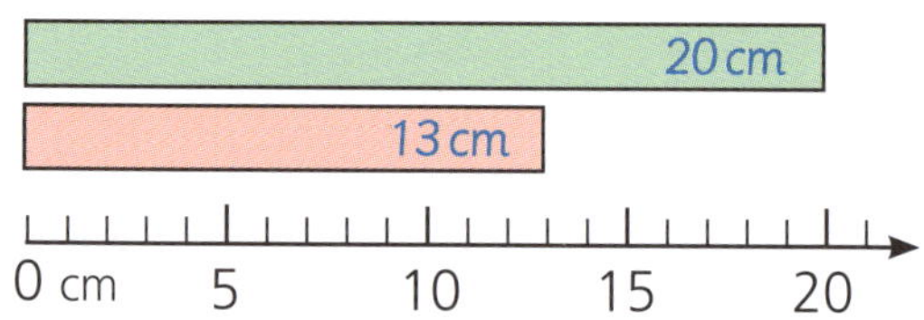


20 – 13 = ☐ so 13 + ☐ = 20

Difference = ☐ cm

**c**

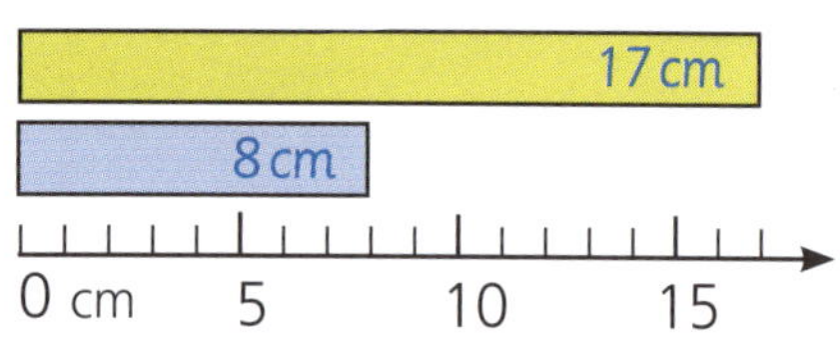


17 – 8 = ☐ so 8 + ☐ = 17

Difference = ☐ cm

**d**

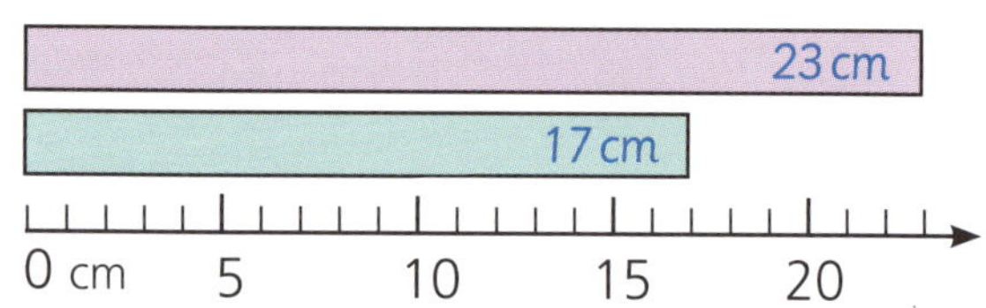


23 – 17 = ☐ so 17 + ☐ = 23

Difference = ☐ cm

**e**

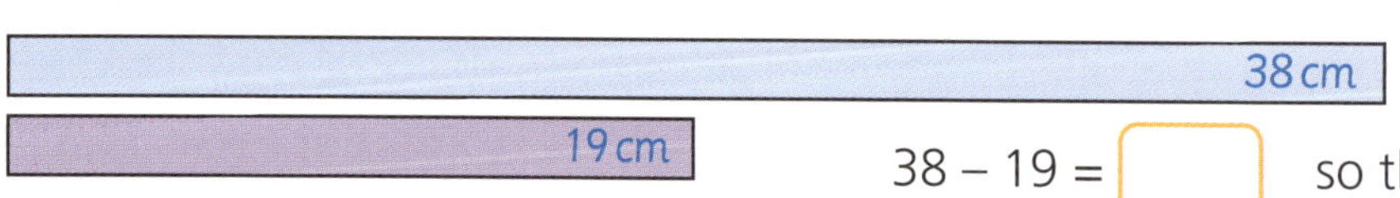


38 – 19 = ☐ so the difference = ☐ cm

**f**

44 – 16 = ☐ so the difference = ☐ cm

ACTIVITY

- Compare pieces of paper cut to different centimetre lengths, or draw lines of different lengths.
- Write your own addition and subtraction number sentences to match the lengths.

# 2:23 Inverse Operations, + and –

CONCEPT

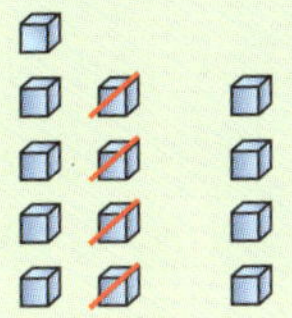

9 – 4 + 4 = 9

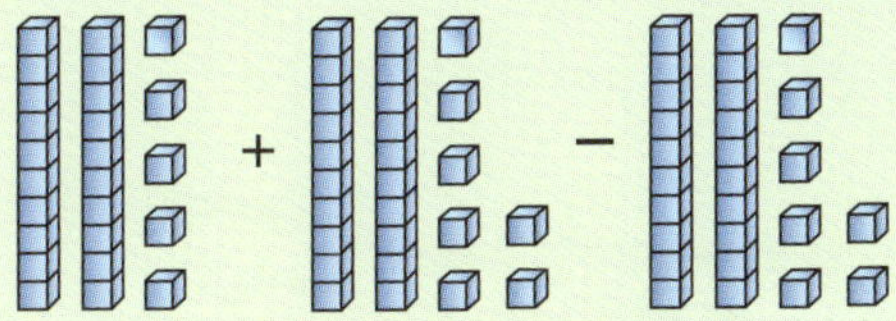

25 + 27 – 27 = 25

27 – 27 = 0

**1** a 5 + 3 – 3 = ☐  b 8 – 2 + 2 = ☐  c 5 + 7 – 7 = ☐

d 10 + 8 – 8 = ☐  e 10 – 7 + 7 = ☐  f 16 – 8 + 8 = ☐

**2** a Craig had $26.
He was given another $57.
He spent $57.

$26 + ____ – ____ = ____

He had ☐ left.

b Chi had 75 balloons.
She was given 68 more.
68 popped.

☐

She has ☐ balloons left.

c Sulin had 35 fish.
16 died.
She bought 16 more.

35 + ____ – ____ = ____

She has ☐ fish.

d Mark had $83.
He bought a toy for $37.
He returned the toy and received $37 back.

☐

He has ☐.

**3** a 27 + 34 – 34 + 1 = ☐  b 47 – 2 + 29 – 29 = ☐

c 93 – 76 + 76 – 2 = ☐  d 39 + 84 + 1 – 84 = ☐

**4** a 32 + 9 – 9 + 27 – 27 = ☐  b 92 – 5 + 5 – 2 + 2 = ☐

c 29 + 15 – 15 + 3 – 3 = ☐  d 89 – 27 + 27 – 9 + 9 = ☐

e 7 + 34 – 2 – 34 + 2 = ☐  f 29 + 6 – 24 – 6 + 24 = ☐

FUN SPOT

Make up your own inverse number sentences.

37 + 64 – 29 + 42 – 64 + 29 – 42 = 37

☐

# 2:24 Sharing and Grouping

**1** Use the pictures to find how many in each share.

**a** 12 are shared by 3 ☐

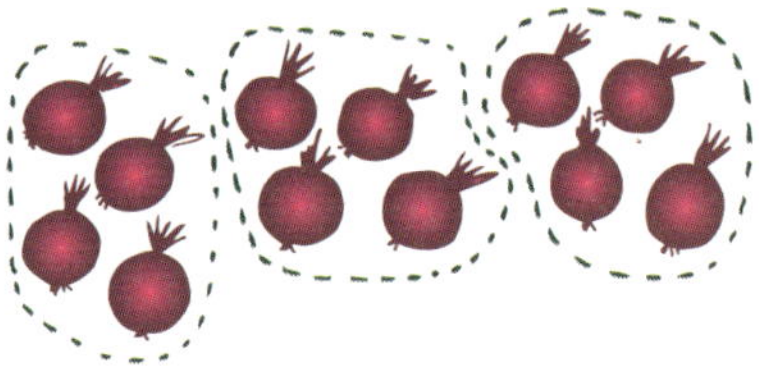

**b** 12 are shared by 4 ☐

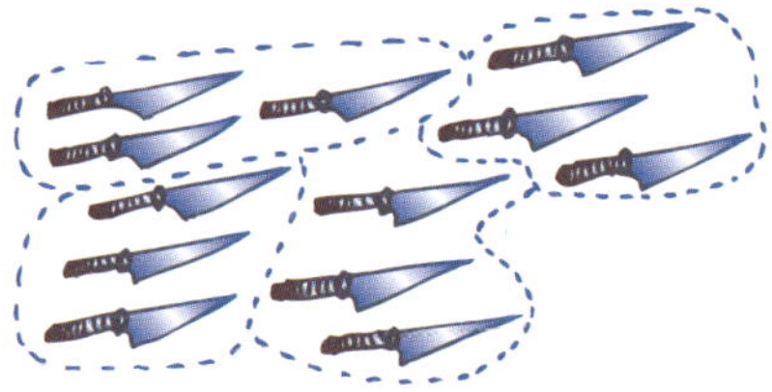

**c** 18 are shared by 9 ☐

**d** 9 are shared by 3 ☐

**e** 16 are shared by 4 ☐

**f** 28 are shared by 4 ☐

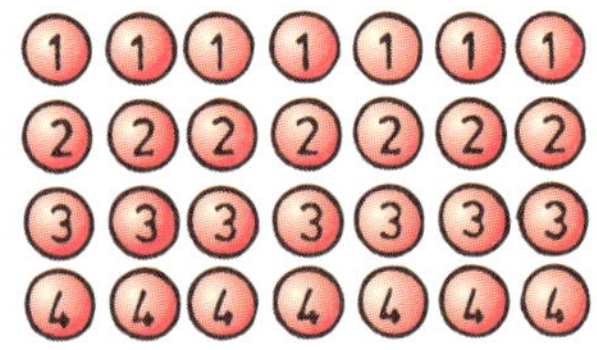


**g** 14 are shared by 2 ☐

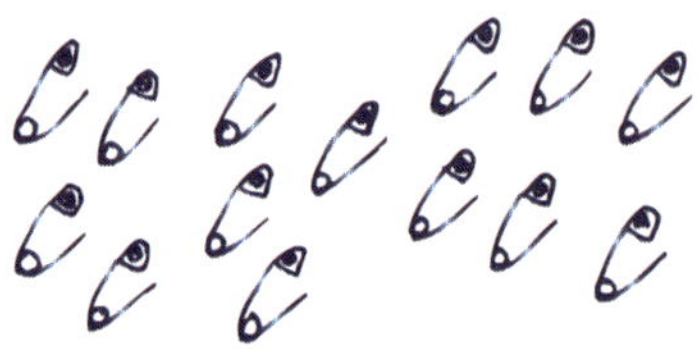

**h** 24 are shared by 6 ☐

**i** 15 are shared by 3 ☐

**2** Use the pictures to find how many groups there are.

**a** Groups of 2 ☐

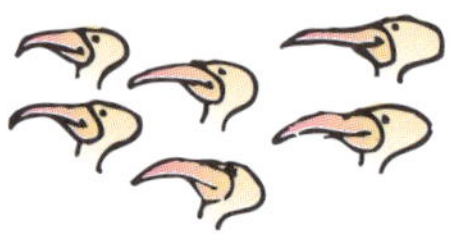

**b** Groups of 6 ☐

**c** Groups of 3 ☐

**d** Groups of 7 ☐

**e** Groups of 4 ☐

**f** Groups of 5 ☐

ACTIVITY

Use counters to find how many in each share.

**3** 12 pieces of toast:

**a** shared by 2 ☐

**b** shared by 3 ☐

**c** shared by 4 ☐

**4** 10 mugs

**a** shared by 3 ☐

**b** shared by 4 ☐

**c** shared by 5 ☐

Discuss what you could do with those left over.

# 2:25 Modelling Division

CONCEPT

÷ means **shared among**
means **in groups of**
means **divided by**

15 shared among 3 ...

Groups of 3 in 15 ...

15 ÷ 3 = ☐

**1** **a**

21 shared among 3 = ☐

Groups of 3 in 21 = ☐

21 ÷ 3 = ☐

**b**

15 shared among 5 = ☐

Groups of 5 in 15 = ☐

15 ÷ 5 = ☐

**c**

18 shared among 6 = ☐

Groups of 6 in 18 = ☐

18 ÷ 6 = ☐

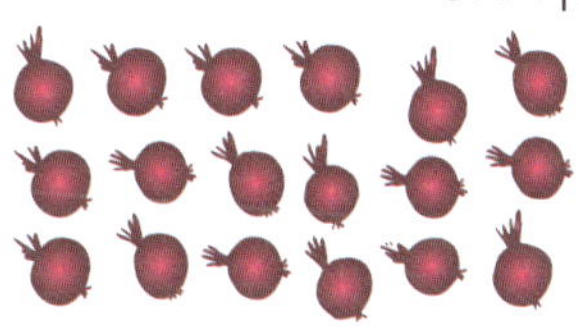

**d**

12 shared among 4 = ☐

Groups of 4 in 12 = ☐

12 ÷ 4 = ☐

**2** **a**

Columns of 4 in 20 = ☐

20 ÷ 4 = ☐

Rows of 5 in 20 = ☐

20 ÷ 5 = ☐

**b**

Rows of 4 in 16 = ☐

16 ÷ 4 = ☐

Groups of 8 in 16 = ☐

16 ÷ 8 = ☐

**c**

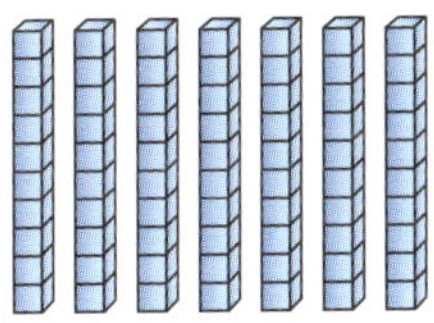

Groups of 10 in 70 = ☐

70 ÷ 10 = ☐

7 groups of 10 = ☐

**d**

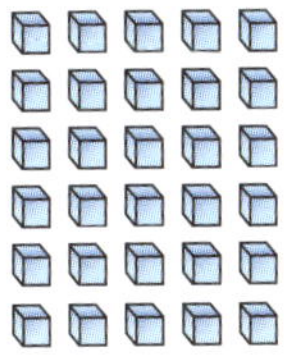

Rows of 5 in 30 = ☐

30 ÷ 5 = ☐

30 ÷ 6 = ☐

**3** Use the addition sequence to solve the division problem.

**a** 8 + 8 + 8 = 24 How many groups of 8 apples in 24? ☐

**b** 9 + 9 + 9 + 9 = 36 How many rows of 9 buttons in 36? ☐

**c** 7 + 7 + 7 + 7 = 28 4 boys share 28 pens. One share = ☐

 • *AUSTRALIAN SIGNPOST MATHS 3* • ISBN 978 1 4886 2184 0

# 2:26 Change from $2

CONCEPT

To give change, we count on from the price paid.

**Example**: What is the change from $2 if the cost is 35c?

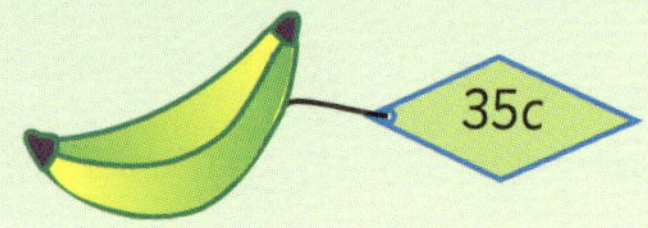


- Count on from 35c.
- 35c and 5c makes 40c, and 10c makes 50c, and 50c makes $1, and $1 makes $2.
- The change is 5c + 10c + 50c + $1 = **$1.65**.

**1** Colour the coins to show the change from $2.

**a**

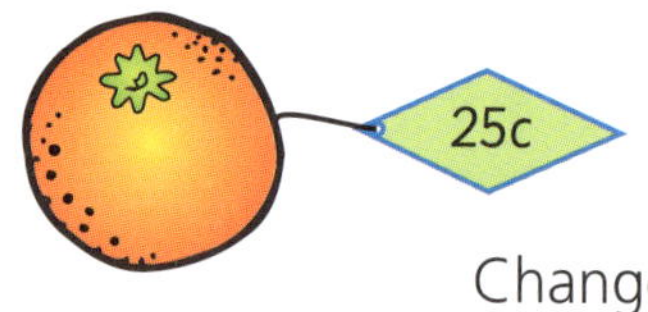


Change = ☐

**b**

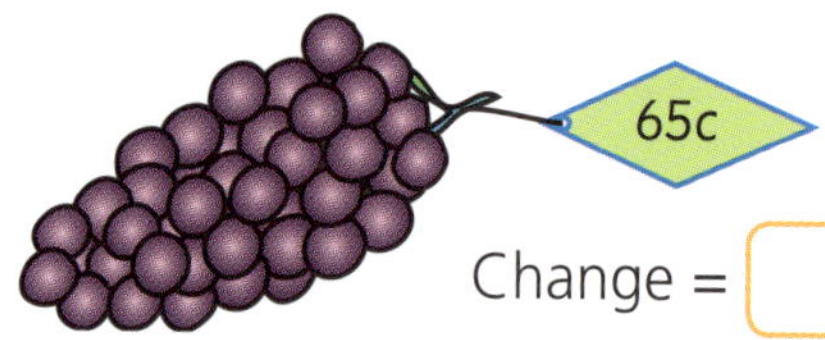


Change = ☐

**c**

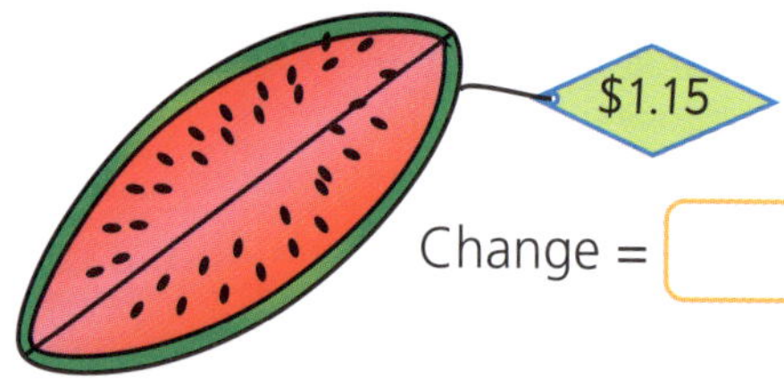


Change = ☐

**2** Use the coins to count on and find the change.

| | Cost | Money Given | Coins Counted | Total Change | Coins |
|---|---|---|---|---|---|
| a | 45c | 60c | 5c, 10c | 15c | |
| b | 65c | 80c | | | |
| c | 55c | $1 | | | |
| d | $1.75 | $2 | | | |

 • *AUSTRALIAN SIGNPOST MATHS 3* • ISBN 978 1 4886 2184 0

# 2:27 Money

1 Write the total value of each group of banknotes.

a

b

c

d

2 Circle the larger amount.

a

b

c

3 Write the amounts in order from smallest to largest.

a

b

c

# 2:28 Addition to 99, No Trading

CONCEPT

| I show | I think | I write |
|---|---|---|
| 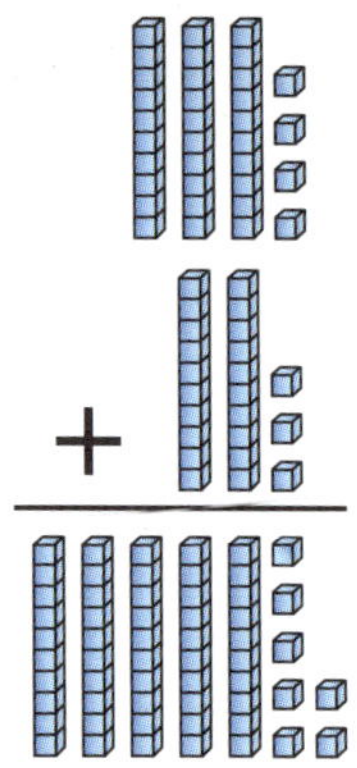 | 3 tens and 4 ones + 2 tens and 3 ones = 5 tens and 7 ones | Tens \| Ones: 3 4 + 2 3 = 5 7 |

**1** Use the split strategy or place-value blocks to answer these.

**a**

| Tens | Ones |
|---|---|
| 3 | 2 |
| + 2 | 6 |
| | |

**b**

| Tens | Ones |
|---|---|
| 4 | 3 |
| + 3 | 5 |
| | |

**c**

| Tens | Ones |
|---|---|
| 3 | 7 |
| + 4 | 0 |
| | |

**d**

| Tens | Ones |
|---|---|
| 6 | 4 |
| + 2 | 5 |
| | |

**e**

| Tens | Ones |
|---|---|
| 5 | 6 |
| + 4 | 0 |

**f**

| Tens | Ones |
|---|---|
| 4 | 2 |
| + 3 | 5 |

**g**

| Tens | Ones |
|---|---|
| 5 | 3 |
| + 3 | 6 |

**h**

| Tens | Ones |
|---|---|
| 7 | 3 |
| + 2 | 5 |

**i**

| Tens | Ones |
|---|---|
| $6 | 6 |
| + $2 | 3 |

**j**

| Tens | Ones |
|---|---|
| $7 | 5 |
| + $2 | 2 |

**k**

| Tens | Ones |
|---|---|
| $4 | 8 |
| + $3 | 1 |

**l**

| Tens | Ones |
|---|---|
| $5 | 4 |
| + $3 | 5 |

**2**

**a** 4 tens and 6 ones
+ 4 tens and 2 ones

**b** 5 tens and 4 ones
+ 3 tens and 3 ones

**c** 2 tens and 4 ones
+ 5 tens and 5 ones

**d** 7 tens and 3 ones
+ 1 ten and 6 ones

**e** 6 tens and 0 ones
+ 2 tens and 9 ones

**f** 3 tens and 1 one
+ 4 tens and 7 ones

**3**

**a** I paid $35 for a shirt and $61 for pants.
How much did I spend? ☐

**b** There are 11 girls and 18 boys in our class.
How many are in our class? ☐

**c** 15 horses and 42 cows are on our farm.
How many animals altogether? ☐

**d** I saved $53. Alana saved $24.
How much did we save? ☐

# 2:29 Subtraction, No Trading

CONCEPT

**1** Model the word problem using place-value blocks. Write a number sentence.

**a** 36 badges, 4 sold.
How many are left?

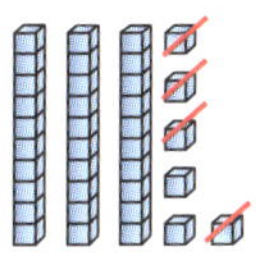

**b** 55 birds, 23 fly away.
How many remain?

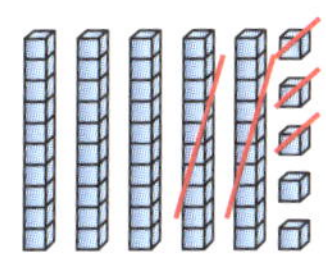

**c** 47 girls, 5 boys.
How many more girls than boys?

**d** 68 needed, 34 collected.
How many more to collect?

**e** 76 books, 42 covered.
How many more to cover?

**f** 59 grapes, 46 eaten.
How many are left?

**2**

**a**

| Tens | Ones |
|---|---|
| 4 | 9 |
| – 1 | 3 |
| | |

**b**

| Tens | Ones |
|---|---|
| 7 | 5 |
| – 4 | 4 |
| | |

**c**

| Tens | Ones |
|---|---|
| 5 | 7 |
| – 3 | 4 |
| | |

**d**

| Tens | Ones |
|---|---|
| 6 | 9 |
| – 4 | 2 |
| | |

**e**

| Tens | Ones |
|---|---|
| 5 | 6 |
| – 2 | 1 |

**f**

| Tens | Ones |
|---|---|
| 4 | 8 |
| – 1 | 4 |

**g**

| Tens | Ones |
|---|---|
| 6 | 7 |
| – 2 | 5 |

**h**

| Tens | Ones |
|---|---|
| 7 | 4 |
| – 4 | 0 |

**i**

| Tens | Ones |
|---|---|
| $9 | 4 |
| – $3 | 1 |

**j**

| Tens | Ones |
|---|---|
| $6 | 3 |
| – $2 | 3 |

**k**

| Tens | Ones |
|---|---|
| $8 | 5 |
| – $8 | 1 |

**l**

| Tens | Ones |
|---|---|
| $7 | 4 |
| – $2 | 2 |

● Use a calculator to check your answers in Question 2.

 • *AUSTRALIAN SIGNPOST MATHS 3* • ISBN 978 1 4886 2184 0

# 2:30 Multiplication

CONCEPT

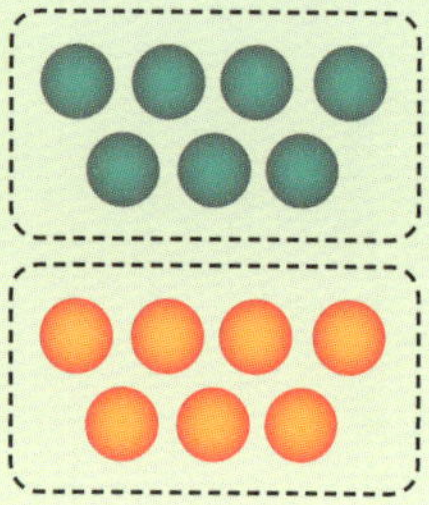

Double 7 = 2 × 7 = 14

Half 14 = 14 ÷ 2 = 7

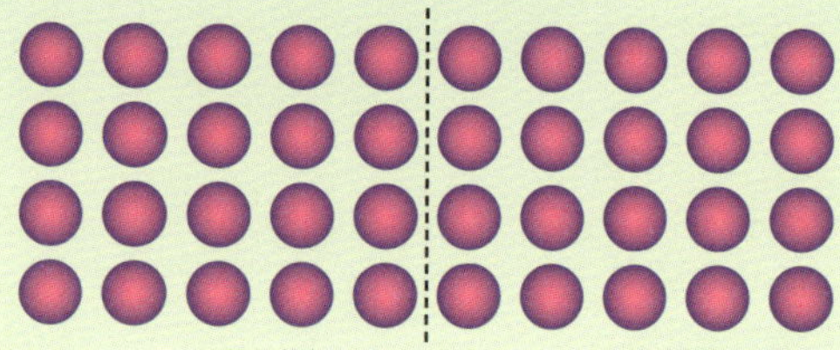

Half of 4 × 10 = 4 × 5

**1** Fill in the multiplication table. Compare the × 5 and × 10 answers.

| | 2 | 5 | 1 | 10 | 3 | 7 | 4 | 9 | 6 | 8 |
|---|---|---|---|---|---|---|---|---|---|---|
| × 5 | | | | | | | | | | |
| × 10 | | | | | | | | | | |

**a** Is 10 × 5 equal to 5 × 10? ☐

**b** Is 3 × 5 half of 3 × 10? ☐

**c** Is 3 × 10 twice 3 × 5? ☐

**d** Is 4 × 5 half of 8 × 10? ☐

**e** Is 4 × 10 twice 4 × 5? ☐

**f** Is 4 × 5 half of 4 × 10? ☐

**g** Are the × 5 answers half of the × 10 answers? ☐ Why is this?

**2** Do what is inside the brackets first.

**a** 5 × 2 = (4 × 2) + (1 × 2) = ☐ + ☐ = ☐

**b** 8 × 5 = (4 × 5) + (4 × 5) = ☐ + ☐ = ☐

**c** 13 × 10 = (10 × 10) + (3 × 10) = ☐ + ☐ = ☐

**d** 11 × 5 = (10 × 5) + (1 × 5) = ☐ + ☐ = ☐

**3** Fill in the multiplication table by doubling.

| | 7 | 9 | 8 | 11 | 14 | 21 | 12 | 6 | 13 | 16 |
|---|---|---|---|---|---|---|---|---|---|---|
| × 2 | | | | | | | | | | |

**a** 16 × 2 = (10 × 2) + (6 × 2) = ☐ + ☐ = ☐

**b** 34 × 2 = (30 × 2) + (4 × 2) = ☐ + ☐ = ☐

**c** 201 × 2 = (200 × 2) + (1 × 2) = ☐ + ☐ = ☐

# 2:31 Linking Multiplication and Division

CONCEPT

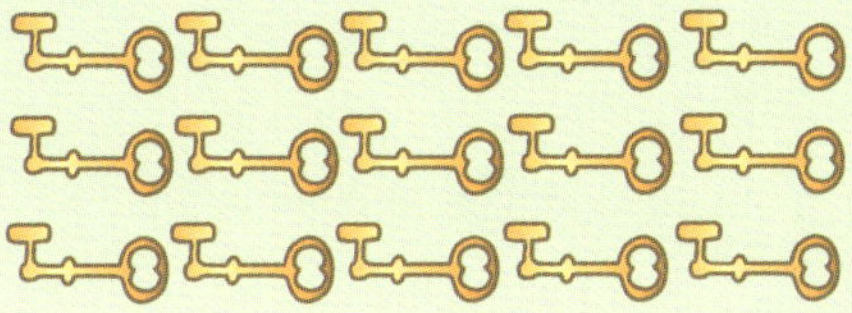

For 15 ÷ 5, ask:
**What** × 5 = 15?

**1** a What is 4 groups of 5? ____ 4 × 5 = ____
b How many 5s in 20? ____ 20 ÷ 5 = ____
c Share 20 among 4. ____ 20 ÷ 4 = ____

**2** a What is 5 groups of 4? ____ 5 × 4 = ____
b How many 4s in 20? ____ 20 ÷ 4 = ____
c Share 20 among 5. ____ 20 ÷ 5 = ____

**3** a What is 3 groups of 8? ____ 3 × 8 = ____
b How many 8s in 24? ____ 24 ÷ 8 = ____
c Share 24 among 3. ____ 24 ÷ 3 = ____

**4** a What is 9 groups of 2? ____ 9 × 2 = ____
b How many 2s in 18? ____ 18 ÷ 2 = ____
c Share 18 among 9. ____ 18 ÷ 9 = ____

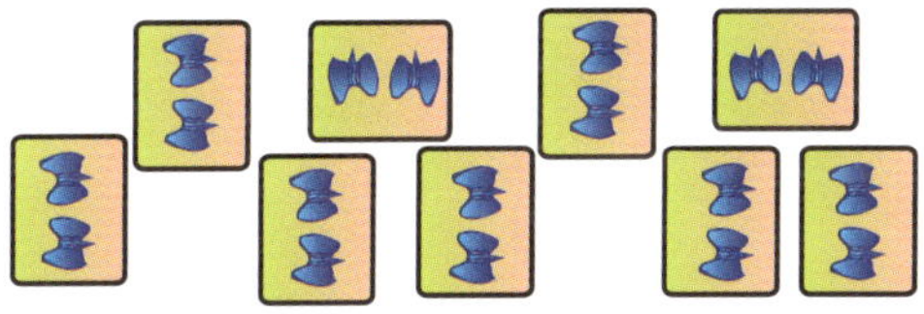

CONCEPT

10 × 5 = 50 ↗ 50 ÷ 5 = 10
↘ 50 ÷ 10 = 5

For 50 ÷ 5 = ?, ask: ↗ ____ × 5 = 50
↘ 5 × ____ = 50

**5** Use the first number sentence to complete the others.

a 5 × 2 = 10
10 ÷ 2 = ____
10 ÷ 5 = ____

b 10 × 3 = 30
30 ÷ 3 = ____
30 ÷ 10 = ____

c 7 × 2 = 14
14 ÷ 2 = ____
14 ÷ 7 = ____

d 9 × 3 = 27
27 ÷ 3 = ____
27 ÷ 9 = ____

**6** Use the first number sentence to complete the others.

a 12 ÷ 3 = 4
____ × 3 = 12
3 × ____ = 12

b 35 ÷ 5 = 7
____ × 5 = 35
5 × ____ = 35

c 50 ÷ 10 = 5
____ × 10 = 50
10 × ____ = 50

d 80 ÷ 10 = 8
____ × 10 = 80
10 × ____ = 80

# 2:32 Multiples

CONCEPT

When we count by 2s, we get the **multiples** of 2.

- 2, 4, 6, 8, 10, 12, … are the multiples of 2.
- 5, 10, 15, 20, 25, … are the multiples of 5.
- 10, 20, 30, 40, 50, … are the multiples of 10.

35 is a multiple of 5. 18 is a multiple of 2.

**1** **a** Skip count by 2s to write the multiples of 2.

2, 4, 6, 8, ____, ____, ____, ____, ____, ____, ____, ____

**b** Skip count by 5s to write the multiples of 5.

5, 10, 15, ____, ____, ____, ____, ____, ____, ____, ____

**c** Skip count by 3s to write the multiples of 3.

3, 6, 9, ____, ____, ____, ____, ____, ____, ____, ____

**2** Write the numbers 2, 5, 10, 12, 15, 16 and 20 that are:

**a** Multiples of 2 ____

**b** Multiples of 5 ____

**c** Multiples of 10 ____

**3** **a** If $2 \times 10 = 20$ then 20 is a multiple of ____ and ____.

**b** If $5 \times 2 = 10$ then 10 is a multiple of ____ and ____.

**c** If $3 \times 4 = 12$ then 12 is a multiple of ____ and ____.

**4** This array has 4 rows and 9 columns.
There are 36 circles altogether,
so 36 is a multiple of ____ and ____.

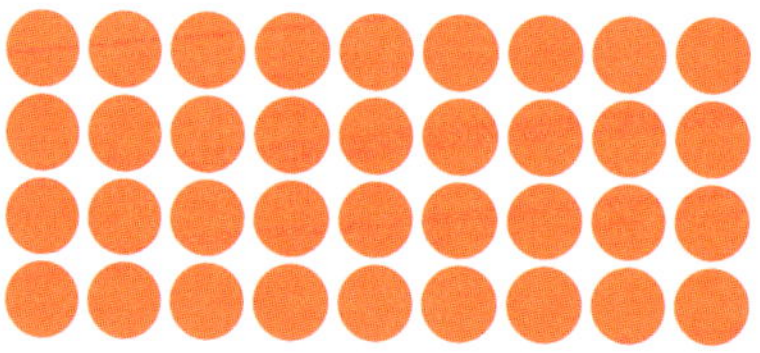

ACTIVITY

Use counters to make arrays like the one in Question 4.
For each array, count the counters and complete this sentence:

____ is a multiple of ____ and ____.


 ISBN 978 1 4886 2184 0

# 2:33 Division Facts from Multiplication Facts

CONCEPT

To find how many 5s in 15, ask:
What multiplied by 5 gives 15?

Groups of 5 in 15:

Answer: 3

**1**
- a 8 ÷ 2 = ?
  ☐ × 2 = 8
  Answer = ☐
- b 50 ÷ 5 = ?
  ☐ × 5 = 50
  Answer = ☐
- c 50 ÷ 10 = ?
  ☐ × 10 = 50
  Answer = ☐
- d 20 ÷ 5 = ?
  ☐ × 5 = 20
  Answer = ☐
- e 45 ÷ 5 = ?
  ☐ × 5 = 45
  Answer = ☐
- f 18 ÷ 3 = ?
  ☐ × 3 = 18
  Answer = ☐
- g 35 ÷ 5 = ?
  ☐ × 5 = 35
  Answer = ☐
- h 80 ÷ 10 = ?
  ☐ × 10 = 80
  Answer = ☐

**2** Use the first number sentence to complete the others.
- a 2 × 10 = 20
  20 ÷ 10 = ☐
  20 ÷ 2 = ☐
- b 6 × 2 = 12
  12 ÷ 2 = ☐
  12 ÷ 6 = ☐
- c 5 × 3 = 15
  15 ÷ 3 = ☐
  15 ÷ 5 = ☐
- d 8 × 5 = 40
  40 ÷ 5 = ☐
  40 ÷ 8 = ☐

**3**
- a ☐ × 10 = 10
- b ☐ × 2 = 20
- c ☐ × 10 = 60
- d ☐ × 3 = 15
- e 5 × ☐ = 5
- f 3 × ☐ = 30
- g 4 × ☐ = 8
- h 2 × ☐ = 16

**4**
- a 10 ÷ 2 = ☐
- b 8 ÷ 2 = ☐
- c 14 ÷ 2 = ☐
- d 12 ÷ 2 = ☐
- e 25 ÷ 5 = ☐
- f 30 ÷ 5 = ☐
- g 40 ÷ 5 = ☐
- h 45 ÷ 5 = ☐
- i 20 ÷ 10 = ☐
- j 80 ÷ 10 = ☐
- k 90 ÷ 10 = ☐
- l 70 ÷ 10 = ☐

**5**
- a How many groups of 2 are in 10? ☐ 10 ÷ 2 = ☐
- b How many groups of 10 are in 2? ☐
- c Is 10 ÷ 2 equal to 2 ÷ 10? ☐

## Number Line Jumps

ACTIVITY

- One person draws a number line, starting at zero, that shows several equal jumps.
- The other person writes multiplication and division number sentences to match the picture.

This shows 6 × 5 = 30 and 30 ÷ 5 = 6.

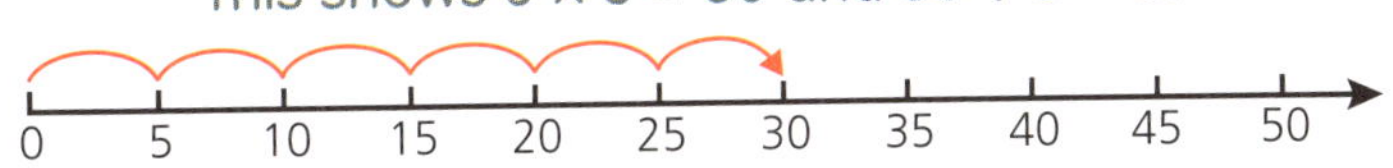

# 2:34 Addition to 99 with Trading

CONCEPT

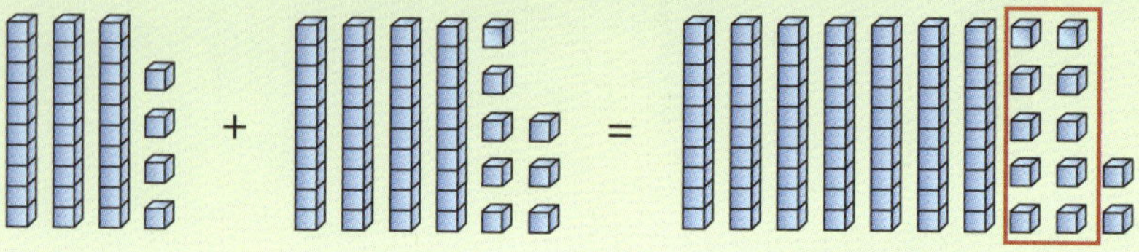

34 + 48 = 82

**1** Use the models to complete the number sentences.

**a**

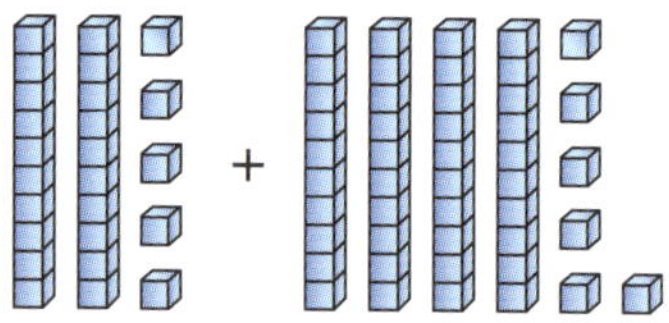

☐ + ☐ = ☐

**b**

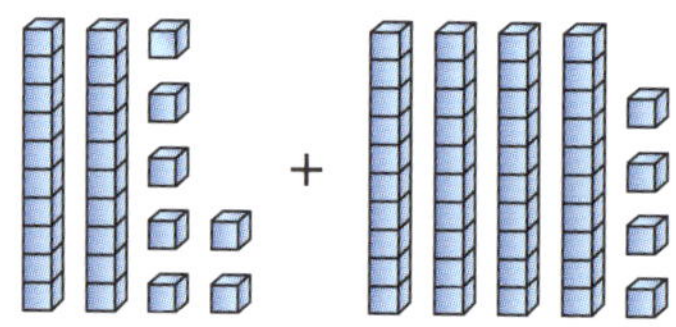

☐ + ☐ = ☐

**c**

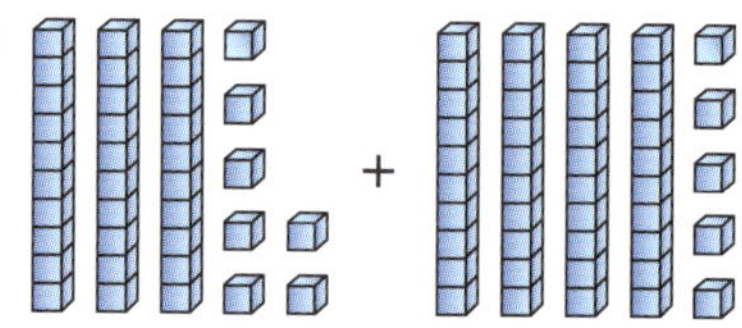

☐ + ☐ = ☐

**d**

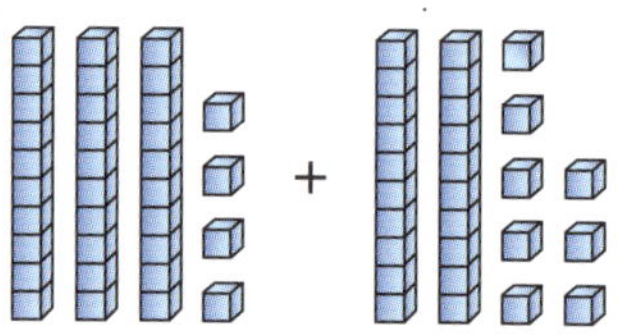

☐ + ☐ = ☐

**e**

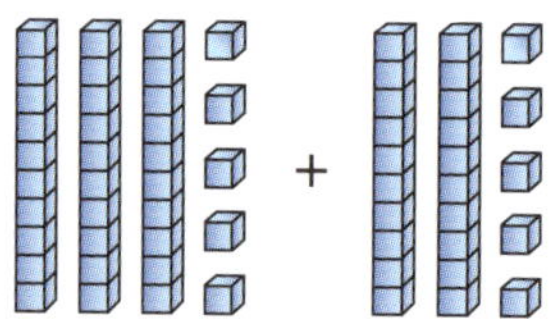

☐ + ☐ = ☐

**f**

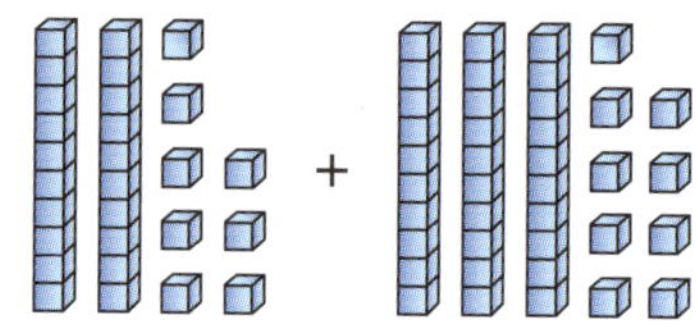

☐ + ☐ = ☐

**2** Use place-value blocks to complete the number sentences.

**a** 27 + 35 = ☐ **b** 56 + 25 = ☐ **c** 46 + 28 = ☐

**d** 66 + 26 = ☐ **e** 35 + 47 = ☐ **f** 28 + 52 = ☐

**g** 53 + 37 = ☐ **h** 44 + 39 = ☐ **i** 37 + 49 = ☐

**3** Use the jump strategy to complete the number sentences.

**a** 27 + 26 = ☐

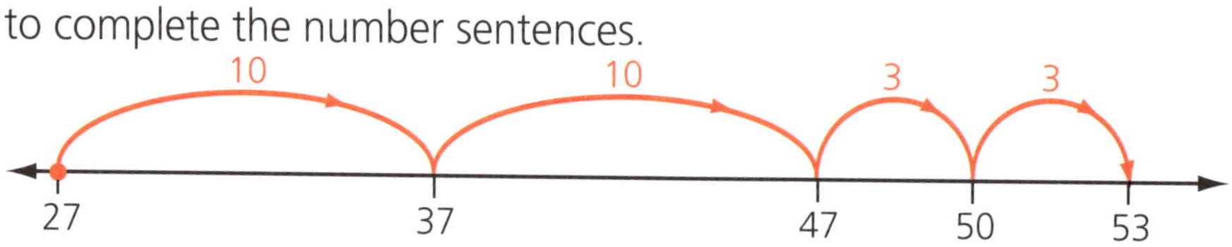


**b** 58 + 24 = ☐

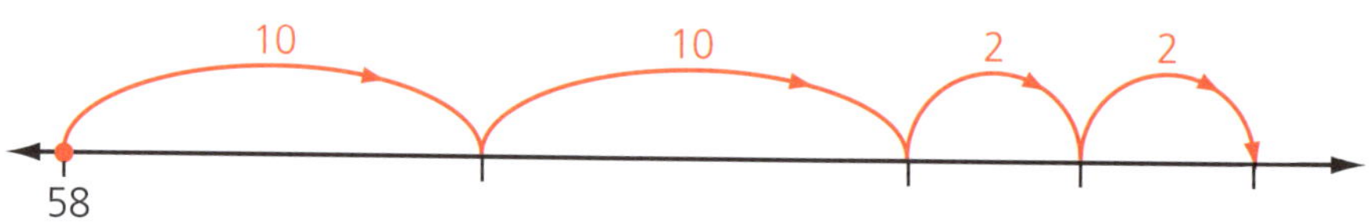


**c** 48 + 19 = ☐

**d** 67 + 24 = ☐

Use a calculator to check your answers.

# 2:35 Addition to 99 with Trading

**1**

| + | 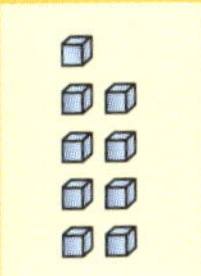 | 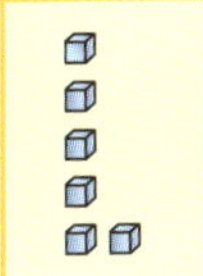 | 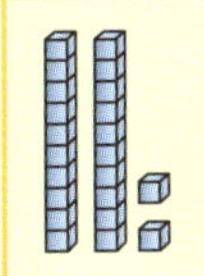 | 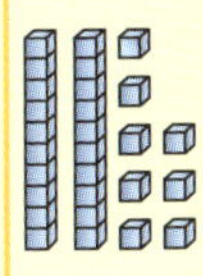 | 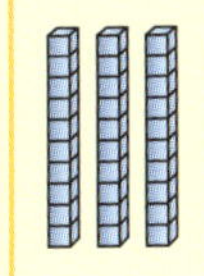 | 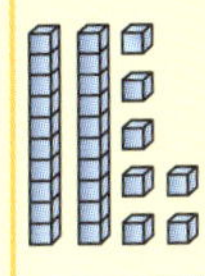 | 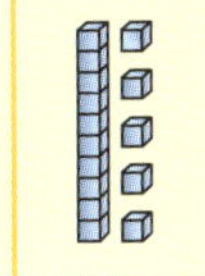 | 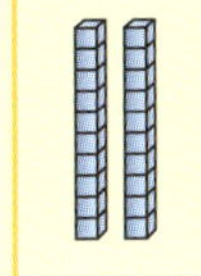 | 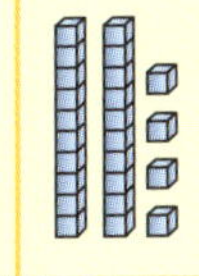 |
|---|---|---|---|---|---|---|---|---|---|
| 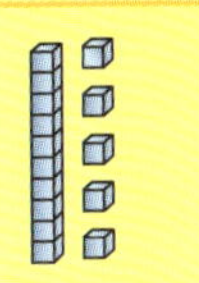 | 24 | | | | | | | | |
| 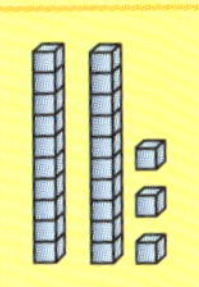 | | | | | | | | | |
| 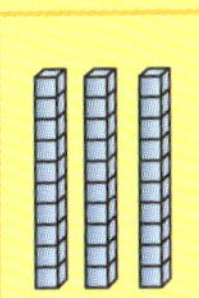 | | | | | | | | | |

**2** Use craft sticks to complete the number sentences.

**a** 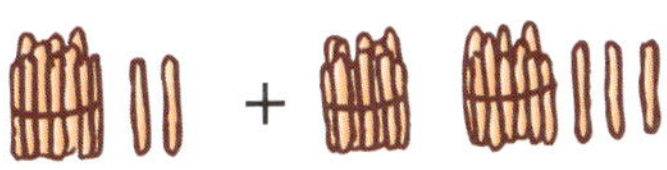

☐ + ☐ = ☐

**b** 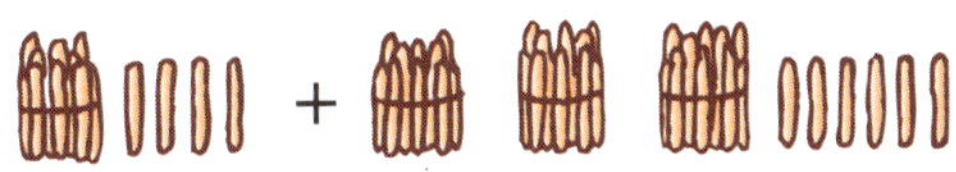

☐ + ☐ = ☐

**c** 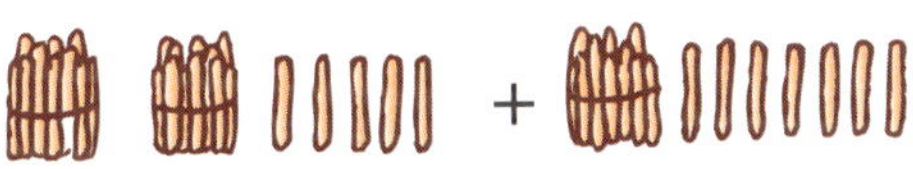

☐ + ☐ = ☐

**d** 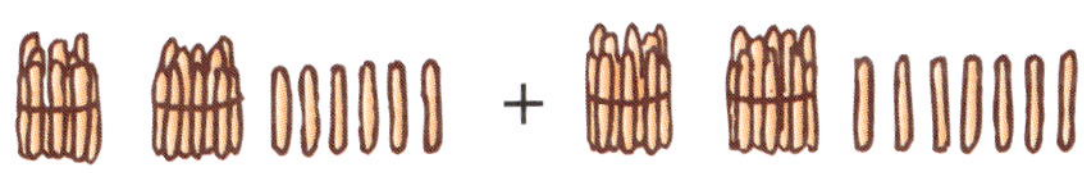

☐ + ☐ = ☐

**3** Use craft sticks or place-value blocks to complete the number sentences.

**a** 24 + 67 = ☐ **b** 46 + 34 = ☐ **c** 26 + 45 = ☐

**d** 75 + 18 = ☐ **e** 35 + 24 = ☐ **f** 62 + 15 = ☐

**g** 47 + 25 = ☐ **h** 38 + 46 = ☐ **i** 51 + 24 = ☐

**j** 28 + 48 = ☐ **k** 53 + 15 = ☐ **l** 34 + 37 = ☐

- Use a calculator to check your answers.

## Adding Craft Sticks

ACTIVITY

- Write the numbers 1 to 50 on craft sticks.
- Take turns to pick two sticks without looking.
- Add the two numbers on your sticks.
- Check your addition with a calculator. If you were correct, keep the sticks.
- The person with the most sticks at the end is the winner.

Try to use the most efficient strategy.

# 2:36 Addition to 99 with Trading

CONCEPT

- 5 ones + 7 ones = 12 ones
- Trade 10 ones for 1 ten
- 4 tens + 1 ten = 5 tens

45 + 7

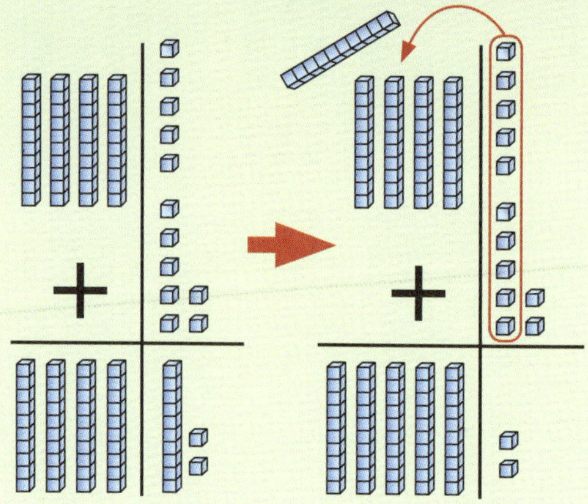

We write it like this.

| Tens | Ones |
|---|---|
| 1 | |
| 4 | 5 |
| + | 7 |
| 5 | 2 |

The 10 ones are shown as 1 ten.

**1** Trade ten ones for one ten to complete each addition.

**a**

| Tens | Ones |
|---|---|
| | |
| 2 | 4 |
| + | 7 |
| | |

**b**

| Tens | Ones |
|---|---|
| | |
| 1 | 8 |
| + | 5 |
| | |

**c**

| Tens | Ones |
|---|---|
| | |
| 3 | 5 |
| + | 6 |
| | |

**d**

| Tens | Ones |
|---|---|
| | |
| 4 | 6 |
| + | 9 |
| | |

**e** Tens Ones

```
     3
+ 5  8
```

**f** Tens Ones

```
     7
+ 4  7
```

**g** Tens Ones

```
     4
+ 7  6
```

**h** Tens Ones

```
     8
+ 3  4
```

**i** Tens Ones

```
  $   5
+ $8  9
```

**j** Tens Ones

```
  $5  4
+ $   9
```

**k** Tens Ones

```
  $2  7
+ $   3
```

**l** Tens Ones

```
  $6  9
+ $   8
```

**2** Write a number sentence for each number story.

**a** Jan has 4 cats and 29 mice. How many animals has she altogether?

**b** Mark has 6 canaries and 37 budgies. How many birds does he have?

**c** Lyn has 65 ants and 8 bees. How many insects has she altogether?

**d** 57 five-cent coins and 6 ten-cent coins. How many coins?



 • *AUSTRALIAN SIGNPOST MATHS 3* • ISBN 978 1 4886 2184 0

# 2:37 Addition to 99 with Trading

CONCEPT

Thirty-eight ducks and twenty-six chickens were in the yard. How many birds were there altogether?

- **Find:** How many birds?
- **Number sentence:** 38 + 26 = ☐
- **Answer:** 64 birds were in the yard.

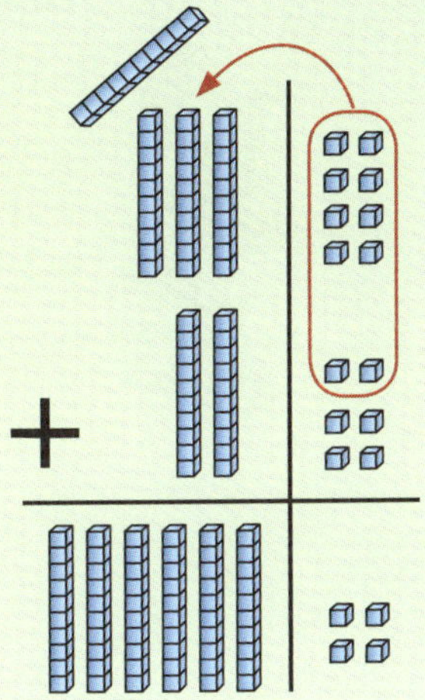

| | Tens | Ones |
|---|---|---|
| | 1 | |
| | 3 | 8 |
| + | 2 | 6 |
| | 6 | 4 |

**1** Complete the additions and show the trading.

**a**

| | Tens | Ones |
|---|---|---|
| | | |
| | 1 | 2 |
| + | 2 | 8 |
| | | |

**b**

| | Tens | Ones |
|---|---|---|
| | | |
| | 3 | 3 |
| + | 3 | 9 |
| | | |

**c**

| | Tens | Ones |
|---|---|---|
| | | |
| | 4 | 5 |
| + | 1 | 6 |
| | | |

**d**

| | Tens | Ones |
|---|---|---|
| | | |
| | 2 | 7 |
| + | 3 | 4 |
| | | |

**e**

| | Tens | Ones |
|---|---|---|
| | 6 | 5 |
| + | 2 | 8 |

**f**

| | Tens | Ones |
|---|---|---|
| | 1 | 1 |
| + | 5 | 9 |

**g**

| | Tens | Ones |
|---|---|---|
| | 3 | 6 |
| + | 5 | 7 |

**h**

| | Tens | Ones |
|---|---|---|
| | 2 | 4 |
| + | 2 | 8 |

**i**

| | Tens | Ones |
|---|---|---|
| | 7 | 7 |
| + | 1 | 7 |

**j**

| | Tens | Ones |
|---|---|---|
| | 6 | 6 |
| + | 2 | 9 |

**k**

| | Tens | Ones |
|---|---|---|
| | 4 | 5 |
| + | 3 | 7 |

**l**

| | Tens | Ones |
|---|---|---|
| | 2 | 8 |
| + | 5 | 4 |

**2**

**a**

| | Tens | Ones |
|---|---|---|
| | $4 | 6 |
| + | $ | 8 |

**b**

| | Tens | Ones |
|---|---|---|
| | $ | 7 |
| + | $8 | 9 |

**c**

| | Tens | Ones |
|---|---|---|
| | $2 | 6 |
| + | $6 | 4 |

**d**

| | Tens | Ones |
|---|---|---|
| | $4 | 9 |
| + | $3 | 1 |

 • *AUSTRALIAN SIGNPOST MATHS 3* • ISBN 978 1 4886 2184 0

# 2:38 Problem Solving

Solve these problems.

1. Davis has 100 bones. He gave his dog 40 of them. How many were left?

   Find: How many bones were left?

   Number sentence: ______ = ______

   Answer: ______

Find: ______

Number sentence: ______ = ______

Answer: ______

2. Ten pies are needed to fill one tray. Three trays are filled. How many pies are there?

   Find: ______

   Number sentence: ______ = ______

   Answer: ______

3. Corina had 24 apricots. She gave 2 to each friend until no apricots were left. How many friends were given apricots?

   Find: ______

   Number sentence: ______ = ______

   Answer: ______

4. In a library, one shelf holds 24 books. Another shelf holds 55. How many books do these shelves hold altogether?

   Find: ______

   Number sentence: ______ = ______

   Answer: ______

5. At the first stop, 15 people got off the bus. At the second stop, 10 people got off. At the third, 8 got off. How many people got off altogether?

   Find: ______

   Number sentence: ______ = ______

   Answer: ______

6. In a shop, bottles are kept in rows of 5. There are 3 full rows and 2 bottles left over. How many bottles are there altogether?

   Find: ______

   Number sentence: ______ = ______

   Answer: ______

7. How many rows of bottles could you make if you had 45 bottles and you put 5 bottles in each row?

   Find: ______

   Number sentence: ______ = ______

   Answer: ______

8. Discuss ways of solving these problems in your head.

   **a** 1000 – 101 = ______ **b** 1000 – 110 = ______ **c** 1000 – 410 = ______

   **d** 1000 – 510 = ______ **e** 1000 – 302 = ______ **f** 1000 – 802 = ______

   **g** Could you use place value to find your answers quickly? ______



 *AUSTRALIAN SIGNPOST MATHS 3* • ISBN 978 1 4886 2184 0

# 2:39 Problem Solving

Problem solving is easy when you set out your work like this.

Find:

Number sentence:

Answer:

Working

Solve these problems.

1. Holly had 43 football cards and bought 25 more. How many altogether?

   Find:

   Number sentence:

   Answer:

   Working

2. 5 boys shared 45 books equally. How many did each boy have?

   Find:

   Number sentence:

   Answer:

   Working

3. A farmer had 96 sheep but sold 28. How many did she have left?

   Find:

   Number sentence:

   Answer:

   Working

4. Marnie collected 54 aluminium cans before lunch and 37 in the afternoon. How many cans did she collect?

   Find:

   Number sentence:

   Answer:

   Working

5. Lee planted seeds in rows of 5. He planted 6 full rows and had 4 seeds left. How many seeds were there altogether?

   Find:

   Number sentence:

   Answer:

   Working

INVESTIGATION

6. Lauren and Justin want to play handball. For a handball court they need two squares next to each other.

   **a** How many different courts could they choose?

   **b** How many courts could all be used at the same time?

   **c** If they needed three squares in a row to make a court, how many courts could be used at the same time?

# 2:40 Addition Involving Hundreds

**I show**

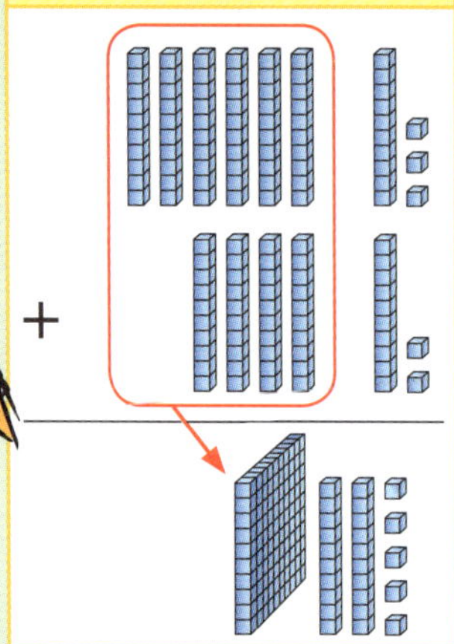


**I write**

| | Tens | Ones |
|---|---|---|
| | 7 | 3 |
| + | 5 | 2 |
| 1 | 2 | 5 |

CONCEPT

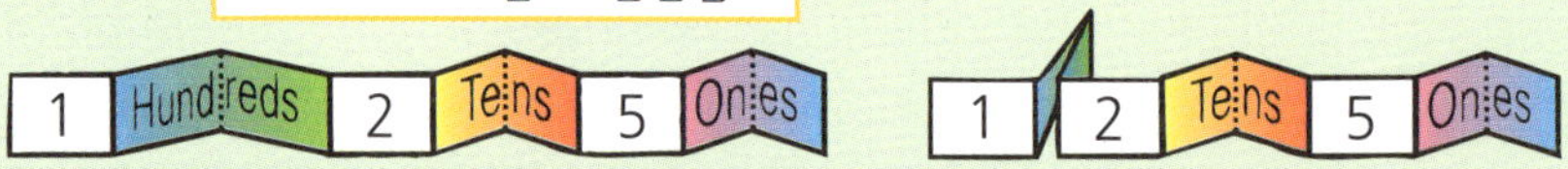


**1** Model these problems with place-value blocks. Fill in the answers.

**a** 72 books,
34 books. ☐ + ☐ = ☐
How many books altogether? ☐

**b** 92 stickers,
45 stickers. ☐ + ☐ = ☐
How many stickers altogether? ☐

**c** 74 cars,
43 cars. ☐ + ☐ = ☐
How many cars altogether? ☐

**d** 82 girls,
54 boys. ☐ + ☐ = ☐
How many children altogether? ☐

**2**

**a**

| | Tens | Ones |
|---|---|---|
| | 7 | 6 |
| + | 3 | 1 |
| | | |

**b**

| | Tens | Ones |
|---|---|---|
| | 8 | 2 |
| + | 5 | 4 |
| | | |

**c**

| | Tens | Ones |
|---|---|---|
| | 6 | 4 |
| + | 6 | 3 |
| | | |

**d**

| | Tens | Ones |
|---|---|---|
| | 8 | 0 |
| + | 3 | 9 |
| | | |

**e**

| | Tens | Ones |
|---|---|---|
| | 9 | 2 |
| + | 9 | 6 |
| | | |

**f**

| | Tens | Ones |
|---|---|---|
| | 5 | 3 |
| + | 7 | 2 |
| | | |

**g**

| | Tens | Ones |
|---|---|---|
| | 2 | 9 |
| + | 9 | 0 |
| | | |

**h**

| | Tens | Ones |
|---|---|---|
| | 3 | 7 |
| + | 8 | 1 |
| | | |

**i**

| | Tens | Ones |
|---|---|---|
| | 9 | 4 |
| + | 4 | 3 |
| | | |


 • *AUSTRALIAN SIGNPOST MATHS 3* • ISBN 978 1 4886 2184 0

# 2:41 Addition Involving Hundreds

CONCEPT

**I show**

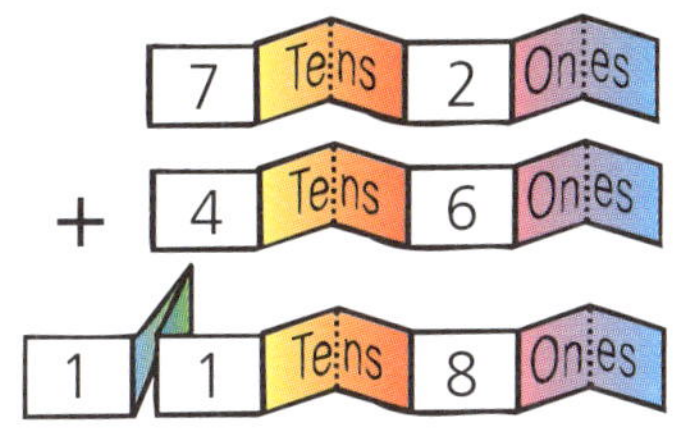


**I write**

$72 + 46 = 118$

| | Tens | Ones |
|---|---|---|
| 1 | 7 | 2 |
| + | 4 | 6 |
| 1 | 1 | 8 |

118 shells were found altogether.

**1** Use the numeral expanders to complete the additions.

**a** 62 balls, 71 balls.

☐ balls altogether.

**b** 32 trees, 75 trees.

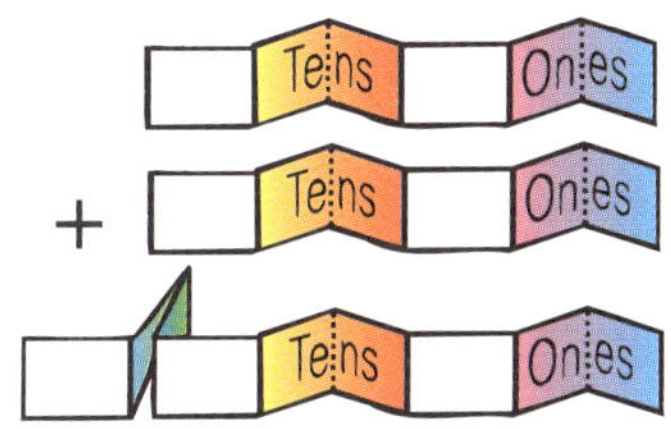


☐ trees altogether.

**c** 64 boats, 82 boats.

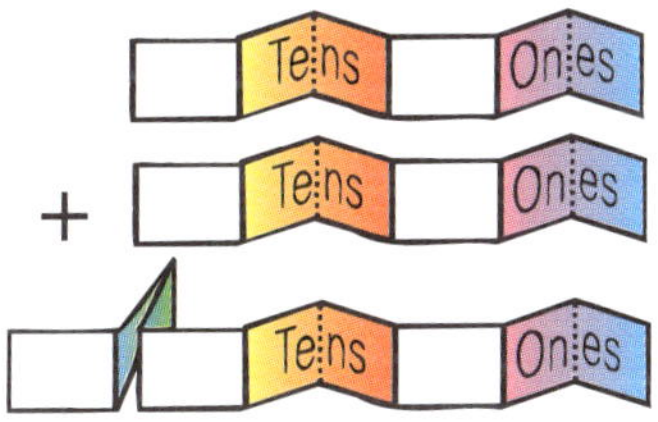


☐ boats altogether.

**d** 96 cards, 83 cards.

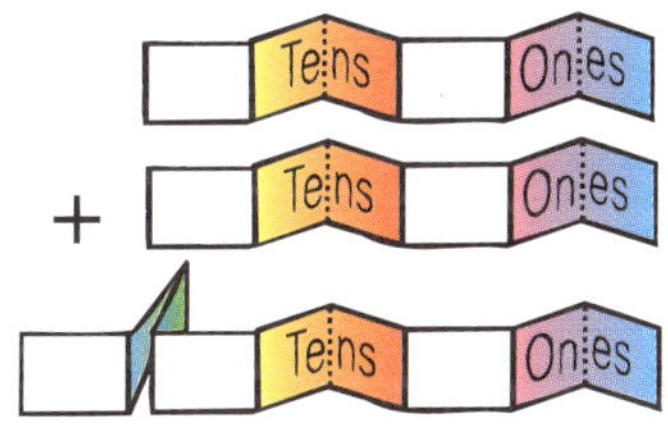


☐ cards altogether.

**2**

**a** 6 tens and 3 ones
+ 8 tens and 4 ones

**b** 5 tens and 5 ones
+ 7 tens and 4 ones

**c** 9 tens and 4 ones
+ 8 tens and 4 ones

**d** 5 tens and 1 ones
+ 7 ten and 3 ones

**e** 8 tens and 3 ones
+ 4 tens and 5 ones

**f** 2 tens and 7 one
+ 9 tens and 0 ones

**3** Do these sums in your head using the method above. Write the answers.

**a** $83 + 42 =$ ☐ **b** $48 + 90 =$ ☐ **c** $74 + 93 =$ ☐

**d** $70 + 84 =$ ☐ **e** $56 + 71 =$ ☐ **f** $77 + 31 =$ ☐

**g** $92 + 41 =$ ☐ **h** $29 + 90 =$ ☐ **i** $38 + 81 =$ ☐

# 2:42 Subtraction to 999

CONCEPT

**I show**

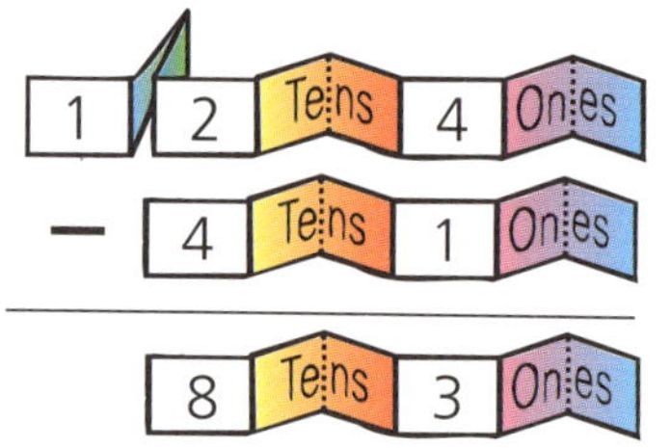


**I write**

124 – 41 = 83

| | Tens | Ones |
|---|---|---|
| 1 | 2 | 4 |
| – | 4 | 1 |
| | 8 | 3 |

There are 83 balloons left.

**1** Use the numeral expanders to complete the subtractions.

**a** 116 grapes, 51 eaten.

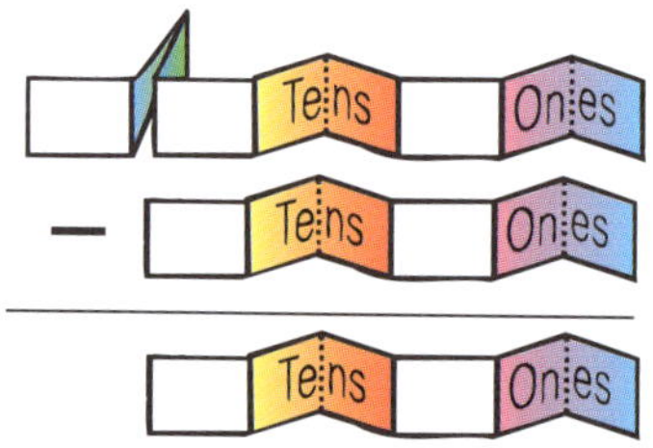


☐ left.

**b** $225, $34 spent.

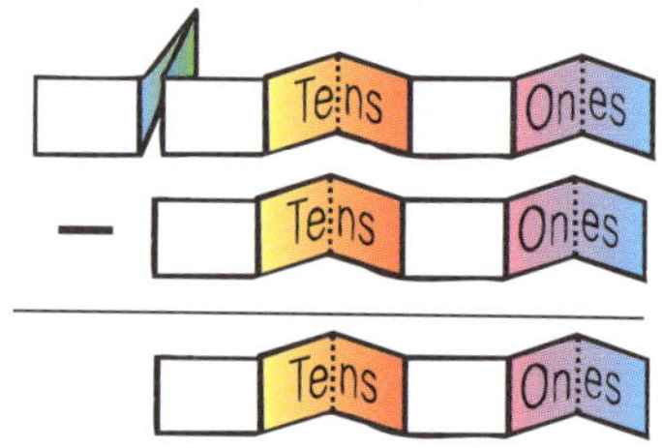


☐ left.

**c** 342 beads, 61 lost.

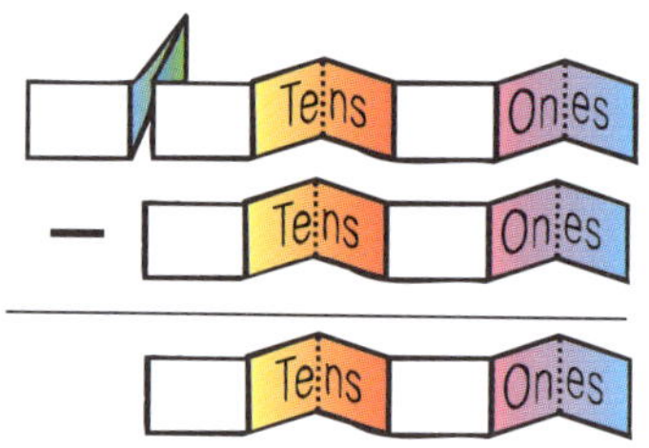


☐ left.

**d** $408, $64 spent.

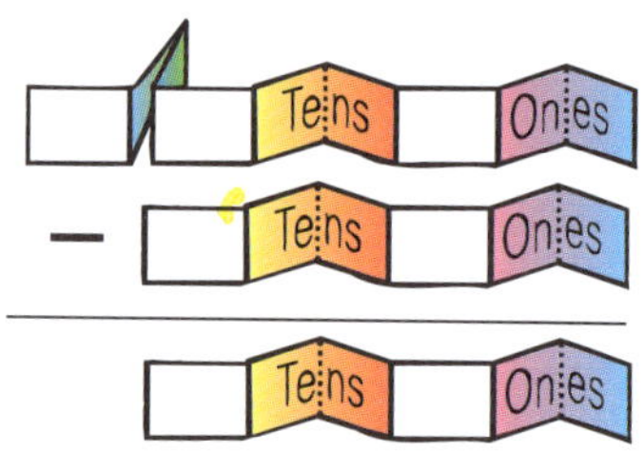


☐ left.

**2**

**a** 12 tens and 9 ones
– 5 tens and 3 ones

**b** 24 tens and 3 ones
– 8 tens and 1 one

**c** 31 tens and 5 ones
– 6 tens and 4 ones

**d** 53 tens and 8 ones
– 4 tens and 5 ones

**e** 65 tens and 7 ones
– 6 tens and 5 ones

**f** 48 tens and 3 ones
– 9 tens and 2 ones

**3** Do these sums in your head using the method above. Write the answers.

**a** 143 – 52 = ☐
**b** 329 – 48 = ☐
**c** 208 – 82 = ☐
**d** 612 – 31 = ☐
**e** 427 – 47 = ☐
**f** 743 – 70 = ☐
**g** 507 – 52 = ☐
**h** 833 – 53 = ☐
**i** 959 – 73 = ☐

● Do you think this strategy is useful? ☐ Discuss.

# 2:43 Addition to 999

CONCEPT

**I show**

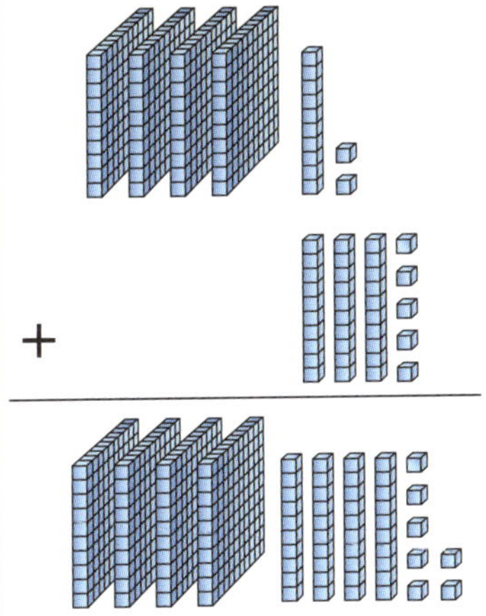

**I write**

| | Tens | Ones |
|---|---|---|
| 4 | 1 | 2 |
| + | 3 | 5 |
| 4 | 4 | 7 |

Use place-value blocks, numeral expanders or mental strategies to complete the additions. Judge for yourself whether this is useful.

**1** **a** 162 eggs, 41 eggs. ☐ + ☐ = ☐

How many eggs altogether? ☐

**b** 171 stamps, 28 stamps. ☐ + ☐ = ☐

How many stamps altogether? ☐

**2**

**a**

| | Tens | Ones |
|---|---|---|
| 2 | 3 | 1 |
| + | 5 | 6 |
| | | |

**b**

| | Tens | Ones |
|---|---|---|
| 3 | 7 | 4 |
| + | 6 | 3 |
| | | |

**c**

| | Tens | Ones |
|---|---|---|
| 1 | 9 | 4 |
| + | 3 | 2 |
| | | |

**d**

| | Tens | Ones |
|---|---|---|
| 2 | 8 | 6 |
| + | 3 | 1 |
| | | |

**e**

| | Tens | Ones |
|---|---|---|
| 4 | 5 | 4 |
| + | 7 | 0 |
| | | |

**f**

| | Tens | Ones |
|---|---|---|
| | 4 | 5 |
| + 4 | 6 | 2 |
| | | |

**g**

| | Tens | Ones |
|---|---|---|
| 5 | 6 | 3 |
| + | 5 | 2 |
| | | |

**h**

| | Tens | Ones |
|---|---|---|
| 6 | 7 | 2 |
| + | 7 | 3 |
| | | |

**i**

| | Tens | Ones |
|---|---|---|
| 6 | 4 | 1 |
| + | 8 | 3 |
| | | |

**j**

| | Tens | Ones |
|---|---|---|
| 8 | 3 | 5 |
| + | 9 | 2 |
| | | |

**k**

| | Tens | Ones |
|---|---|---|
| 5 | 7 | 6 |
| + | 5 | 2 |
| | | |

**l**

| | Tens | Ones |
|---|---|---|
| 4 | 4 | 6 |
| + | 8 | 3 |
| | | |

**3** **a** 762 + 60 = ☐ **b** 391 + 24 = ☐ **c** 280 + 54 = ☐

**d** 584 + 33 = ☐ **e** 762 + 80 = ☐ **f** 545 + 62 = ☐

# 2:44 Checking Subtraction by Addition

Here is the problem.

Diane had 57 books. She gave 31 to the library. How many did she have left?

Here is the working.

| Tens | Ones |
|---|---|
| 5 | 7 |
| – 3 | 1 |
| 2 | 6 |

Answer

This is how you check the answer.

| Tens | Ones |
|---|---|
| 2 | 6 |
| + 3 | 1 |
| 5 | 7 |

Check

Add the second number to your answer. You should get the first number back.

CONCEPT

Dianne has 26 books left.

**1** Complete each problem and check the answer.

**a** The teacher had 36 frogs. 14 got away. How many were left?

Tens Ones: 3 6 − 1 4 → Tens Ones: ☐ + 1 4

☐ frogs were left.

**b** James had $87. He spent $42. How much did he have left?

Tens Ones: 8 7 − 4 2 → Tens Ones: ☐ + 4 2

James had $ ☐ left.

Addition is the inverse of subtraction.

**c** 47 flowers were given to Kaylee. She gave away 9. How many were left?

Tens Ones: 4 7 − 9 → Tens Ones: ☐ + 9

Kaylee had ☐ flowers left.

**d** The zoo has 24 emus and 8 koalas. How many more emus than koalas?

Tens Ones: 2 4 − 8 → Tens Ones: ☐ +

There are ☐ more emus.

To find the distance, you subtract.

**2** Find the distance between the places on each sign.

**a** Blaxland 25 km / Lawson 6 km — Distance = ☐ km

**b** Mackay 41 km / Clairview 7 km — Distance = ☐ km

**c** Horsham 8 km / Nhill 36 km — Distance = ☐ km

**3** Use the signpost to find these distances:

**a** Dunnart to Bellus ☐

**b** Dunnart to Caves ☐

**c** Bellus to Caves ☐

**d** Gumnut to Caves ☐

**e** Dunnart to Gumnut ☐

**f** Bellus to Gumnut ☐

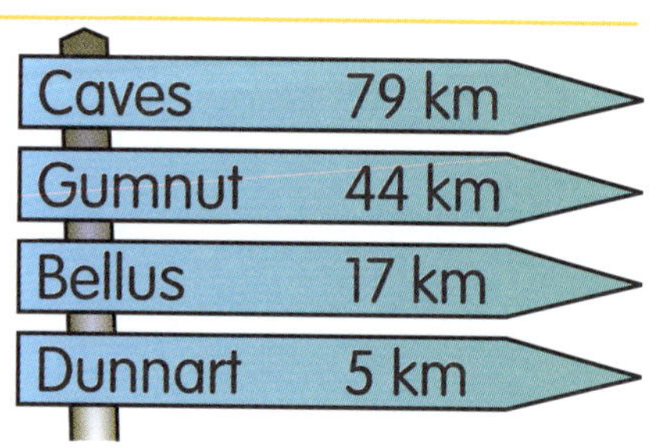

# 2:45 Inverse Operations, × and ÷

CONCEPT

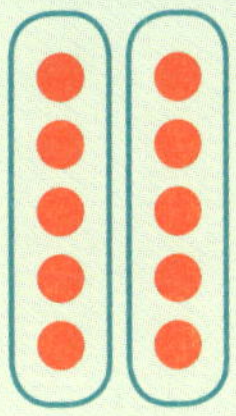

Double 5 = 10
Half 10 = 5
So,
5 (× 2)(÷ 2) = 5

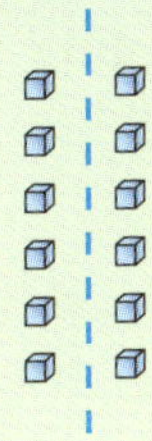

12 ÷ 2 = 6
6 × 2 = 12
So,
12 (÷ 2)(× 2) = 12

**1** a 20 ÷ 10 × 10 =
b 10 × 5 ÷ 5 =
c 3 × 2 ÷ 2 =
d 30 ÷ 5 × 5 =
e 8 × 2 ÷ 2 =
f 60 ÷ 10 × 10 =

**2** a 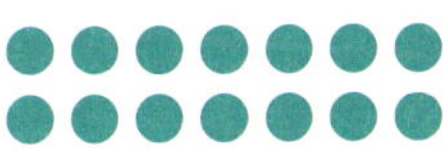

Half 14 =
Double 7 =
So, 14 ÷ 2 × 2 =

b 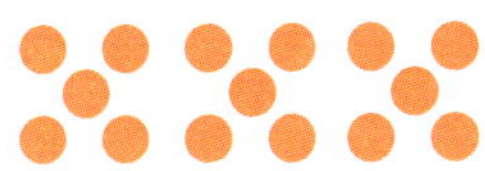

15 ÷ 5 =
3 × 5 =
So, 15 ÷ 5 × 5 =

c 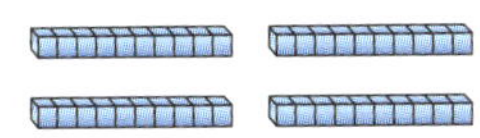

4 × 10 =
40 ÷ 10 =
So, 4 × 10 ÷ 10 =

d 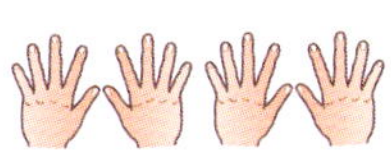

4 × 5 =
20 ÷ 5 =
So, 4 × 5 ÷ 5 =

e 

5 × 2 =
10 ÷ 2 =
So, 5 × 2 ÷ 2 =

f 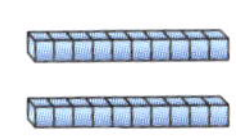

20 ÷ 2 =
10 × 2 =
So, 20 ÷ 2 × 2 =

**3** a 6 × 2 ÷ 2 =
b 10 ÷ 5 × 5 =
c 30 ÷ 10 × 10 =
d 4 ÷ 2 × 2 =
e 8 × 5 ÷ 5 =
f 50 ÷ 10 × 10 =
g 9 × 2 ÷ 2 =
h 7 × 5 ÷ 5 =
i 45 ÷ 5 × 5 =
j 6 × 10 ÷ 10 =

Circle parts that undo each other.
5 (× 2)(÷ 2)

FUN SPOT

Make up your own inverse number sentences.
Example: 8 (× 10)(÷ 10)(× 3)(÷ 3)(× 42)(× 2)(÷ 42)(÷ 2) = 8

# 2:46 Counting Change

To give change, we count on from the price paid.

**Example**: What is the change from $50 if the cost is $9.40?

- **Count on from $9.40.**
- $9.40 → $9.50 → $10.00 → $30 → $50
  10c | 50c | $20 | $20
- The change is 10c + 50c + $20 + $20 = **$40.60**.

CONCEPT

**1** Count on and write the change in coins and notes.

| | Cost | Money given | Counting on to find the change | Total change |
|---|---|---|---|---|
| a | $18.60 | $50 | → $18.80 → $19 → $20 → $40 → $50 | |
| b | $7.15 | $20 | → $7.20 → $7.40 → $7.50 → $8 → $10 → $20 | |
| c | $1.35 | $10 | → $1.40 → $1.50 → $2 → $4 → $5 → $10 | |
| d | $14.50 | $100 | → $15 → $20 → $40 → $50 → $100 | |
| e | $1.30 | $10 | | |
| f | $2.55 | $20 | | |

**2** Find the change from $20. Use only the boxes you need.

a $14.40 Change =

b $9.90 Change =

c $5.25 Change =

The square edges are for notes.

**3** Show three different ways of giving change from $10 if the cost is $1.95. Practise giving change using play money.

ACTIVITY



 • *AUSTRALIAN SIGNPOST MATHS 3* • ISBN 978 1 4886 2184 0

# 2:47 Rounding to the Nearest 5c

CONCEPT

Because we don't have 1 cent pieces, we always round the total bill to the nearest 5 cents.

20c: 18c, 19c, 20c, 21c, 22c

25c: 23c, 24c, 25c, 26c, 27c

30c: 28c, 29c, 30c, 31c, 32c

**1** Choose which answer is the cost, rounded to the nearest 5 cents.

a SALE $1.38 — $1.35 or $1.40 — $ ____

b SALE $2.53 — $2.50 or $2.55 — $ ____

**2** Round to the nearest 5 cents to find the cost.

a SALE $6.21 — $ ____
b SALE $5.36 — $ ____
c SALE $7.86 — $ ____

d SALE $4.77 — $ ____
e SALE $9.12 — $ ____
f SALE $3.99 — $ ____

g SALE $8.70 — $ ____
h SALE $0.54 — $ ____
i SALE $9.99 — $ ____

**3** Round each amount to the nearest 5c.

a $3.74 ____  b $1.87 ____  c $2.11 ____

d $6.73 ____  e $4.02 ____  f $8.26 ____

g $1.59 ____  h $3.55 ____  i $2.99 ____

**4** a Jodie bought a book for $8.92 and scissors for $3.02.
What was the rounded total? ____

b Rajiv bought two bunches of flowers.
Each bunch was $15.44. What was the rounded total? ____

c Peter bought a calculator for $15.37.
He paid with a $20 note. How much change did he get? ____

 • *AUSTRALIAN SIGNPOST MATHS 3* • ISBN 978 1 4886 2184 0

# 2:48 Problem Solving

1 Darren paid $5 for each light globe. How much would he pay for this many globes?

a 2 globes ☐ b 4 globes ☐ c 5 globes ☐
d 8 globes ☐ e 6 globes ☐ f 7 globes ☐

2 In each packet there are 10 times as many pencils as pens.
With this many pens, how many pencils?

a 3 pens ☐ b 5 pens ☐
c 9 pens ☐ d 10 pens ☐

3 Suzy has won 4 times as many games as Tom.
Tom has won 5 games. How many has Suzy won? ☐

4 Trinh earns 50c for each cake she sells. How much does she earn if she sells this many cakes?

a 10 cakes ☐ b 8 cakes ☐ c 12 cakes ☐

5 a $65 \div 5 = (50 \div 5) + (15 \div 5)$ = ☐ + ☐ = ☐
b $28 \div 2 = (20 \div 2) + (8 \div 2)$ = ☐ + ☐ = ☐
c $4 \times 6 = (5 \times 6) - (1 \times 6)$ = ☐ – ☐ = ☐
d $99 \times 6 = (100 \times 6) - (1 \times 6)$ = ☐ – ☐ = ☐

INVESTIGATION

6 Rachel put 20 chairs in front of the stage in 2 rows of 10 chairs, as shown. Use 20 counters to find other ways to arrange the chairs in equal rows. List all the ways.

| | |
|---|---|
| 2 × 10 | |
| | |
| | |

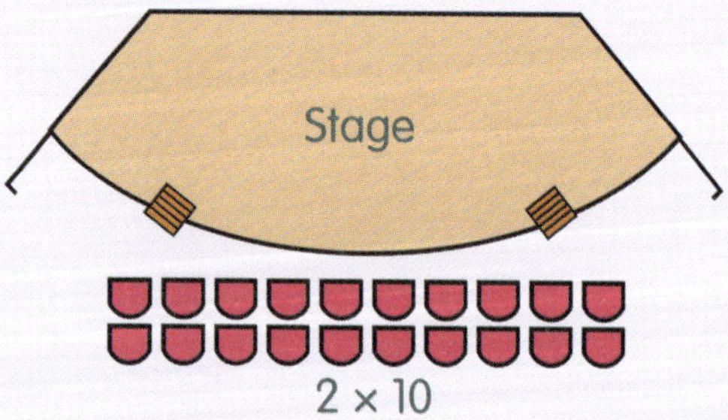


7 List all the ways that 30 chairs can be arranged in equal rows.

| | | | |
|---|---|---|---|
| | | | |
| | | | |


 • *AUSTRALIAN SIGNPOST MATHS 3* • ISBN 978 1 4886 2184 0

# 2:49 Subtraction with Trading

## Trading Down

| | | |
|---|---|---|
| **Problem**<br>23 cups,<br>9 used.<br>How many left?<br>You can't take 9 ones from 3 ones. | 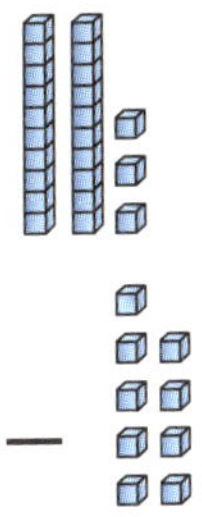 | Tens Ones<br>2 3<br>− 9 |
| **Trading**<br>Trade to get more ones.<br>You now have 1 ten and 13 ones. | 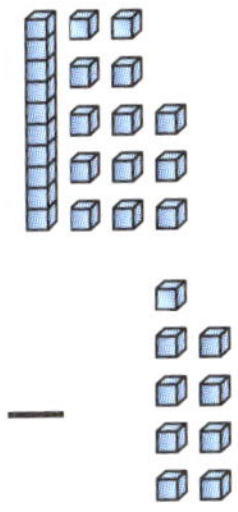 | Tens Ones<br>1 13<br>~~2~~ ~~3~~<br>− 9 |
| **Ones**<br>13 ones take away 9 ones equals 4 ones. | 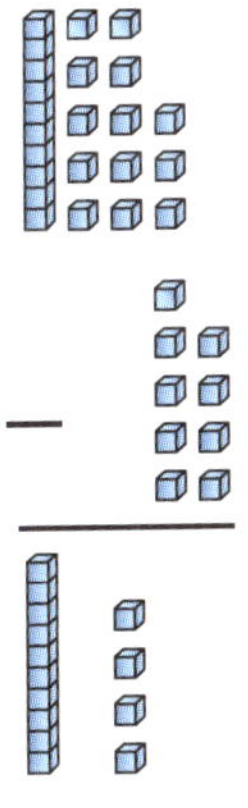 | Tens Ones<br>1 13<br>~~2~~ ~~3~~<br>− 9<br>4 |
| **Tens**<br>1 ten take away 0 tens equals 1 ten. | | Tens Ones<br>1 13<br>~~2~~ ~~3~~<br>− 9<br>1 4 |

**ACTIVITY**

Make up number stories for Question 2 parts **g** and **h**.

**1** Show trading down 1 ten.

| | Tens | Ones |
|---|---|---|
| **a** | 3 | 4 |
| **b** | 5 | 7 |
| **c** | 4 | 6 |
| **d** | 7 | 2 |
| **e** | 6 | 3 |
| **f** | 8 | 5 |

**2** Use place-value blocks to do these. Estimate your answer first.

| | Tens | Ones |
|---|---|---|
| **a** | 2 | 4 |
| | − | 6 |
| **b** | 2 | 5 |
| | − | 9 |
| **c** | 3 | 7 |
| | − | 8 |
| **d** | 5 | 6 |
| | − | 7 |
| **e** | 7 | 3 |
| | − | 5 |
| **f** | 6 | 1 |
| | − | 4 |
| **g** | 3 | 5 |
| | − | 7 |
| **h** | 4 | 2 |
| | − | 5 |

# 2:50 Subtraction with Trading

| Trading Down | | |
|---|---|---|
| **Problem**<br>34 birds,<br>18 leave.<br>How many stay?<br>You can't take 8 ones from 4 ones. | 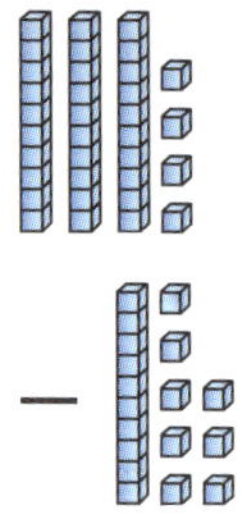 | Tens \| Ones<br>3 \| 4<br>− 1 \| 8 |
| **Trading**<br>Trade to get more ones.<br>You now have 2 tens and 14 ones. | 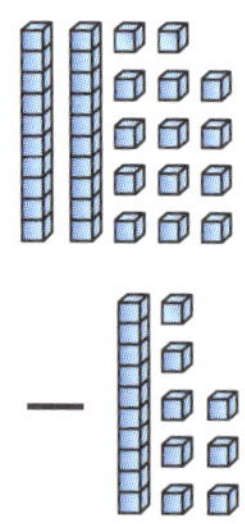 | Tens \| Ones<br>2 \| 14<br>~~3~~ \| ~~4~~<br>− 1 \| 8 |
| **Ones**<br>14 ones take away 8 ones equals 6 ones. | 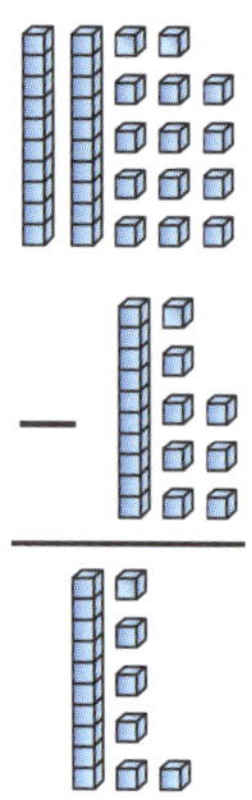 | Tens \| Ones<br>2 \| 14<br>~~3~~ \| ~~4~~<br>− 1 \| 8<br>\| 6 |
| **Tens**<br>2 tens take away 1 ten equals 1 ten. | | Tens \| Ones<br>2 \| 14<br>~~3~~ \| ~~4~~<br>− 1 \| 8<br>1 \| 6 |

**1** Show trading down 1 ten.

a Tens | Ones: 4 | 5

b Tens | Ones: 2 | 7

c Tens | Ones: 3 | 1

d Tens | Ones: 7 | 2

**2** Use place-value blocks to do these. Estimate your answer first.

a Tens | Ones: 5 6 − 1 8

b Tens | Ones: 8 3 − 2 9

c Tens Ones: 6 2 − 3 4

d Tens Ones: 3 7 − 1 8

e Tens Ones: 4 1 − 2 7

f Tens Ones: 6 5 − 3 6

g Tens Ones: 7 4 − 4 9

h Tens Ones: 9 2 − 3 5

© PEARSON AUSTRALIA 2018 • *AUSTRALIAN SIGNPOST MATHS 3* • ISBN 978 1 4886 2184 0

# 3:01 Direct Comparison of Length

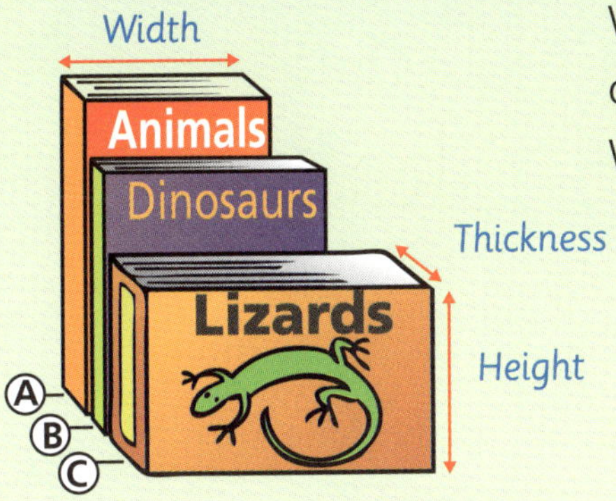


CONCEPT

We can compare the height, width and thickness of books by placing them side by side.
When we do this, we are using **direct comparison**.

1. Use the labels A, B and C on the picture above to answer these questions.
   - **a** Which book is the widest?
   - **b** Which book is the thickest?
   - **c** Which book is the tallest?

2. List the books A, B and C in order from least to most, according to:
   - **a** width
   - **b** thickness
   - **c** height

3. 
   - **a** Which is longer, G or R?
   - **b** Which is longer, B or Y?
   - **c** Which is longer, G or B?
   - **d** Which pencil is the longest?
   - **e** Which pencil is the shortest?
   - **f** List the pencils in order from shortest to longest.

4. 
   - **a** Which is longer, A or B?
   - **b** Which is longer, B or C?
   - **c** Is it easier to compare lengths if the ends are in line?

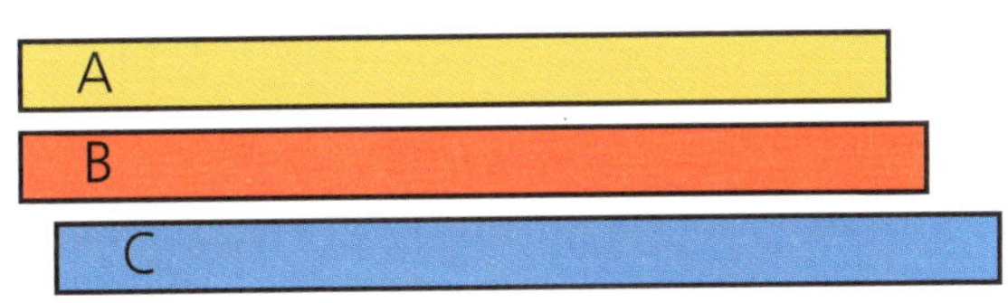


ACTIVITY

- Use direct comparison to compare the height, width and thickness of books.
- Compare the lengths of 4 objects. List them in order from shortest to longest.

# 3:02 Time Revision

CONCEPT

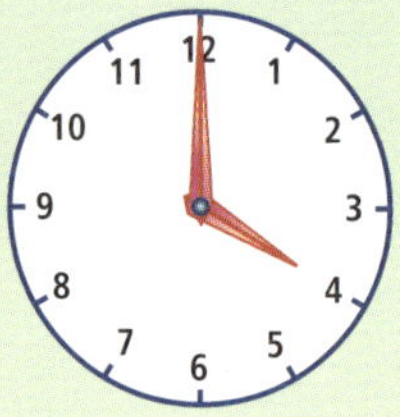
4 o'clock

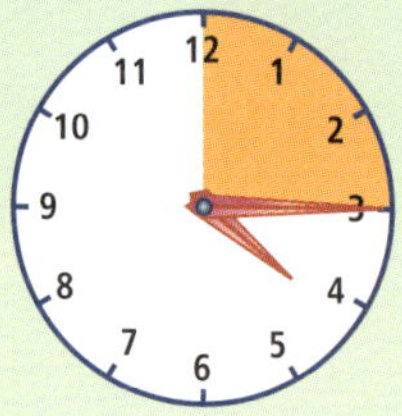
a quarter past 4

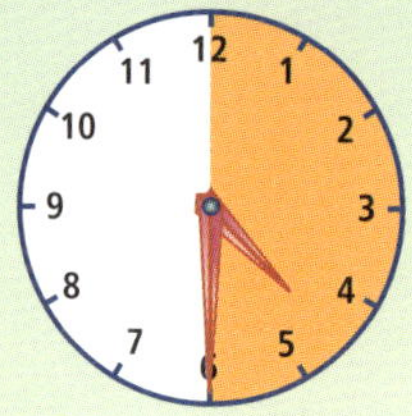
half past 4

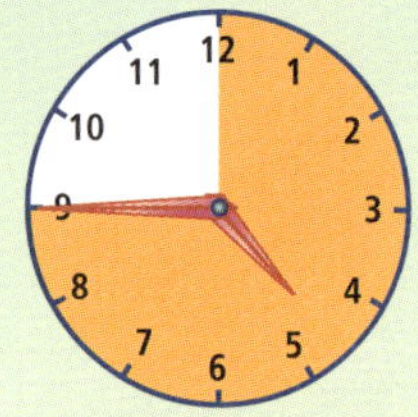
a quarter to 5

**1** Write the time shown.

a
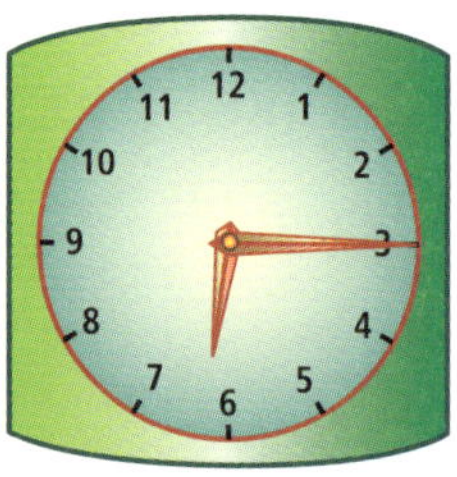

b

c

d

e

f

g

h

**2** Show the time given.

a

a quarter past 10

b
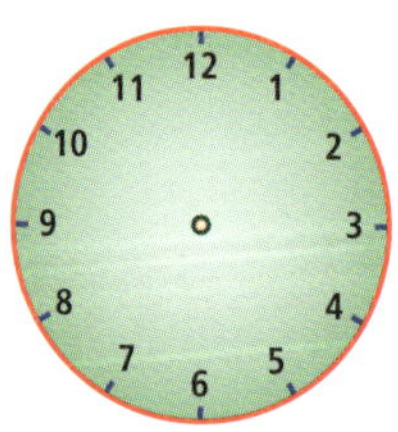
a quarter to 8

c
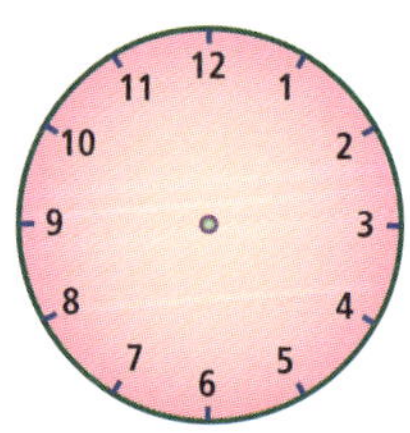
4 o'clock

d
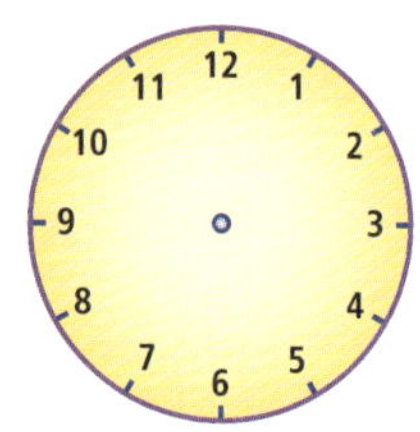
half past 7

e
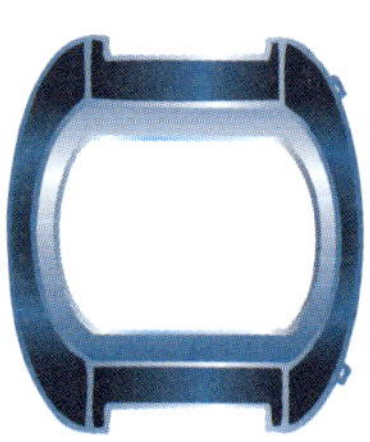
a quarter past 8

f
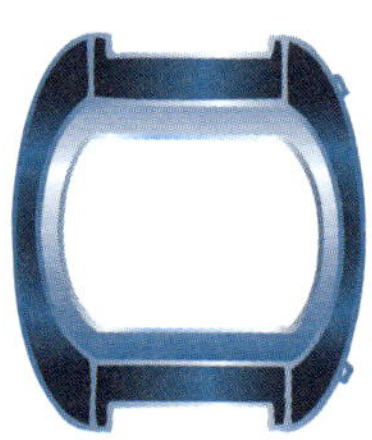
half past 2

g
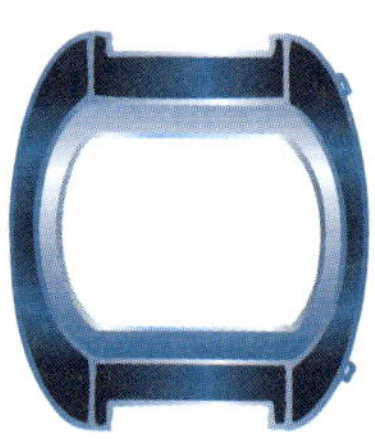
a quarter to 1

h

7 o'clock

# 3:03 Analogue Time

CONCEPT

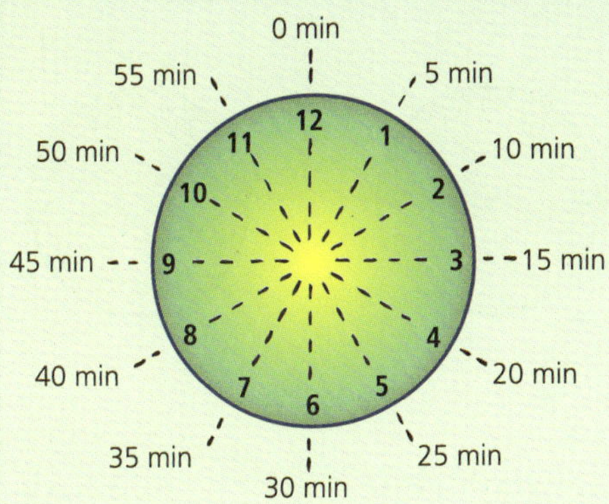


It takes 5 minutes for the minute hand to move from one numeral to the next.

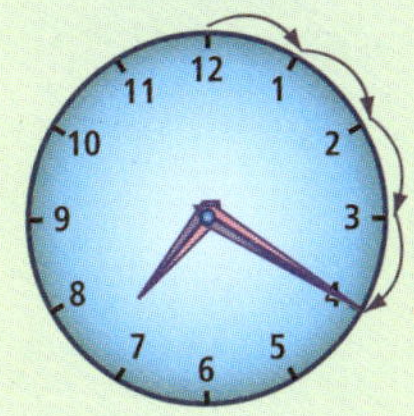

20 past 7

- The hour hand has passed 7.
- The minute hand has moved 5, 10, 15, 20 minutes from 12.
- The time is **20 past 7**.

**1** Write the time shown.

a

☐ past ☐

b
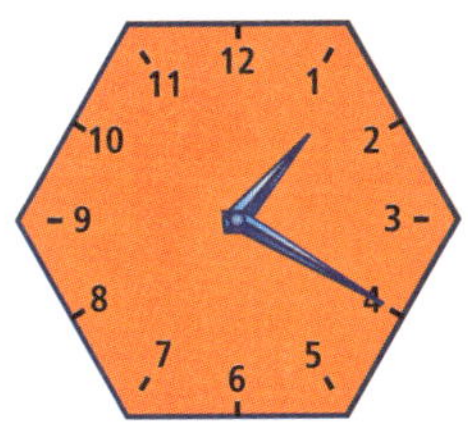
☐ past ☐

c

☐ past ☐

d
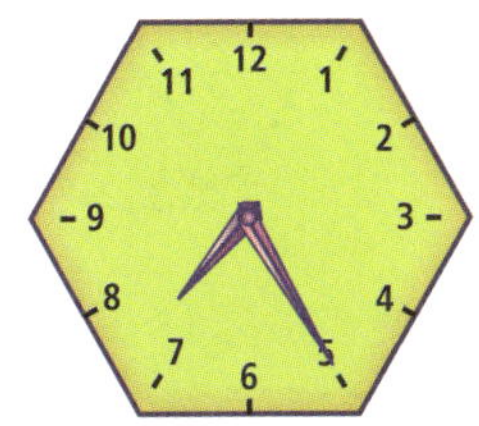
☐ past ☐

e
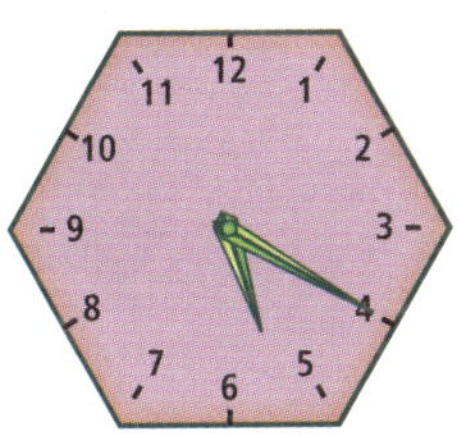
☐ past ☐

f

☐ past ☐

g

☐ past ☐

h

☐ past ☐

CONCEPT

50 minutes past 2

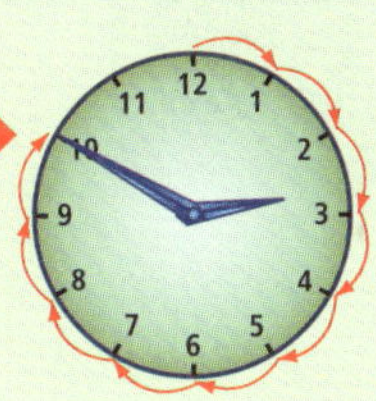

- The hour hand has passed 2.
- The minute hand has moved 5, 10, 15, … 50 minutes from 12.
- The time is **50 minutes past 2**.

**2** Write the time shown.

a

☐ past ☐
☐ to ☐

b

☐ past ☐
☐ to ☐

c

☐ past ☐
☐ to ☐

d
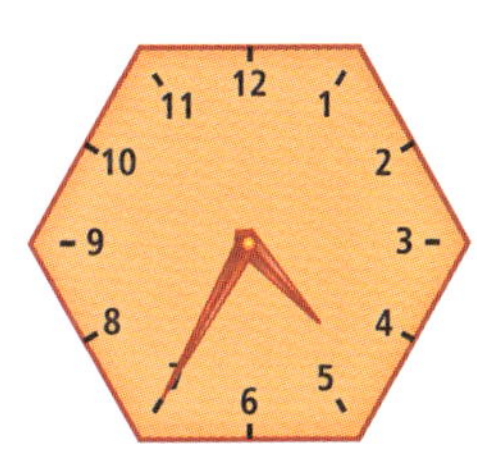
☐ past ☐
☐ to ☐

 • *AUSTRALIAN SIGNPOST MATHS 3* • ISBN 978 1 4886 2184 0

# 3:04 Analogue and Digital Time

1 Look carefully at each clock face and complete the labels.

**a** 

4: ____

____ past 4

**b** 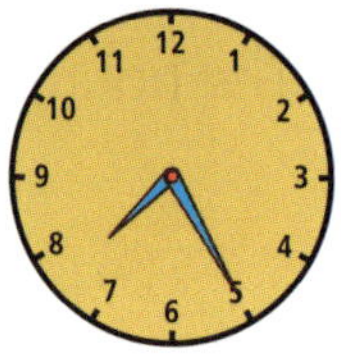

7: ____

____ past 7

**c** 

12: ____

____ past 12

**d** 

4: ____

____ past 4

**e** 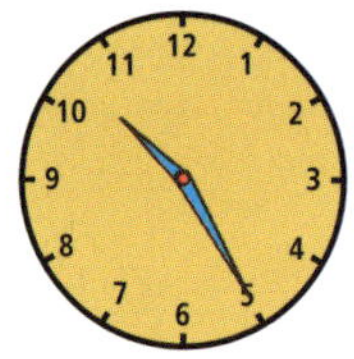

10: ____

____ past 10

**f** 

2: ____

____ past 2

**g** 

11: ____

____ past 11

**h** 

8: ____

____ past 8

**i** 

3: ____

____ past 3

**j** 

4: ____

____ past 4

**k** 

____ past ____

**l** 

____ past ____

**m** 

____ past ____

**n** 

____ past ____

**o** 

____ past ____

**p** 

____ past ____

**q** 

____ past ____

**r** 

____ past ____

**s** 

____ past ____

**t** 

____ past ____

ACTIVITY

- Collect advertisements, timetables and TV program guides that show times. Why is digital time usually used?
- Collect pictures of different watches and clocks. How are they different?

# 3:05 Analogue and Digital Time

CONCEPT

- The hour hand has passed 1.
- The minute hand has moved 5, 10, 15, 20 minutes from 12.
- The time is **1:20** or **20 past 1**.

**1** Look carefully at each clock face and complete the labels.

**a**

10: ___
___ past 10

**b**

11: ___
___ past 11

**c**

2: ___
___ past 2

**d**

9: ___
___ past 9

**e**

4: ___
___ past 4

**f**

5: ___
___ past 5

**g**

2: ___
___ past 2

**h**

4: ___
___ past 4

**i**

7: ___
___ past 7

**j**

1: ___
___ past 1

**k**

___ past ___

**l**

___ past ___

**m**

___ past ___

**n**

___ past ___

**o**

___ past ___

**2** Draw lines to join the same times together.

**a**

| | |
|---|---|
| 11:35 | 10 minutes past 6 |
| 6:10 | 50 minutes past 9 |
| 9:50 | eleven thirty-five |

**b**

| | |
|---|---|
| 1:40 | 25 minutes past 8 |
| 8:25 | five-oh-five |
| 5:05 | 40 minutes past 1 |

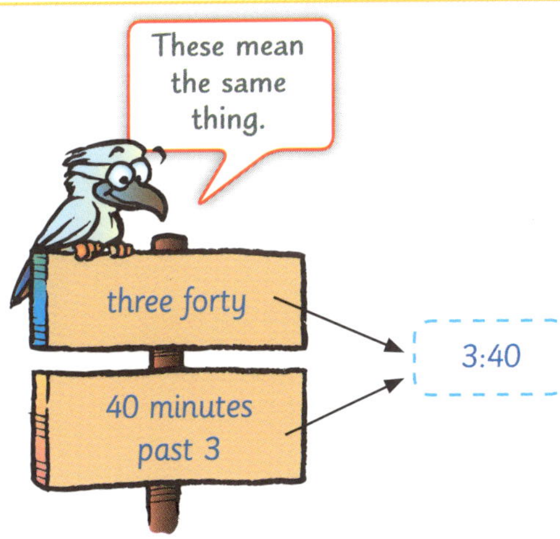

# 3:06 The Metre

CONCEPT

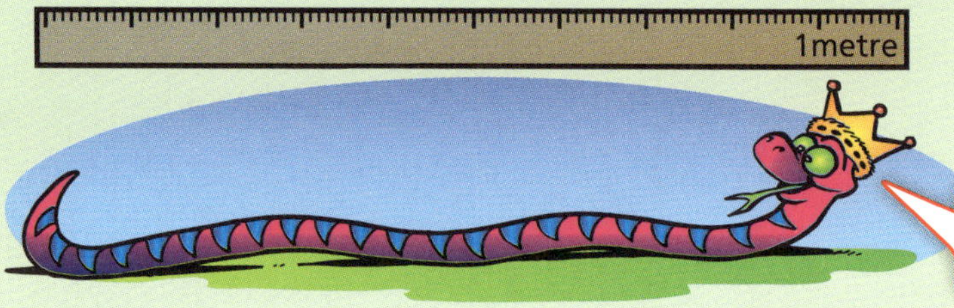


1 metre

This ruler is 1 metre long.

❶ Make a list of objects that are about 1 metre long.

______________________________

______________________________

______________________________

❷ Cut a strip of paper that is 1 metre long. How many of the items below would it take to match this length?

| Item | | Estimate | Measurement |
|---|---|---|---|
|  | How many shoes? | | |
|  | How many books? | | |
|  | How many hands? | | |

❸ Match each length with the best answer.

- height of a tree
- width of a door
- length of a shoe
- length of a desk
- width of a path
- height of your teacher

- less than 1 metre
- about 1 metre
- more than 1 metre

# 3:07 Using the Metre

- The metre is a standard unit of the metric system.
- One metre is usually about the length of your arm span, or two of your steps.
- **m** is short for **metre**.

CONCEPT

ACTIVITY

1. Estimate and measure each length to the nearest metre.

| Length | Estimate | Measurement |
|---|---|---|
| length of 10 steps | m | m |
| width of a gate | m | m |
| length of the classroom | m | m |
| distance to the canteen | m | m |

2. Match each length with the best answer.

| | | |
|---|---|---|
| width of a door | 1 metre | height of yourself |
| length of a whiteboard | 2 metres | length of a car |
| height of a door | 5 metres | width of your desk |
| width of a window | | length of ten steps |

3. Measure the following. Show whether each length is less than, about or more than 1 metre.

| | Less than 1 m | About 1 m | More than 1 m |
|---|---|---|---|
| shoulder to toe | | | |
| standing jump | | | |
| height of your teacher | | | |

 • *AUSTRALIAN SIGNPOST MATHS 3* • ISBN 978 1 4886 2184 0

# 3:08 Centimetres

- **cm** is short for **centimetre**.
- 100 centimetres is the same as 1 metre.

100 cm = 1 m

CONCEPT

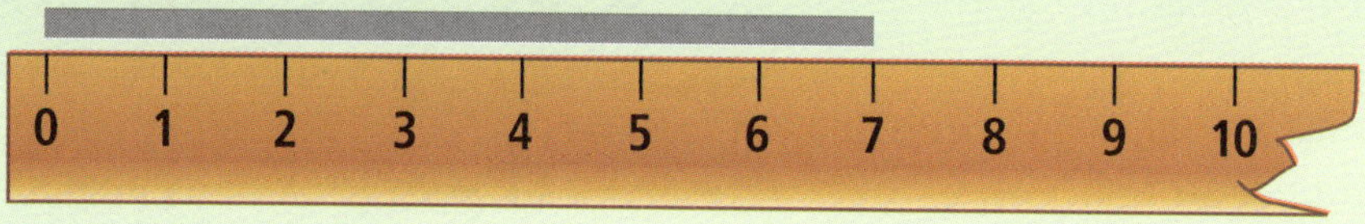


The grey line is 7 cm long.

**1** How long is each line?

a
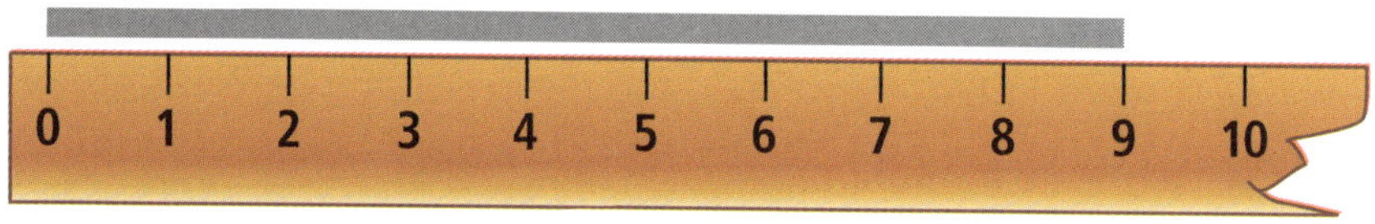


☐ cm

b
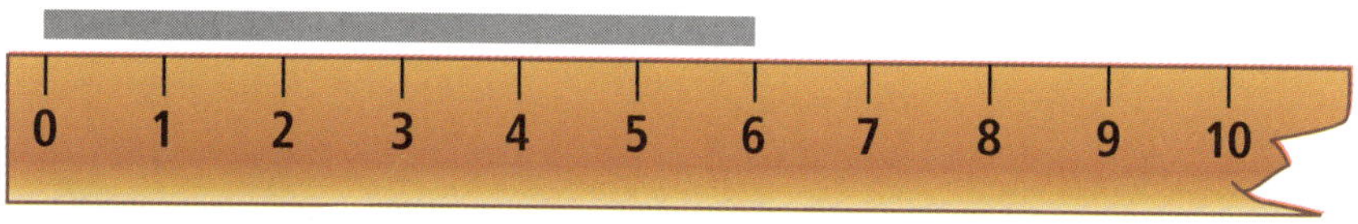


☐ cm

c
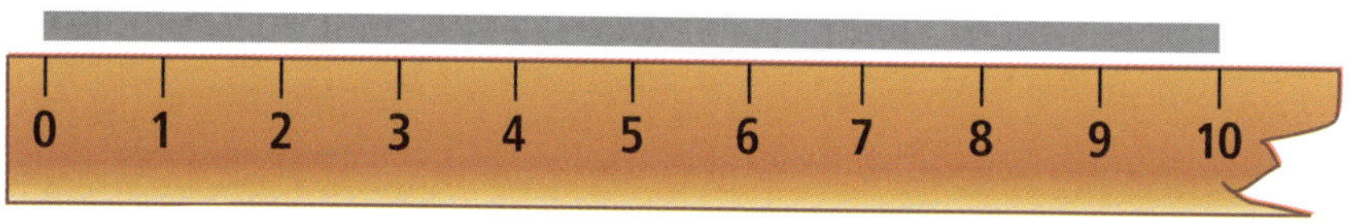


☐ cm

**2** Estimate and then measure the length of each item. Write your measurement.

a

☐ cm

b

☐ cm

c

☐ cm

d
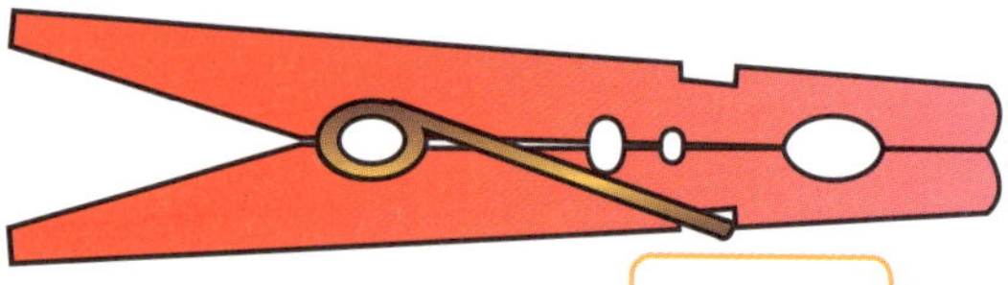

☐ cm

**3** Estimate and then measure the length of each line. Write your measurement.

a
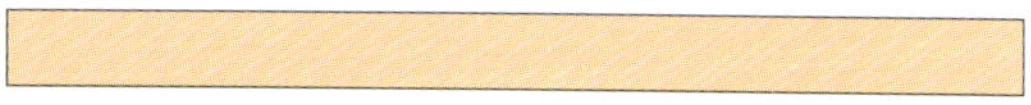

☐ cm

b

☐ cm

c

☐ cm

d

☐ cm

e

☐ cm


 *AUSTRALIAN SIGNPOST MATHS 3* • ISBN 978 1 4886 2184 0

# 3:09 Measuring with Centimetres

CONCEPT

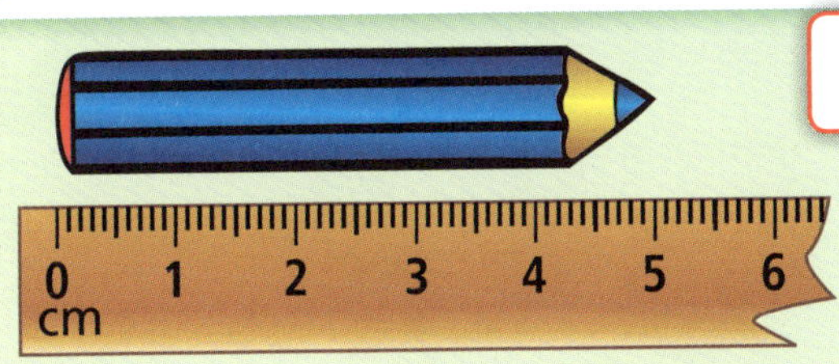

1 cm is about as long as your fingernail.

**1** Estimate and then measure the length of each bar. Write the measurements.

**2** Estimate and then measure the sides of each shape. Write your measurements.

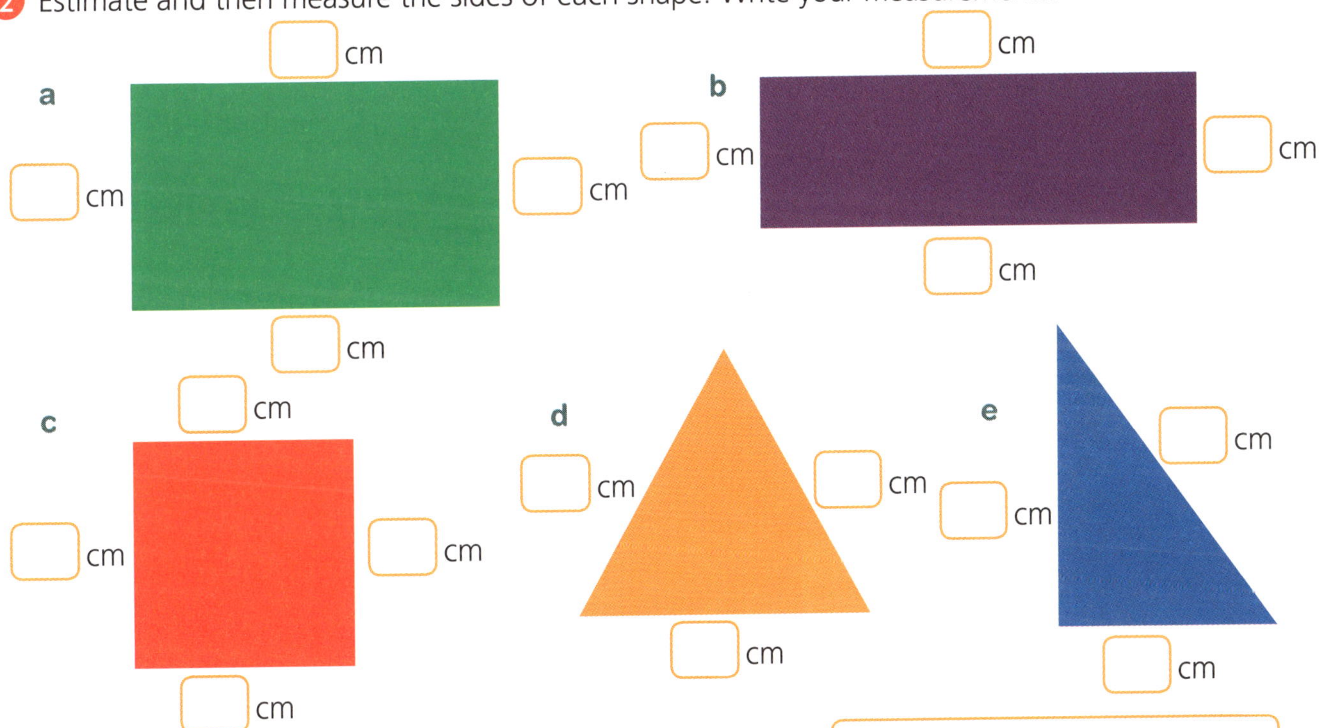


**f** Which of these shapes is the longest all the way around?

ACTIVITY

Measure items in your classroom to complete the table.

| Item | Measurement |
|---|---|
| | Length: cm |
| | Width: cm |
| | Height: cm |

| Item | Measurement |
|---|---|
| | Length: cm |
| | Width: cm |
| | Height: cm |

# 3:10 Comparing Areas

Area is the measure of the amount of surface.

The red triangle has more area than the green triangle.

CONCEPT

1 For each pair of shapes, circle the shape that has more area.

a 

b 

c 

2 For each group of shapes, circle the largest area and colour the smallest area.

a 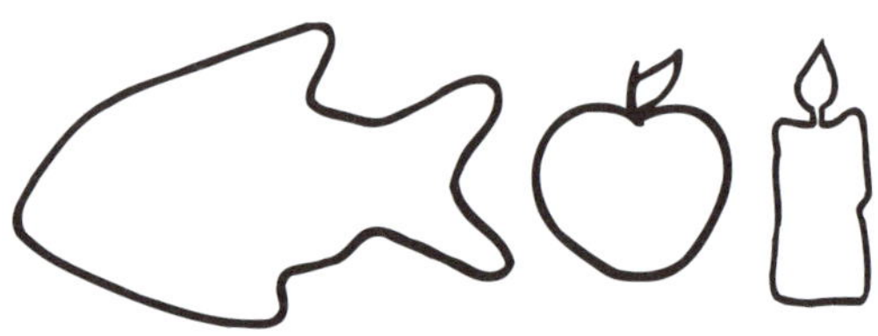

b 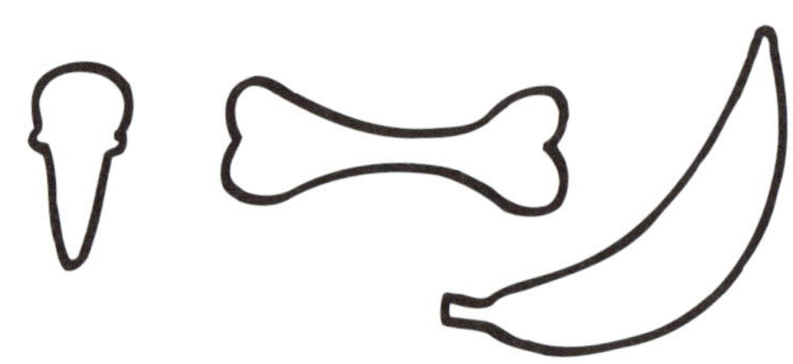

ACTIVITY

Use sheets of newspaper to find the larger area. Count how many sheets it takes to cover one area, then compare this with the next area.

**You will need**:

- newspaper

- sticky tape

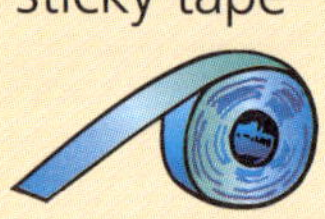

- scissors

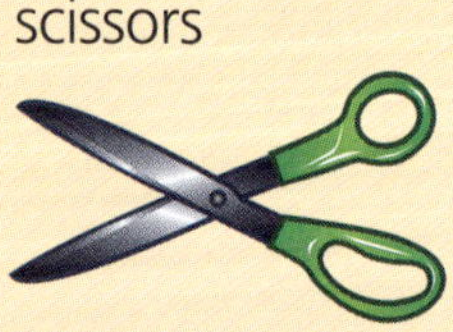

| First Area | Second Area | Comparison |
|---|---|---|
|  Door |  Window | |
| 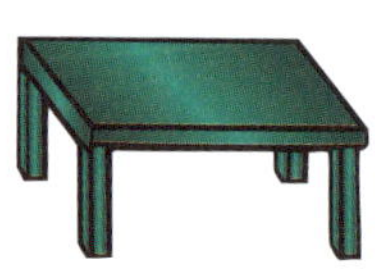 Desk | 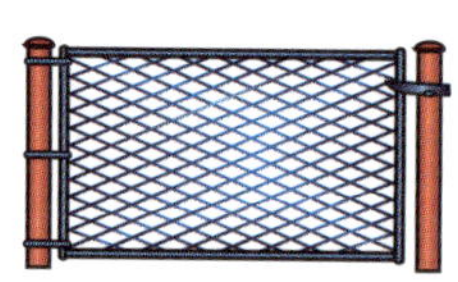 Gate | |
|  Computer |  Television | |

Explain the reasoning you used to find your answers.

# 3:11 Area

**1** Record the area of each shape.

a

☐ diamonds

b
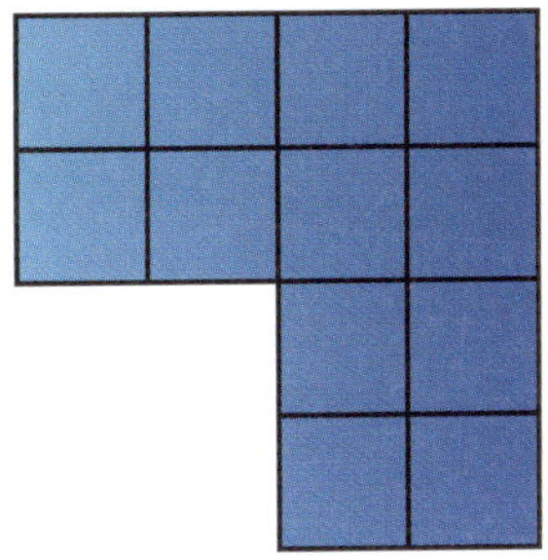
☐ squares

c

☐ triangles

d
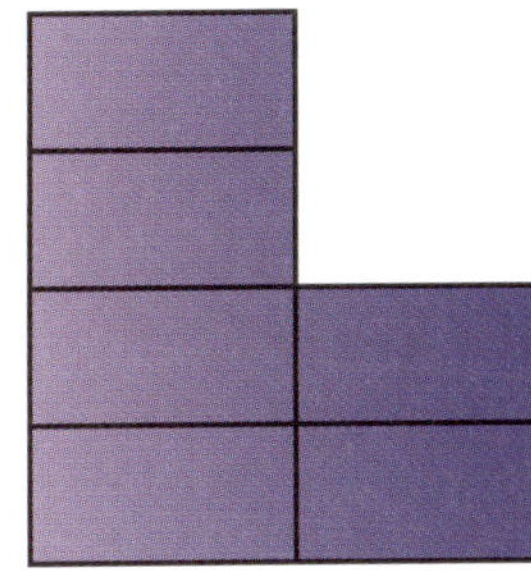
☐ rectangles

**2** On the grid paper, colour shapes with these areas.

a Shape **A**: 10 diamonds

b Shape **B**: 12 diamonds

c Shape **C**: 9 diamonds

d Which shape has the largest area? ☐

e Which shape has the smallest area? ☐

**3** a How many ones place-value blocks are needed to cover each letter?

☐ blocks

☐ blocks

☐ blocks

b Which letter has the largest area? ☐

c Which letter has the smallest area? ☐

d Which letter has an area smaller than F? ☐

e Which letter has an area larger than F? ☐

# 3:12 Estimating the Litre

CONCEPT

- The litre is a standard unit for measuring capacity.
- **L** is short for **litre**.
  We write 1 litre as 1 L.

**1** Colour in blue the containers that hold more than 1 litre.
Colour in red the containers that hold less than 1 litre.

ACTIVITY

**2** How many cups of water does it take to fill a 1 litre container?

Estimate: ☐ cups    Measure: ☐ cups

**3** Use different containers to fill a 1 litre carton.
Write the number of containers needed to completely fill the carton.

| Container | Estimated number to fill 1 L | Measured number to fill 1 L |
|---|---|---|
| eggcup | | |
| | | |
| | | |

 ISBN 978 1 4886 2184 0

# 3:13 The Litre

ACTIVITY

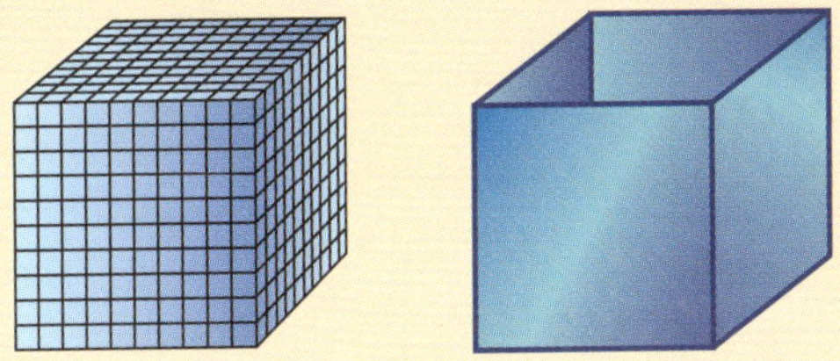

1000 block    1 litre measure

Compare a 1 litre measure with a 1 000 block. Discuss how they are the same.

**1** Find the number of cups needed to fill the 1 litre measure with:

**a** water ☐   **b** sand ☐   **c** rice ☐

**2** Procedure:

- Pour water from a 1 litre measure into a large drink carton.
- Mark the level of the water in the carton.
- To make a new 1 litre measure, cut the carton at the mark. Label your new measure **1 L**.

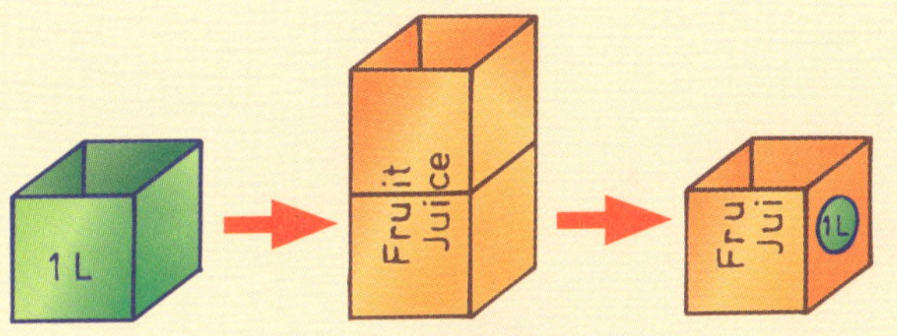


**3** Procedure:

- Collect several containers that hold 1 litre (such as milk cartons, cordial bottles and ice-cream containers).
- Pour 1 litre of water from your 1 litre measure into each of your containers.
- Discuss why the containers are not completely filled.

**4** Use your 1 litre measure to find containers to complete this table.

| | Less than 1 L | About 1 L | More than 1 L |
|---|---|---|---|
| Containers | glue bottle | 1 L milk carton | bucket |
| | | | |
| | | | |
| | | | |
| | | | |
| | | | |

- 1 litre of water has the same volume as a 1 000 block.

# 3:14 The Kilogram

CONCEPT

1 litre of water has a mass of 1 kilogram.

ACTIVITY

**1** **a** On one side of a balance scale, put a 1 kilogram mass. On the other side of the scale, put a bag of sand that you estimate to have the same mass.

**b** Does it balance? Was the bag of sand too heavy or too light?

**c** Repeat this activity with nails or stones instead of sand.

**2** Estimate whether the items listed below are heavier or lighter than 1 kilogram. Use balance scales to check.

| Item | Estimate (compared to 1 kg) | Measurement (balancing with 1 kg) |
|---|---|---|
| 1 feather | lighter | lighter |
| 10 books | | |
| 5 shoes | | |
| 1 brick | | |
| 6 pencil cases | | |
| 1 tape dispenser | | |
| 1litre of water | | |

**3** Circle the items that you estimate have a mass less than 1 kilogram.

**a** peach **b** large dog **c** scissors **d** cat
**e** dessert spoon **f** lamb chop **g** mushroom **h** pony
**i** chair **j** this book **k** car tyre **l** television

**4** Circle the items that are measured in kilograms, according to the number on the label.

**a**

**b**

**c**

**d**
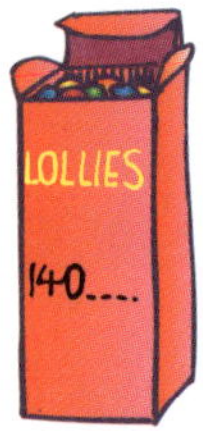


**e**

# 3:15 Using the Kilogram

CONCEPT

**1** Use the short form to write these measurements.

a 3 kilograms ☐ b 10 kilograms ☐ c 8 kilograms ☐

d 5 kilograms ☐ e 7 kilograms ☐ f 11 kilograms ☐

g twelve kilograms ☐ h twenty kilograms ☐ i sixteen kilograms ☐

**2** Circle the items that would have a mass greater than 1 kilogram.

a bicycle b chair c nail d your teacher

e desk f marble g computer h brick

i matchbox j car tyre k large dog l golf ball

**3** Look at the pictures. What is the total mass of the items together?

a grapes and cheese ☐ b meat and grapes ☐

c flour and laundry powder ☐ d onions and meat ☐

e onions and grapes ☐ f cheese and meat ☐

g flour and onions ☐ h cheese and onions ☐

**4** Circle the labels that are written correctly.

a 9kg b 9KG c 7kg d 6kg e 11Kg

f 14kg g 3kg h 17kg i 21kgs j 32kg

## Choose and Check

ACTIVITY

- Choose 10 classroom items that you think are heavier than 1 kg. Use balance scales and a 1 kg mass to check.
- Choose 10 items that you think are lighter than 1 kg. Use balance scales to check.
- Without measuring, arrange six different items in order from lightest to heaviest. Use balance scales to check if you were right.

# 3:16 Analogue Time

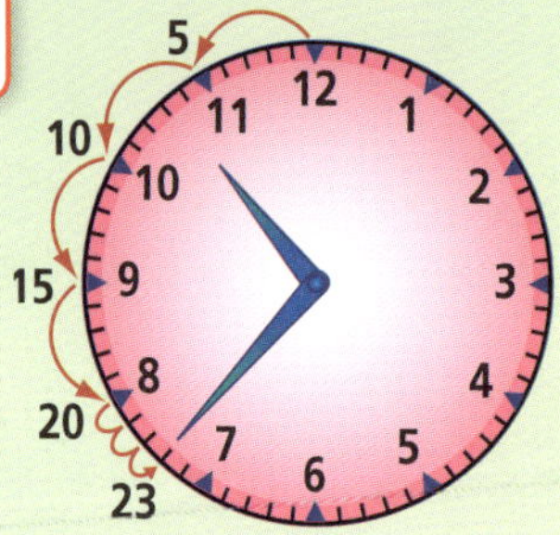


23 minutes to 11

CONCEPT

- The hour hand is approaching 11.
- The minute hand has 23 minutes to go before it reaches the 11.
- The time is **23 to 11**.

**1** Write the time shown.

a  ☐ to ☐

b  ☐ to ☐

c  ☐ to ☐

d  ☐ to ☐

e  ☐ to ☐

f  ☐ to ☐

g  ☐ to ☐

h  ☐ to ☐

**2** Show the time given.

a 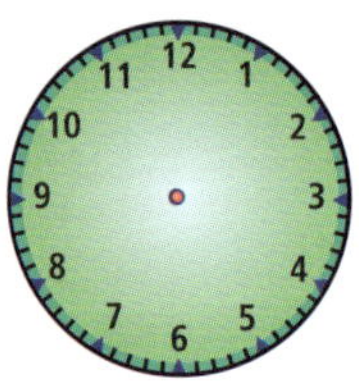 17 to 3

b 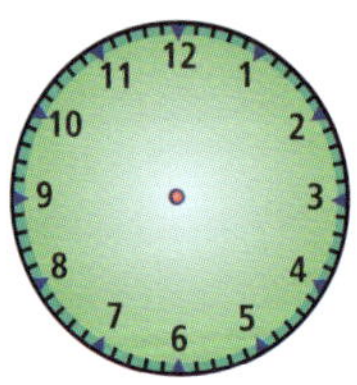 14 to 8

c  28 to 1

d 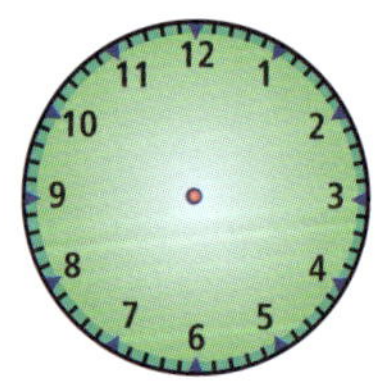 12 to 4

**3** Write the time that is four minutes after.

a 20 to 6 ☐
b 15 to 9 ☐
c 25 to 3 ☐
d 20 to 10 ☐
e 25 to 11 ☐
f 10 to 7 ☐

**4** Write the time that is three minutes before.

a 5 to 6 ☐
b 15 to 11 ☐
c 20 to 9 ☐
d 10 to 12 ☐
e 20 to 8 ☐
f 12 o'clock ☐

# 3:17 Analogue and Digital Time

CONCEPT

22 past 5
5:22

14 to 6
5:46

**1** Write the time shown.

a  ☐ past ☐

b  ☐ to ☐

c  ☐ past ☐

d  ☐ past ☐

e  ☐ to ☐

f  ☐ past ☐

g  ☐ to ☐

h  ☐ to ☐

**2** Write the time shown.

a 

☐ past ☐

b 

☐ past ☐

c 

☐ past ☐

d 

☐ past ☐

e 

☐ past ☐

f 

☐ to ☐

g 

☐ to ☐

h 

☐ to ☐

i 

☐ to ☐

j 

☐ to ☐

 • *AUSTRALIAN SIGNPOST MATHS 3* • ISBN 978 1 4886 2184 0

# 3:18 Recording Length

CONCEPT

**1** Write the lengths in short form.

a 1 metre 70 centimetres ☐ m ☐ cm

b 3 metres 36 centimetres ☐ m ☐ cm

c 4 metres 26 centimetres ☐ m ☐ cm

d 7 metres 13 centimetres ☐ m ☐ cm

e 6 metres 83 centimetres ☐ m ☐ cm

f 5 metres 42 centimetres ☐ m ☐ cm

g 9 metres 59 centimetres ☐ m ☐ cm

h 8 metres 77 centimetres ☐ m ☐ cm

**2** Would you use metres (m) or centimetres (cm) to measure the following things?

a the height of your house ☐

b the length of an eraser ☐

c the length of your hair ☐

d the width of your room ☐

e the width of your thumb ☐

f the length of a pencil ☐

g the length of a path ☐

h the height of your desk ☐

INVESTIGATION

- Find some items that you think would match the lengths in this table. Use a tape measure to measure the items and complete the table.

| Length | Item | Measured Length |
|---|---|---|
| 60 cm | | |
| 1 m 20 cm | | |
| 13 cm | | |
| 2 m 50 cm | | |
| 4 m | | |

- Use a trundle wheel to find these lengths:
  - the length of a fence
  - the width of the school playground
  - the distance from your classroom to the office.
- Measure the length of your steps.
  - How many of your steps make 1 metre? ☐
  
    Now roll a ball along the ground.
  - It rolled ☐ steps.
  - That is about ☐ metres.

# 3:19 Finding Length

CONCEPT

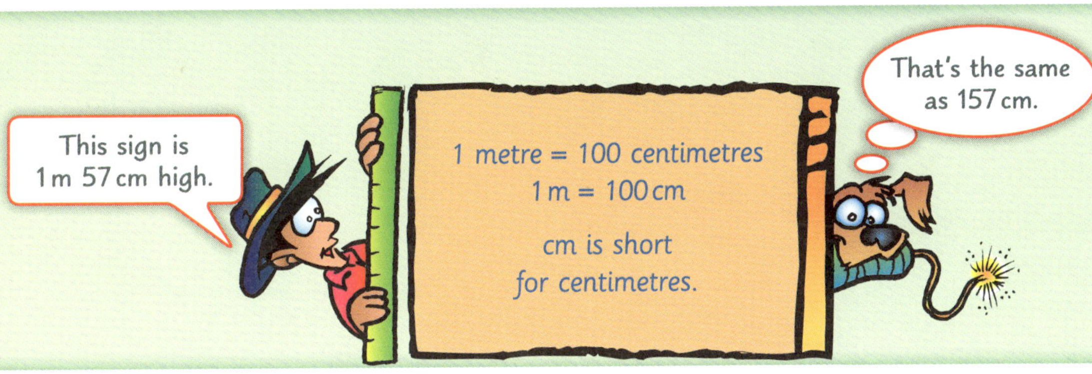


**1** Write the lengths in centimetres.

| | | |
|---|---|---|
| **a** 1 m 26 cm | **b** 1 m 73 cm | **c** 2 m 51 cm |
| **d** 2 m 97 cm | **e** 3 m 16 cm | **f** 3 m 42 cm |
| **g** 1 m 68 cm | **h** 2 m 6 cm | **i** 3 m 30 cm |

**2** Write the lengths as metres and centimetres.

| | |
|---|---|
| **a** 238 cm | **b** 213 cm |
| **c** 358 cm | **d** 307 cm |
| **e** 149 cm | **f** 227 cm |

CONCEPT

**Perimeter** is the total distance around a shape.

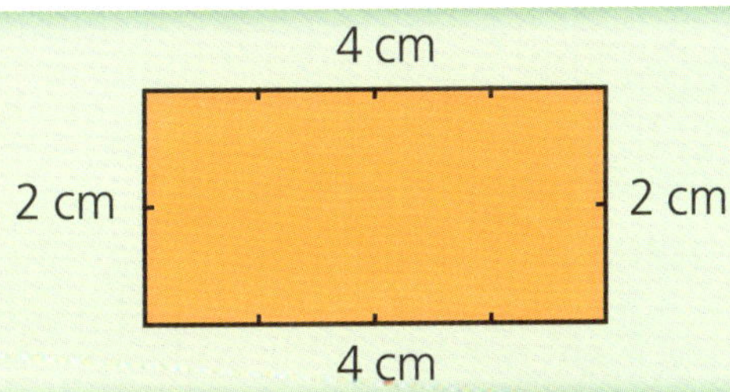


The perimeter of this shape is 12 cm.

**3** Estimate and then measure the perimeter of each shape.

**a**

**b**

**c**

**d**

**e**

**f**

 • *AUSTRALIAN SIGNPOST MATHS 3* • ISBN 978 1 4886 2184 0

# 3:20 The Calendar

A calendar shows how a year is divided into months, weeks and days.

CONCEPT

Use the calendar to answer the questions.

| May 2012 | | | | | | |
|---|---|---|---|---|---|---|
| **Sun** | **Mon** | **Tues** | **Wed** | **Thurs** | **Fri** | **Sat** |
| | | 1 | 2 | 3 | 4 | 5 |
| 6 | 7 | 8 | 9 | 10 | 11 | 12 |
| 13 | 14 | 15 | 16 | 17 | 18 | 19 |
| 20 | 21 | 22 | 23 | 24 | 25 | 26 |
| 27 | 28 | 29 | 30 | 31 | | |

1. How many days are there in May? ______

2. What day of the week are these dates?
   - **a** 1 May ______
   - **b** 12 May ______
   - **c** 9 May ______

3. Write these dates in short form.
   - **a** first Monday in May ______
   - **b** last Friday in May ______
   - **c** last Wednesday in May ______

4. **a** How many school days are there in May? ______
   **b** How many weekend days are there in May? ______

5. **a** What are the dates of the last weekend in May? ______
   **b** What is the day and date of the last day in April? ______
   **c** What is the day and date of the first day in June? ______

## Time Lines

ACTIVITY

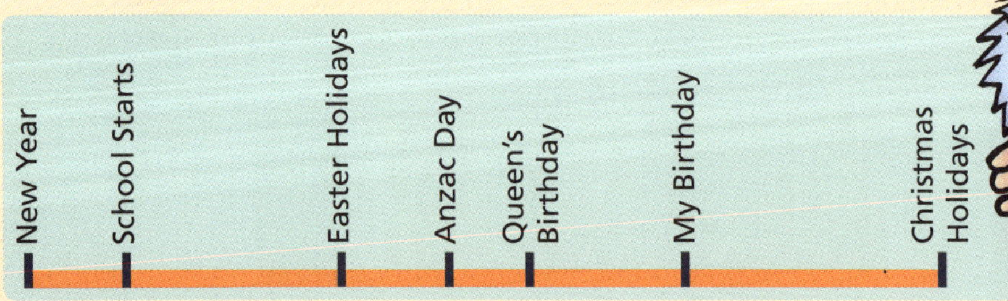

- Make a time line to show the events of last week.

# 3:21 Analogue and Digital Time

CONCEPT

**1** Complete the sentences.

**a** 55 minutes past 2 is the same as 5 to 3, so it is nearly ☐ o'clock.

**b** 52 minutes past 7 is the same as ☐ to ☐, so it is nearly ☐ o'clock.

**c** 48 minutes past 11 is the same as ☐ to ☐, so it is nearly ☐ o'clock.

**2** Circle the activity which takes more time.

**a** cook an egg or cook a sausage

**b** sing a song or write your name

**c** watch a movie or get dressed

**d** brush your hair or make your bed

**3** Order these events 1, 2, 3, 4 from taking the least time to taking the most time.

**a** cook dinner ☐

**b** defrost a slice of bread ☐

**c** run 100 metres ☐

**d** walk to school ☐

**4** Write the times.

| | Start | Finish |
|---|---|---|
| **a** The party starts at three thirty. It finishes two hours later. | | |
| **b** I finished training at half past six. I started thirty minutes earlier. | | |

ACTIVITY

**5** Put these events in order to make a time line of Greg's life.

**A** Went to the Olympic Games in 2000

**B** Went to Fiji in 2010

**C** Born in 1993

**D** Started school in 1998

**E** Completed university in 2014

**F** Started high school in 2005

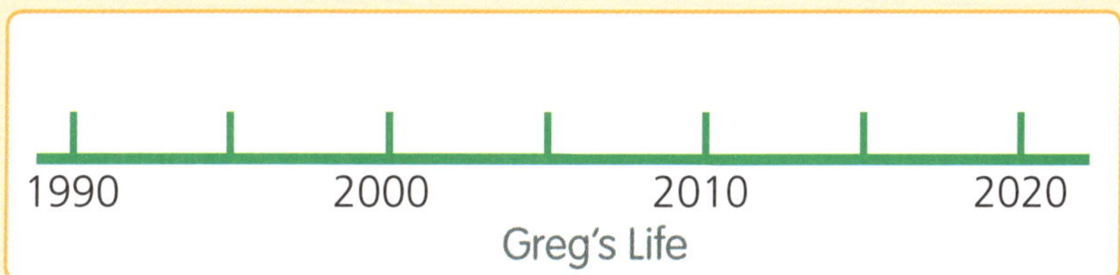

# 3:22 The Millimetre

CONCEPT

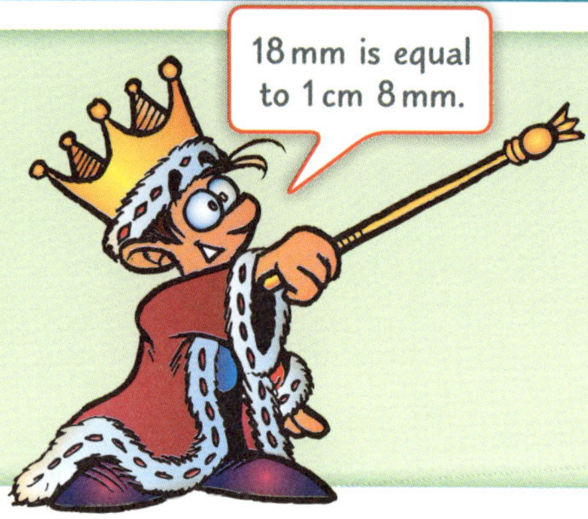


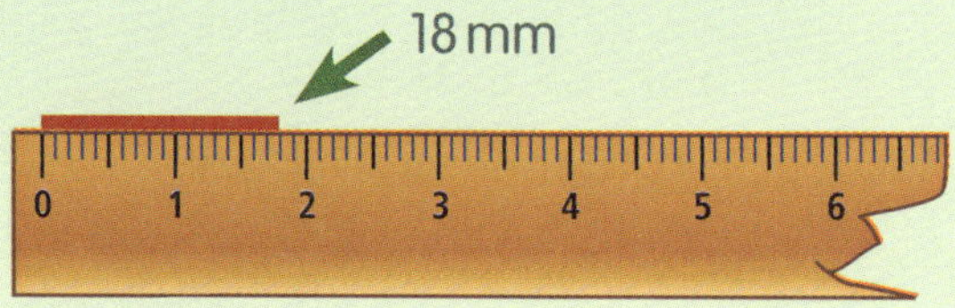


1 centimetre is divided into 10 millimetres.

**mm** is short for millimetres.

**1** Write the length of each interval.

**a**  ☐ mm

**b** 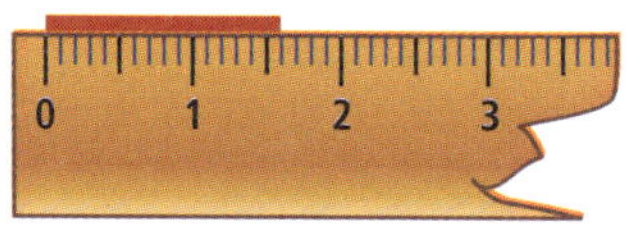 ☐ mm

**c**  ☐ mm

**d** 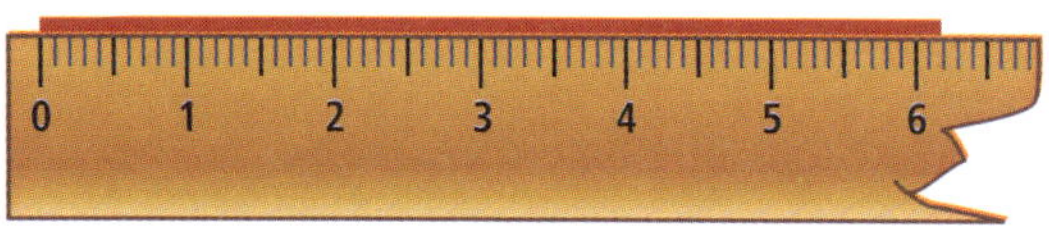 ☐ mm

**e**  ☐ mm

**2** Write the lengths in millimetres.

**a** 1 cm 6 mm ☐ **b** 2 cm 7 mm ☐ **c** 1 cm 2 mm ☐ **d** 4 cm 5 mm ☐

**e** 3 cm 9 mm ☐ **f** 4 cm 3 mm ☐ **g** 3 cm 6 mm ☐ **h** 2 cm 4 mm ☐

**3** Write the lengths as centimetres and millimetres.

**a** 18 mm ☐ **b** 47 mm ☐ **c** 27 mm ☐

**d** 13 mm ☐ **e** 34 mm ☐ **f** 53 mm ☐

**g** 21 mm ☐ **h** 32 mm ☐ **i** 45 mm ☐

ACTIVITY

Estimate and then measure various items to the nearest millimetre. Complete the table below.

| Item | Estimated Length | Measured Length |
|---|---|---|
| length of a pencil | | |
| height of a pencil case | | |
| width of a ruler | | |
| thickness of this book | | |

 • *AUSTRALIAN SIGNPOST MATHS 3* • ISBN 978 1 4886 2184 0

# 3:23 Using Millimetres

**1** Estimate and then measure the length of each interval. Write the measurement.

**a** ______ mm  **b** ______ mm

**c** ______ mm  **d** ______ mm

**e** ______ mm  **f** ______ mm

**2** Write the length of each interval.

**a** 

______ mm

**b** 

______ mm

**c** 

______ mm

**d** 

______ mm

**e** 

______ mm

**3** Write the lengths in millimetres.

**a** 3 cm 4 mm ______  **b** 1 cm 6 mm ______  **c** 6 cm 5 mm ______  **d** 8 cm 8 mm ______

**e** 4 cm 7 mm ______  **f** 9 cm 2 mm ______  **g** 7 cm 3 mm ______  **h** 2 cm 9 mm ______

**4** Write the lengths as centimetres and millimetres.

**a** 19 mm ______  **b** 62 mm ______  **c** 37 mm ______

**d** 71 mm ______  **e** 45 mm ______  **f** 86 mm ______

**g** 58 mm ______  **h** 23 mm ______  **i** 94 mm ______

**j** 99 mm ______  **k** 69 mm ______  **l** 81 mm ______

FUN SPOT

- Play this game with a partner.
  - Take turns to choose an object.
  - Each player estimates the length and width of the object to the nearest millimetre.
  - Measure the object. The closest estimate gets 1 point.
  - The first player to 10 points wins.

- Choose a suitable unit to measure each item. Record the measurements.
  - width of this book ______
  - length of a pen ______
  - length of a ruler ______
  - length of a fingernail ______
  - width of your hand ______
  - width of a ruler ______

Compare your measurements with others. Discuss the results.

 • *AUSTRALIAN SIGNPOST MATHS 3* • ISBN 978 1 4886 2184 0

# 3:24 Area

1 a How many times is each shape used to cover these areas?

A

B

C
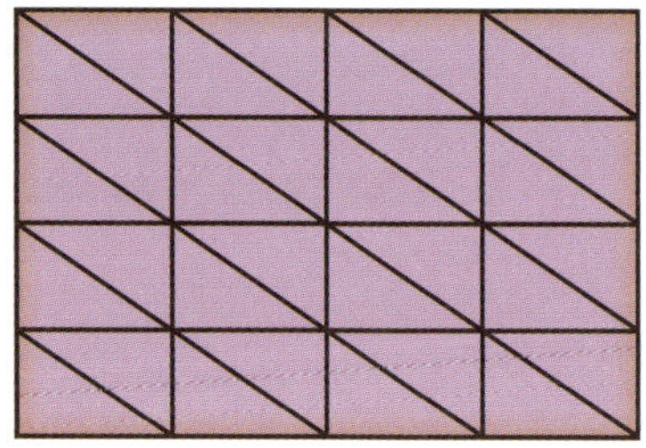

b Do A, B and C have the same area?

2 a How many times does the square cover each shape?

L

M

N
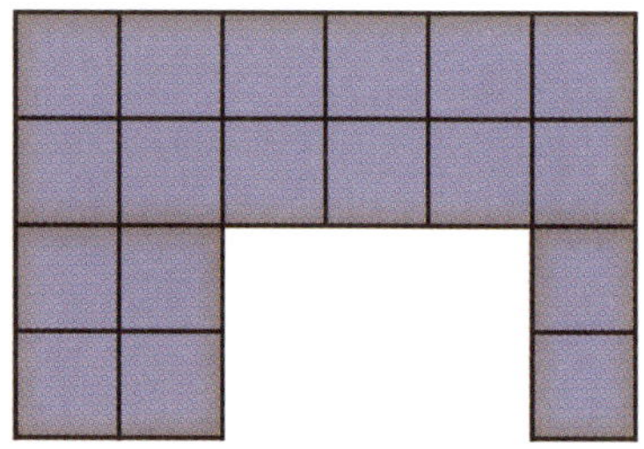

b Write L, M and N in order from least area to most area.

3 Estimate which area is the largest. Count the number of squares and discuss what you find.

X
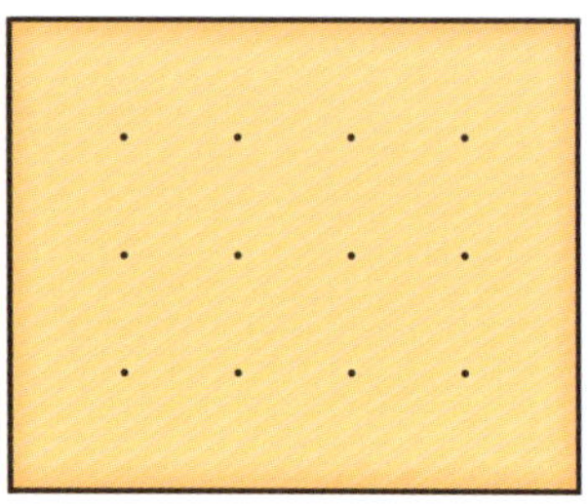

Y
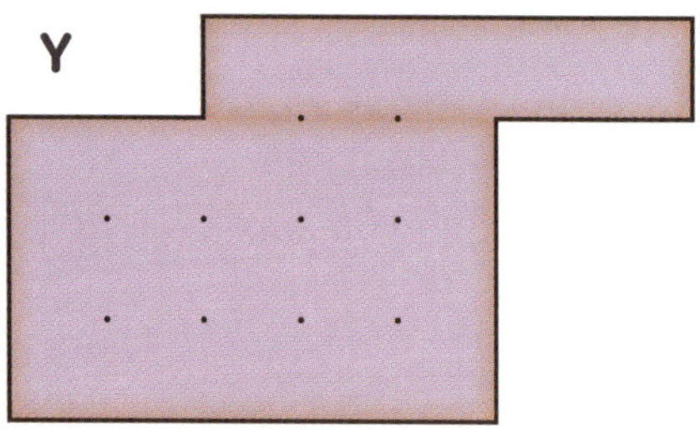

Z
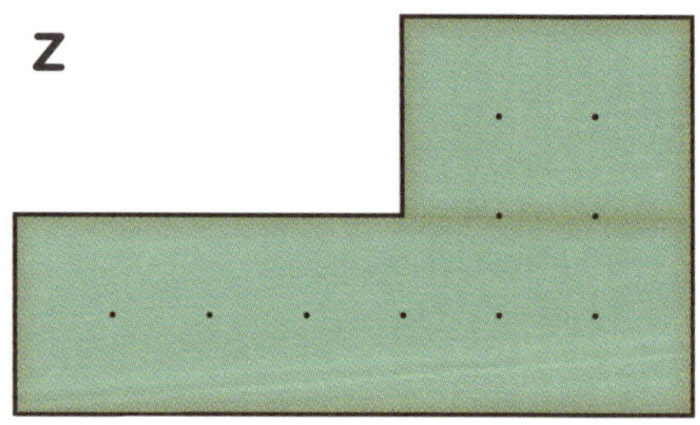

ACTIVITY

- Make some tiles like these shapes from thin cardboard.
- Trace the shape of your hand onto a piece of paper.
- Find the number of each tile needed to cover the shape of your hand.
- Use your tiles to measure other areas.
- Discuss which tile is best for covering areas.

# 3:25 Comparing Areas

**1** Write the area of each shape.

**a**

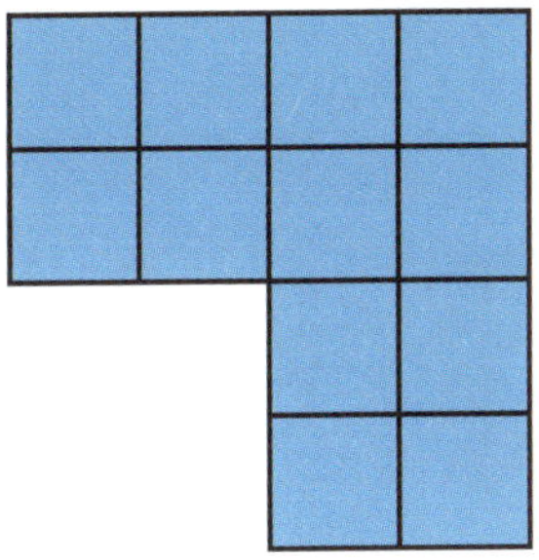

☐ squares

**b**

☐ triangles

**c**

☐ rectangles

**d**

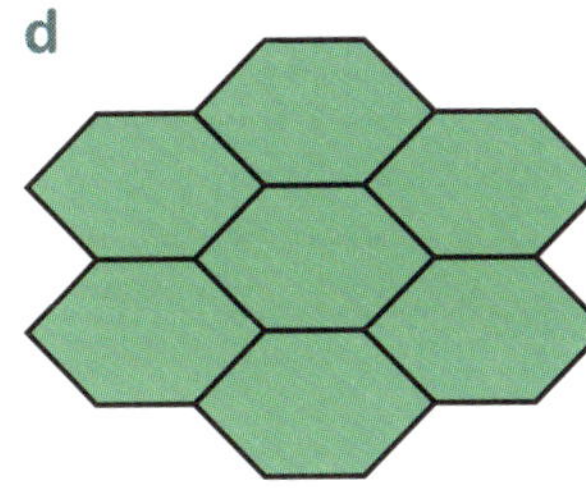

☐ hexagons

**2** Use ones place-value blocks to compare the areas of each pair of shapes. Tick the larger area of each pair.

**a**

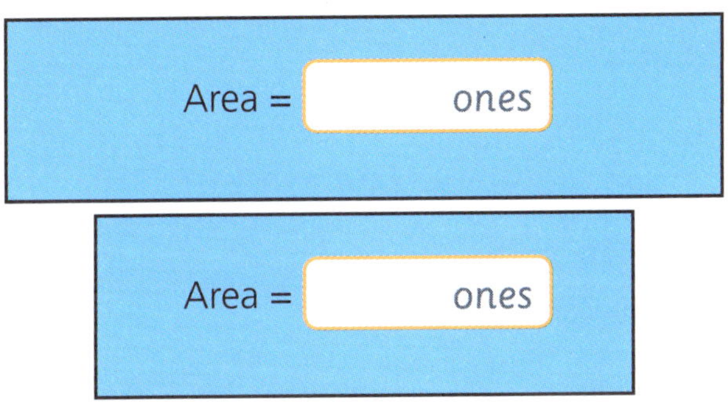


**b**

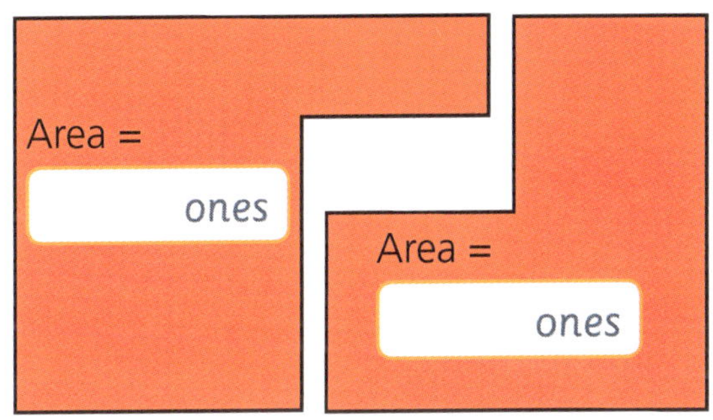


**c**

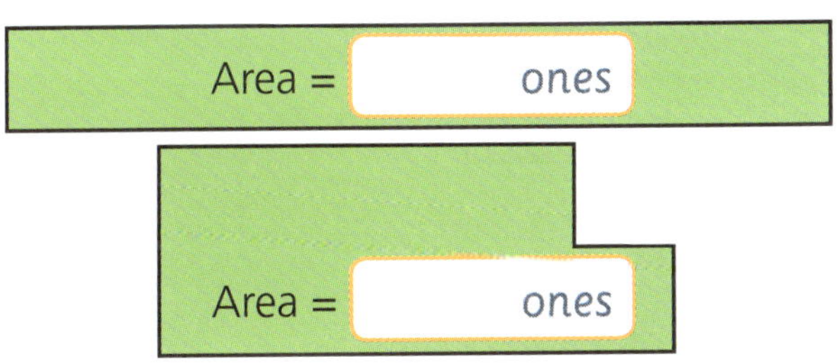


**d**

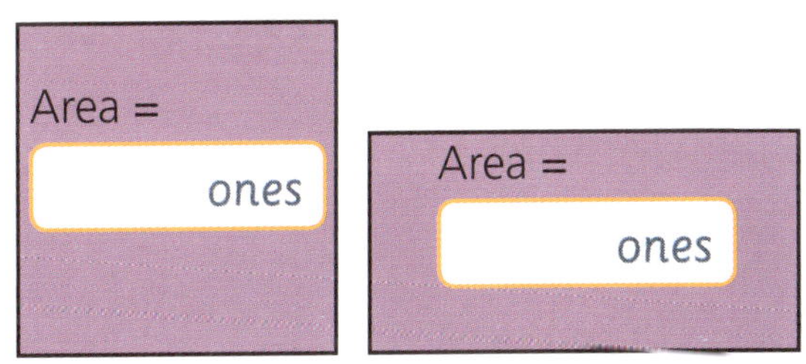


ACTIVITY

- Find two different books. Call them A and B.
  - Estimate: Which book cover has the larger area? ☐
    Compare the areas to check. Were you right? ☐
    Repeat with two different pieces of paper.
  - Of the two books, which has the greater height? ☐
  - Which book is wider? ☐
  - Which book is thicker? ☐

# 3:26 The Gram

CONCEPT

ACTIVITY

- Using balance scales, slowly fill a plastic bag with sand until it balances a 100 g mass. Seal the bag and label it **100 g**. Prepare 11 more bags like this.
- Slowly fill a plastic bag with sand until it balances a 500 g mass. Seal the bag and label it **500 g**. Prepare 2 more bags like this.
- Use balance scales to find:
  - How many 100 g bags balance a 500 g bag? ☐
  - How many 100 g bags balance two 500 g bags? ☐
  - How many 500 g bags balance a 1 kg mass? ☐
  - How many 100 g bags balance a 1 kg mass? ☐
- Use balance scales and your bags of sand to complete the table.

| Mass | Items found in the classroom |
|---|---|
| less than 100 g | |
| between 100 g and 500 g | |
| between 500 g and 1 kg | |

**1** Write the mass in short form.

a forty grams ☐ b one hundred grams ☐
c six hundred grams ☐ d three hundred grams ☐
e eighty grams ☐ f fifty grams ☐
g twenty grams ☐ h seventy grams ☐
i four hundred grams ☐

**2** Look at the pictures. What is the total mass of the items together?

a lollies and orange ☐ b cheese and jam ☐
c chocolates and jam ☐ d orange and chocolate ☐
e jam and lollies ☐ f orange and cheese ☐

# 3:27 Using Grams

ACTIVITY

- Use hefting to order three items from lightest to heaviest.
- Heft with standard masses to find:
  - five items lighter than 500 g
  - five items with a mass between 500 g and 1 kg
  - five items heavier than 1 kg
- Use balance scales and standard masses to complete this table.

| Mass | Items found in the classroom |
|---|---|
| less than 200 g | |
| between 200 g and 400 g | |
| between 400 g and 800 g | |
| between 800 g and 1 kg | |

**1** Write the mass in short form.

| | | | | | |
|---|---|---|---|---|---|
| **a** sixty grams | | **b** one hundred grams | | **c** forty kilograms | |
| **d** thirty grams | | **e** nine kilograms | | **f** seventy grams | |
| **g** eighty grams | | **h** seven kilograms | | **i** two hundred grams | |
| **j** ten kilograms | | **k** sixty kilograms | | **l** fifty kilograms | |

**2** Which unit of measurement (**g** or **kg**) has been left off each package?

**a**

**b**

**c**

**d**

**e**

# 3:28 Using a Ruler

CONCEPT

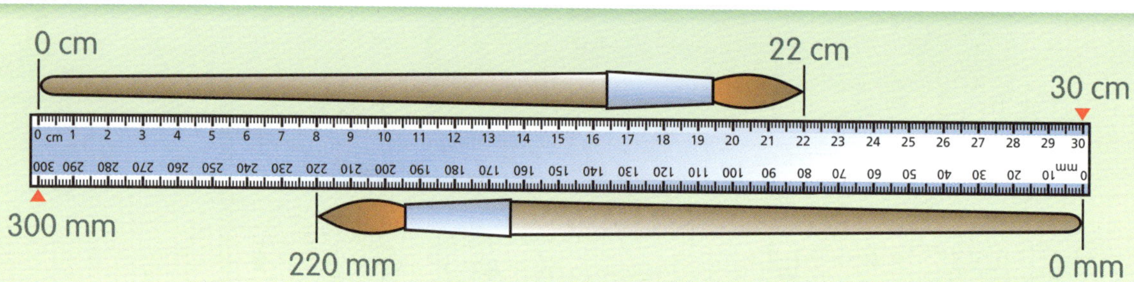


100 centimetres = 1 metre  100 cm = 1 m

10 millimetres = 1 centimetre  10 mm = 1 cm

**1** Use a ruler to measure these lengths.

**a** Place-value ones block

Length in centimetres = ☐ cm

Length in millimetres = ☐ mm

**b** Place-value long block

Length in centimetres = ☐ cm

Length in millimetres = ☐ mm

1000 mm = 1 m

**2** When measuring lengths, why might **centimetres** be better than **hand spans**?

INVESTIGATION

- Use a ruler to measure the lengths of 5 items.

**A** ☐ **B** ☐ **C** ☐

**D** ☐ **E** ☐

- Measure the height of a ball in millimetres. How did you do this?

 • *AUSTRALIAN SIGNPOST MATHS 3* • ISBN 978 1 4886 2184 0

# 3:29 Comparing Masses

CONCEPT

ACTIVITY

- Complete this table. Estimate first, then use balances and a 1 kg mass to check.

| Item | Less than 1 kg | About 1 kg | More than 1 kg |
|---|---|---|---|
| Dictionary | | | |
| 1 L milk carton | | | |
| Telephone book | | | |
| Vase | | | |

- Repeat this activity with other items in the classroom.
  Draw each item on this chart to show where it belongs.

| | | | | | | | |
|---|---|---|---|---|---|---|---|
| Less than 1 kg | | | | | | | |
| About 1 kg |  | | | | | | |
| More than 1 kg | | | | | | | |

- Collect labels from packages that show kilograms. Display the labels on a chart.

**1** Circle the items that would have a mass greater than 1 kilogram.

| | | | |
|---|---|---|---|
| **a** television | **b** ant | **c** pair of socks | **d** car battery |
| **e** football | **f** dog collar | **g** horse | **h** ruler |
| **i** slice of bread | **j** spoon | **k** table | **l** refrigerator |

**2** Circle the items that would be measured in kilograms.

| | | |
|---|---|---|
| **a** packet of jelly beans | **b** bag of potatoes | **c** suitcase of clothes |
| **d** jar of jam | **e** bag of onions | **f** a person |
| **g** packet of sultanas | **h** packet of chips | **i** sack of flour |

# 3:30 The Millilitre

ACTIVITY

- Use different containers to fill a 1 litre jug. Estimate and then record how many of each container is needed to fill the jug.

| Container | Estimated number | Measured number |
| --- | --- | --- |
| | | |
| | | |
| | | |
| | | |
| | | |

The space taken up by an item is called its **volume**.

- Collect labels from small drink containers. Discuss the use of the symbol **mL**.

*mL stands for millilitres. There are 1 000 mL in 1 L.*

**1** Circle the items that would hold less than 1 litre.

**a** bathtub **b** eggcup **c** ice-cream tub
**d** small jam jar **e** drinking glass **f** swimming pool
**g** teacup **h** large bucket **i** teaspoon

**2** Write each amount in short form.

**a** 3 litres ☐ **b** 5 litres ☐ **c** 1 litre ☐
**d** 6 litres ☐ **e** 9 litres ☐ **f** 4 litres ☐
**g** 250 millilitres ☐ **h** 400 millilitres ☐ **i** 750 millilitres ☐
**j** 300 millilitres ☐ **k** 900 millilitres ☐ **l** 500 millilitres ☐

**3** How many millilitres in each amount?

**a** 1 litre ☐ **b** 2 litres ☐ **c** 5 litres ☐
**d** 3 litres ☐ **e** 8 litres ☐ **f** 6 litres ☐

**4** How many litres in each amount?

**a** 1 000 mL ☐ **b** 4 000 mL ☐ **c** 7 000 mL ☐


 • *AUSTRALIAN SIGNPOST MATHS 3* • ISBN 978 1 4886 2184 0

# 3:31 Using Litres

ACTIVITY

- Look at four containers. Estimate the amount of water that each container could hold.
- Use a 1 litre jug to measure how much each container holds, to the nearest litre.
- Order the containers from smallest to largest.
- Find various containers to help you complete this table.

| Container | Estimated | Measured |
|---|---|---|
| | 1 L | |
| | 2 L | |
| | 5 L | |
| | 10 L | |

1 Write each amount in short form.

a 2 litres ☐ b 4 litres ☐ c 1 litre ☐
d 10 litres ☐ e 7 litres ☐ f 5 litres ☐
g 19 litres ☐ h 20 litres ☐ i 14 litres ☐

2

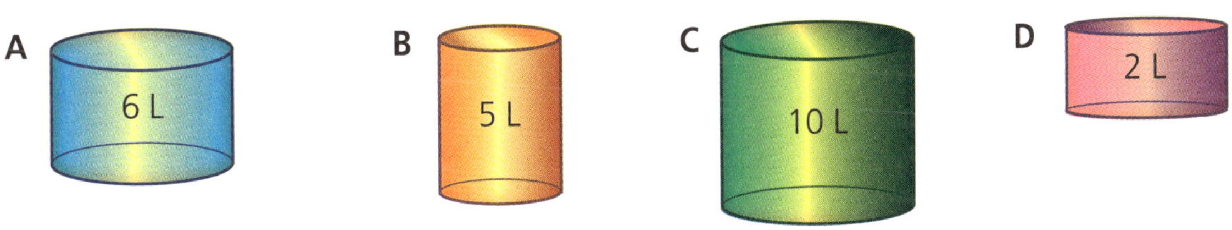


Write which of these containers would hold:

a the most water ☐ b the least water ☐
c more water than A ☐ d less water than B ☐
e more water than B but less than C ☐
f How much water would be needed to fill both A and C? ☐

# 3:32 The Calendar

CONCEPT

There are 12 months in each year. How many days in each month?

30 days has September, April, June and November. All the rest have 31 except for February alone, which has 28 days clear but 29 in each leap year.

**Months of the year**

| | | | |
|---|---|---|---|
| January | (Jan) | July | (Jul) |
| February | (Feb) | August | (Aug) |
| March | (Mar) | September | (Sep) |
| April | (Apr) | October | (Oct) |
| May | (May) | November | (Nov) |
| June | (Jun) | December | (Dec) |

**November 2012**

| Sun | Mon | Tues | Wed | Thurs | Fri | Sat |
|---|---|---|---|---|---|---|
| | | | | 1 | 2 | 3 |
| 4 | 5 | 6 | 7 | 8 | 9 | 10 |
| 11 | 12 | 13 | 14 | 15 | 16 | 17 |
| 18 | 19 | 20 | 21 | 22 | 23 | 24 |
| 25 | 26 | 27 | 28 | 29 | 30 | |

Look at the calendar for November.

1 Study the calendar and answer the questions.

- **a** How many days are in November?
- **b** What day of the week is 1 November?
- **c** What day of the week is 17 November?
- **d** What day of the week is 29 November?
- **e** What is the date of the first Friday?
- **f** What is the date of the first Tuesday?
- **g** What is the date of the last Tuesday?
- **h** What is the date of the last Saturday?
- **i** Write the dates of all the Mondays in November.
- **j** How many Tuesdays are in November?
- **k** On what day of the week does October end?

2 How many days are in the following?

- **a** one week
- **b** one fortnight
- **c** January
- **d** June
- **e** March
- **f** May
- **g** April
- **h** July
- **i** October
- **j** September
- **k** November
- **l** August
- **m** December
- **n** February (not in a leap year)

# 3:33 Standard Metric Units

CONCEPT

We use **standard metric units** so everybody can use the same units.

kilogram | litre | metre

**1** Match the unit to the type of measurement.

kilogram | metre | litre | centimetre | millimetre

capacity — How much does it hold? | mass — How heavy is it? | length — How long is it?

L | kg | cm | mm | m

**2** Which quantity does each tool measure: capacity, mass or length?

a 1 litre container

b Balance scales

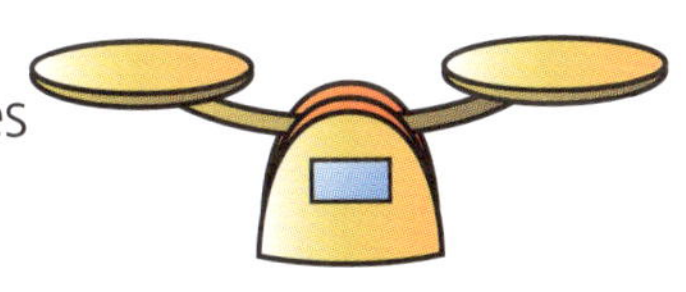

c Rulers

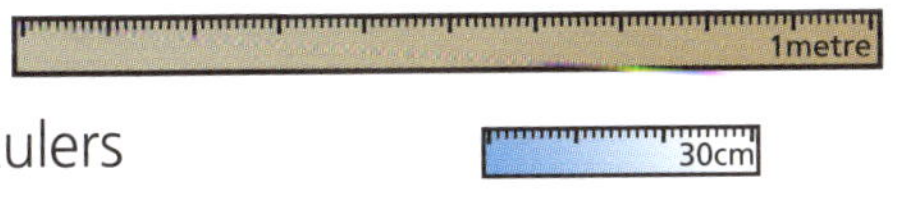

**3** What is one advantage of using standard units to describe and compare objects?

FUN SPOT

How would you compare an apple, a banana and an orange? List as many features as you can.

# 4:01 Properties of 3D Objects

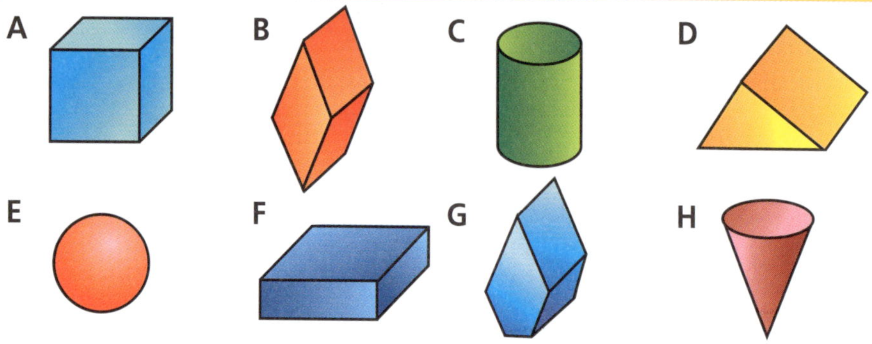

1. This picture shows holes cut in a block of wood.
   Which of the objects A to H would fit through the holes?

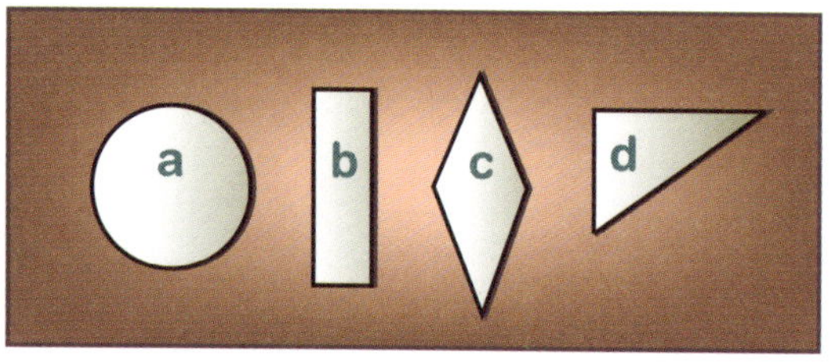


**a** circle hole ______

**b** rectangle hole ______

**c** rhombus hole ______

**d** triangle hole ______

2. Which of the objects A to H can't be easily stacked? ______

3. Which of the objects A to H are the same shape as these?

**a** ______ **b** ______ **c** ______ **d** ______

4. Which of the objects A to H could fit together like this?

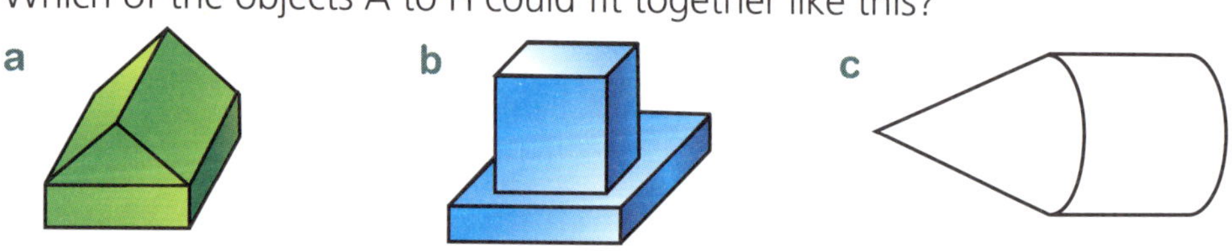


**a** ______ and ______

**b** ______ and ______

**c** ______ and ______

**d** Draw the hidden lines to make the diagram for part **c** look 3D.

5. Draw the missing object in each pattern.

**a**

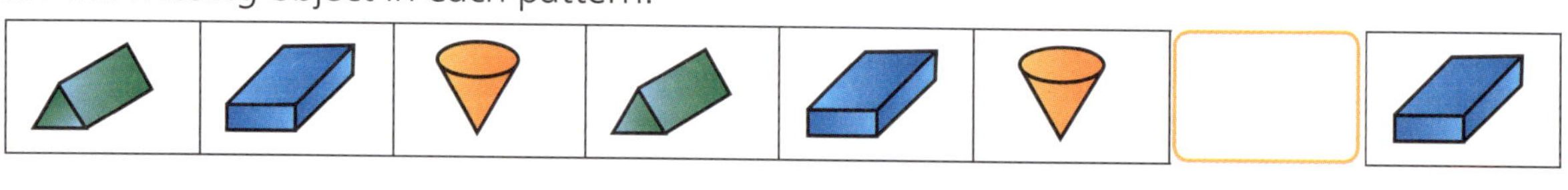

**b**

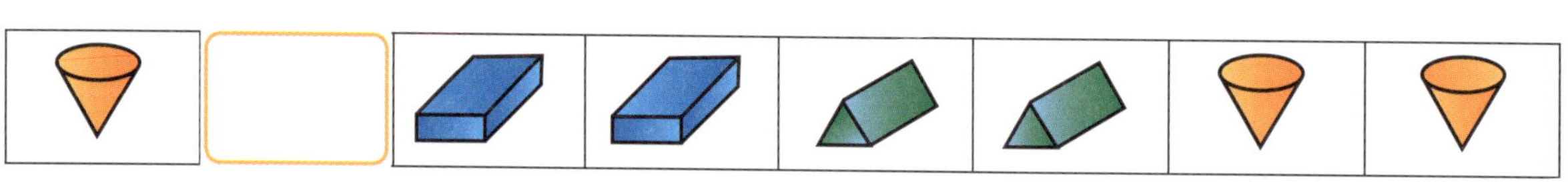



 • *AUSTRALIAN SIGNPOST MATHS 3* • ISBN 978 1 4886 2184 0

# 4:02 Symmetry

These dashed lines are **lines of symmetry**.

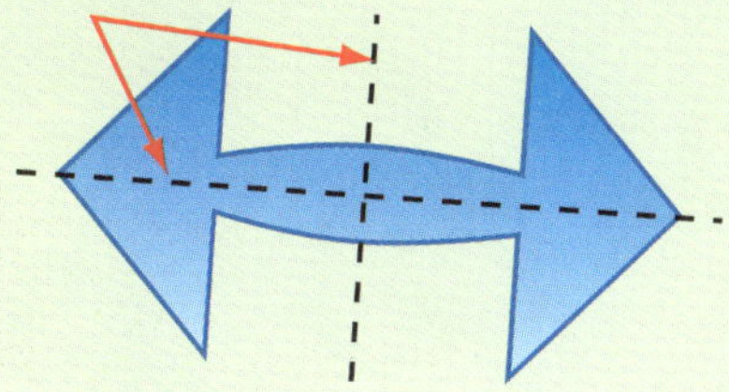

CONCEPT

1. Use a ruler to draw a line of symmetry on each picture.

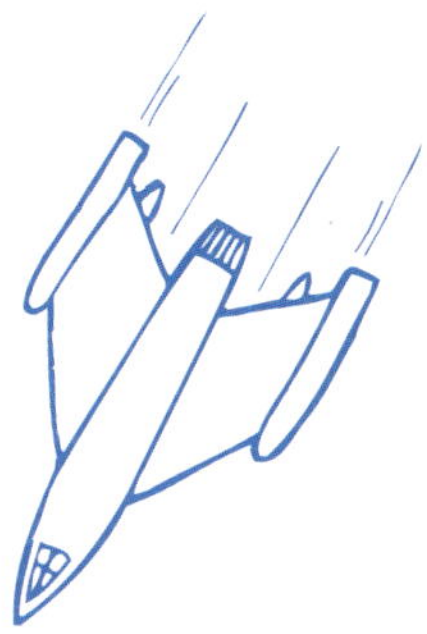

Slide back the edge of this page so that it matches the pictures on page 123 underneath. What do you see?

2. Draw lines of symmetry on each letter.

BTXUHYOV

3. Draw lines of symmetry on each shape.

- Use block shapes to make symmetrical designs.
- Paint a piece of paper. While the paint is still wet, fold the paper onto itself to make a symmetrical painting. What can you see?

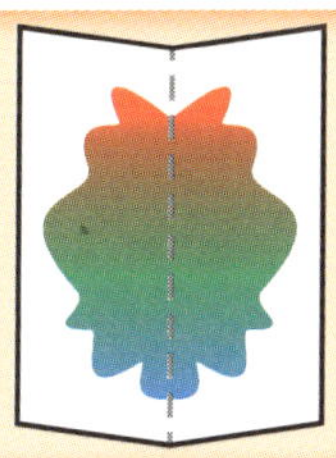

ACTIVITY

# 4:03 Properties of 3D Objects

CONCEPT

- How would you describe the shape of this tent?

**1** How many flat surfaces does each object have? Name each object.

**a** 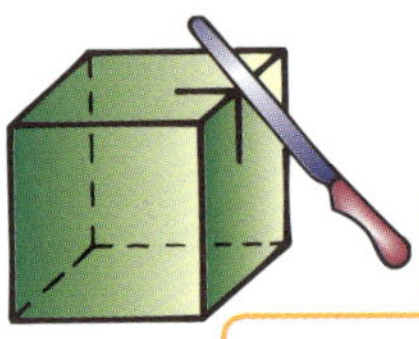

**b** 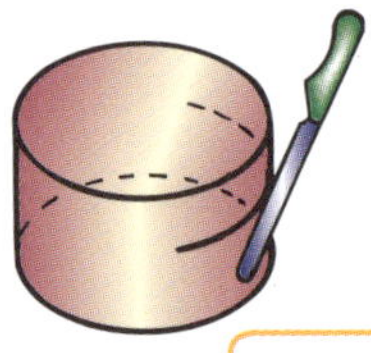

**c** 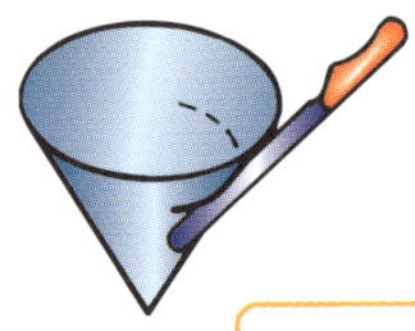

**d** 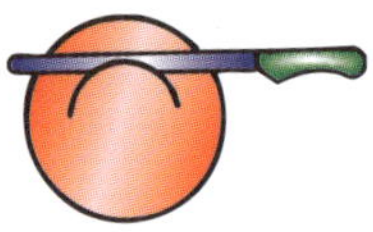

**e** 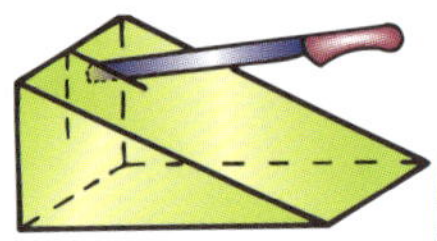

**2** What is the shape of each object's cross-section?

**a** ______ **b** ______

**c** ______ **d** ______ **e** ______

**3** **a** Which of the objects in Question 1 has six corners? ______

**b** Which has 8 corners? ______

**c** Which has 12 edges? ______

**d** Which has 1 corner? ______

**e** Which has 5 faces? ______

**f** Which has 9 edges? ______

**g** Which has only curved surfaces? ______

**h** Which has only 2 flat surfaces? ______

**i** Which have a circular cross-section? ______

**4** Which block would fit into which hole? Write the number.

**a** Hole A ______ **b** Hole B ______

**c** Hole C ______ **d** Hole D ______

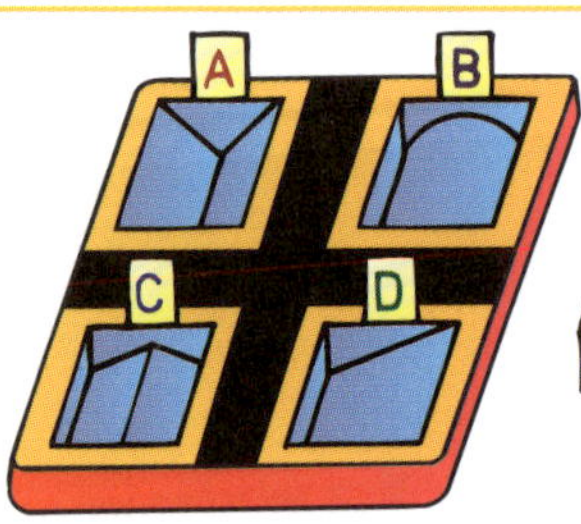

 • *AUSTRALIAN SIGNPOST MATHS 3* • ISBN 978 1 4886 2184 0

# 4:04 Symmetry in Our World

CONCEPT

In our world, there are many things that are almost symmetrical. Talk about some of these things.

Draw the other half of each picture.

1. Draw every line of symmetry.

2. Draw the other half to make each picture symmetrical.

ACTIVITY

- Use block shapes to make symmetrical patterns.

3. Which of **flip**, **slide** or **turn** always makes a symmetrical picture (when put together with its original)? ____________

# 4:05 Parallel and Perpendicular Lines

CONCEPT

**Parallel lines** are always the same distance apart.

- They are straight lines.
- They are in the same direction.
- They never touch.

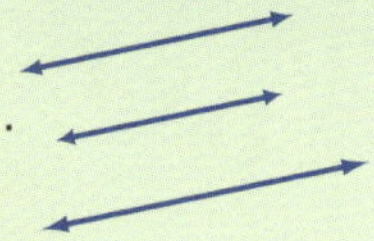

**Perpendicular lines** meet at right angles.

- They are straight lines.
- They cross over at right angles.

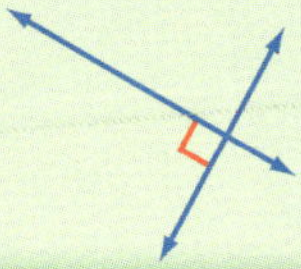

1. At how many points does each group of lines cross?

a  ____

b 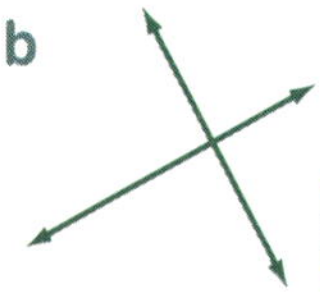 ____

c  ____

d  ____

e 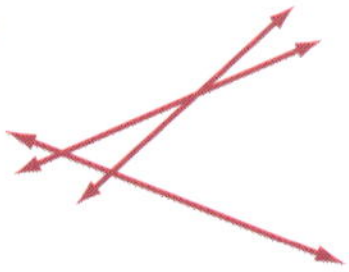 ____

f 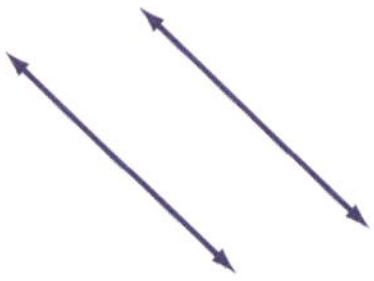 ____

g 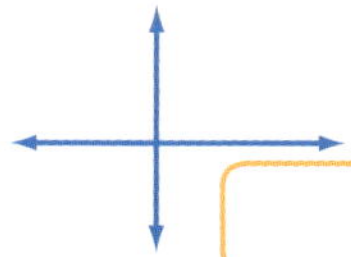 ____

h 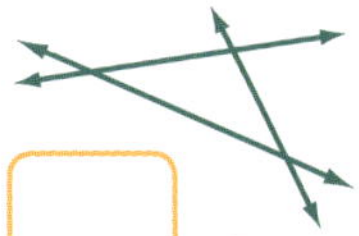 ____

2. Which groups of lines in Question 1 are all parallel? ____

   Why are they parallel? ____

3. Why are the lines in 1**b** perpendicular? ____

4. Name two shapes that have perpendicular sides. ____

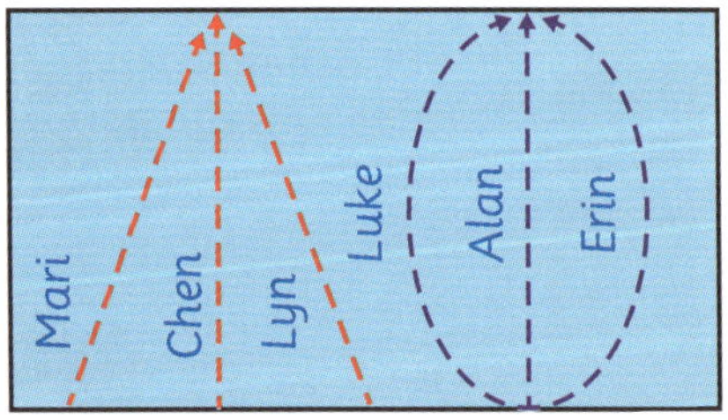


5. Six people swam across a swimming pool.
   - **a** Who swam the shortest path? ____
   - **b** Who swam perpendicular to the sides? ____
   - **c** Who swam parallel paths? ____

6. 
   - **a** List things in this picture that are parallel. ____
   - **b** List things in your classroom that are parallel. ____



 ISBN 978 1 4886 2184 0

# 4:06 Regular and Irregular Shapes

1 Join the dots to complete these regular shapes. Under each shape, write its name.

2 Which shapes in Question 1 have opposite sides parallel?

3 Join the dots to complete these irregular shapes. Under each shape, write its name.

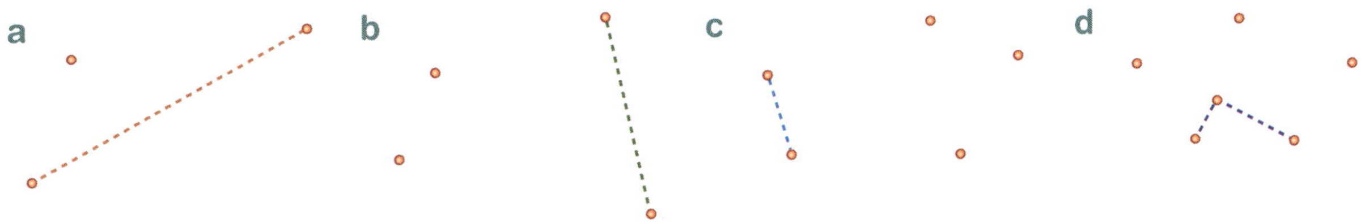


4 Trapeze artists swing on trapeziums. A trapezium has at least one pair of parallel lines. Join the dots to complete each trapezium.

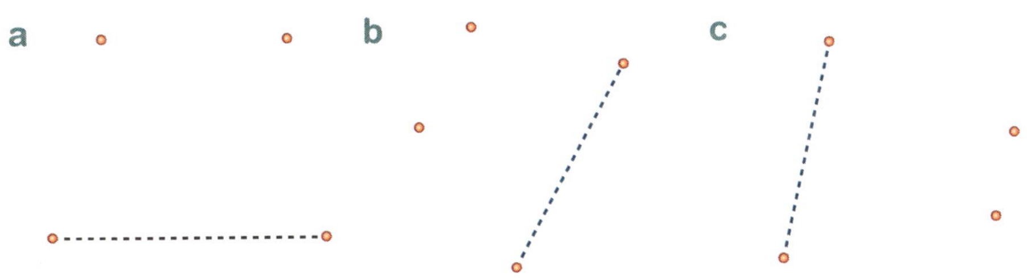

5 a How many lines are used to make the letter H?

b Does the letter H have any parallel lines?

c Write all the capital letters that have parallel lines.

6 To draw a lattice, use diagonal parallel lines to join these dots in both directions.

ACTIVITY

- In the lattice pattern, what shape is repeated?
- When shapes repeat with no gaps between them, the pattern is called a **tessellation**.
- Draw a tessellation of your own.

# 4:07 Position and Giving Directions

A map gives us a view from above.

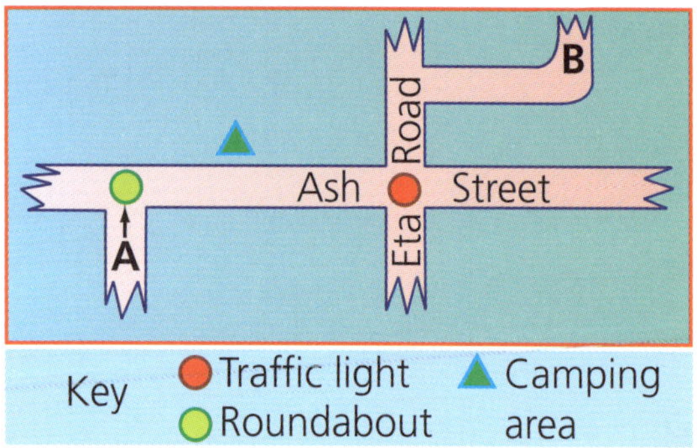

**1** Write directions to get from place to place.

a **A** to **C** ______

b **A** to **D** ______

c **A** to **E** ______

d **A** to **F** ______

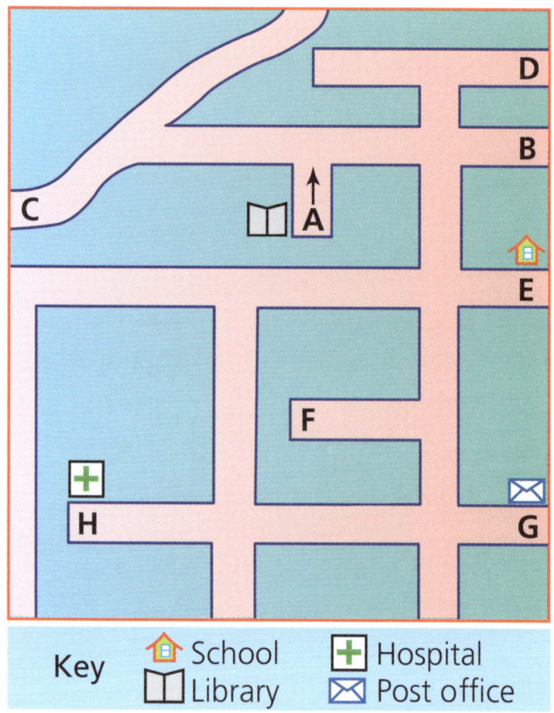


**2** Follow these directions. Where do they lead? Draw each path on the map.

a Start at **A**. Turn right, then the first right, then the second left.

______

b Start at **A**. Turn right, then the first right, then the first right, then the first left, then the first right.

______

**3** a Jenna is sitting in seat **J**. She moves two seats to the right, then one seat forwards. Where is she now? ______

b Patrick is sitting in seat **P**. He moves three seats to the left, then two seats forwards. Where is he now? ______

c Dmitri is sitting in seat **D**. He moves three seats backwards, then two seats to the left. Where is he now? ______

d Erin is sitting in seat **E**. She moves one seat to the right, then two seats backwards, then one seat left, then two seats forwards. Where is she now? ______

© PEARSON AUSTRALIA 2018 • *AUSTRALIAN SIGNPOST MATHS 3* • ISBN 978 1 4886 2184 0

# 4:08 Position and Giving Directions

**1** Follow the spider's path and write the letter for the point where it stops.

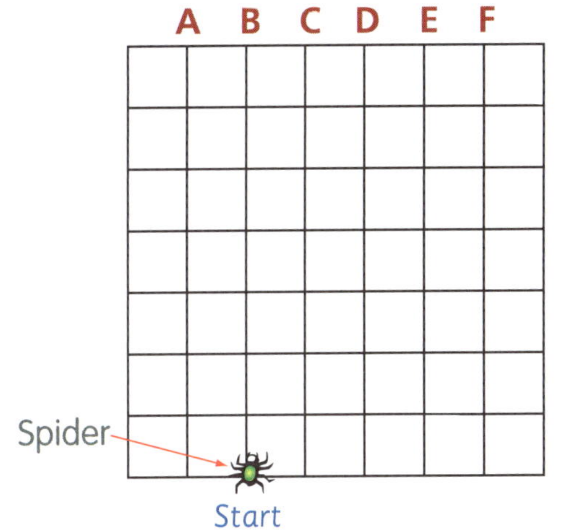


- **a** 4 spaces up, 2 right, 3 up
- **b** 3 spaces up, 1 left, 4 up
- **c** 2 up, 4 right, 3 up, 5 left, 2 up
- **d** 6 up, 2 right, 4 down, 1 left, 5 up
- **e** 1 right, 7 up
- **f** 2 up, 1 left, 2 up, 5 right, 3 up
- **g** Write directions to take the spider to **E**.

**2** Use directions like those above to describe these paths.

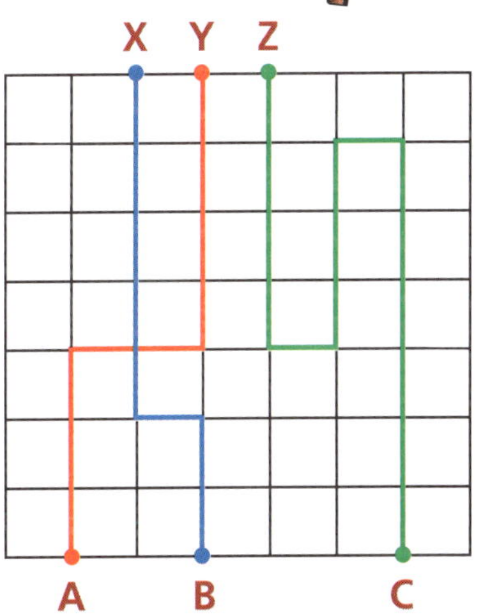


- **a** **A** to **Y**
- **b** **B** to **X**
- **c** **C** to **Z**
- **d** **X** to **B**
- **e** **Y** to **A**

**3** You are in a car at **A**. To get to **E** you would follow these directions: 1 forwards, turn right, 6 forwards, turn left, 4 forwards.
Write directions for these paths:

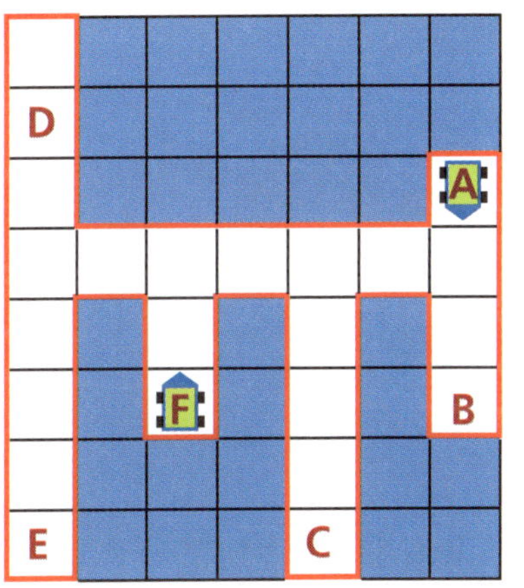


- **a** **A** to **B**
- **b** **A** to **C**
- **c** **A** to **D**
- **d** **F** to **D**
- **e** **F** to **E**

# 4:09 Shapes Revision

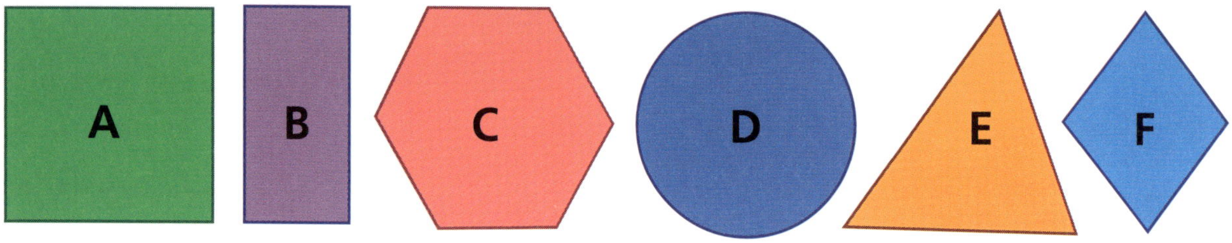


1 Write A, B, C, D, E or F to match the shape name.

a circle ☐ b square ☐ c triangle ☐

d hexagon ☐ e rhombus ☐ f rectangle ☐

2 Write A, B, C, D, E or F to match the description.

a three sides ☐ b four sides ☐

c six sides ☐ d no straight sides ☐

3 a Which of the shapes above have all sides equal? ☐

b Which of the shapes above have opposite sides equal? ☐

4 Write the names of the two shapes used in each picture.

a 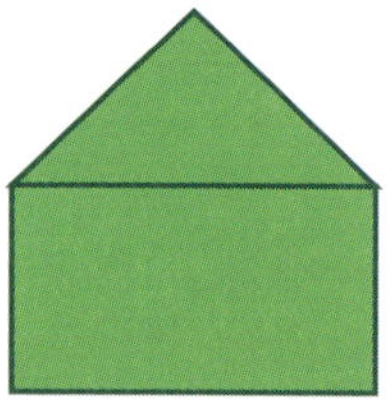

b 

c 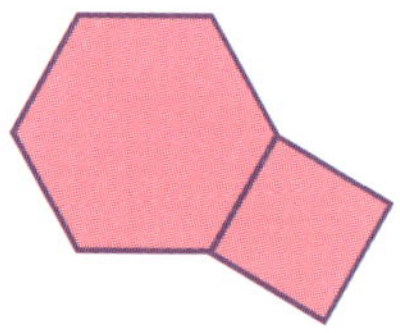

d 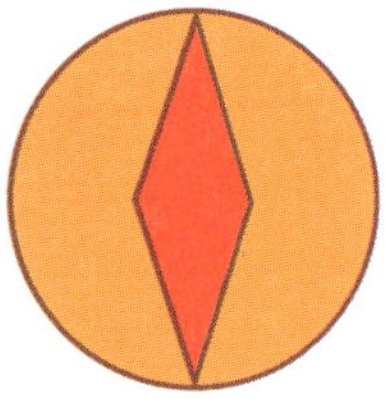

ACTIVITY

5 How many shapes are in the picture?

a squares ☐

b triangles ☐

c rectangles ☐

d circles ☐

- Draw your own picture using shapes.

 • *AUSTRALIAN SIGNPOST MATHS 3* • ISBN 978 1 4886 2184 0

# 4:10 Properties of 2D Shapes

Antonio's teacher made a **geoboard** like this by hammering nails into wood. Antonio used the geoboard to make shapes by stretching a rubber band over the nails.

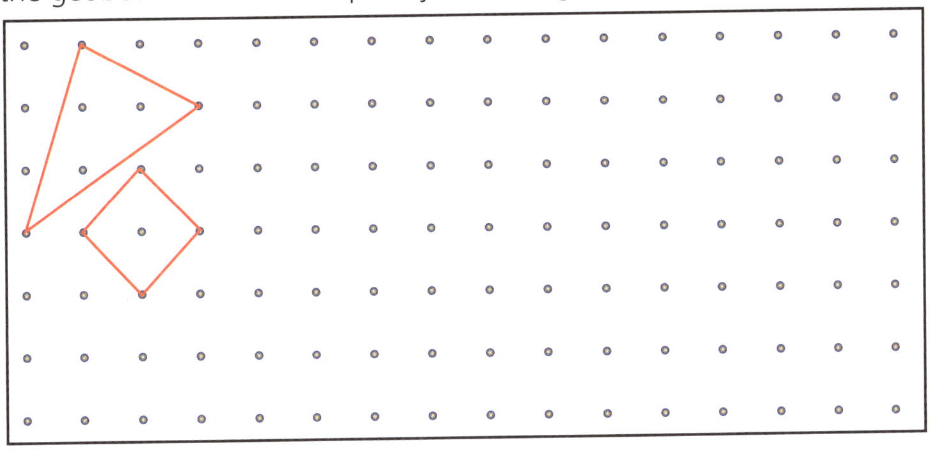

**1** a How many nails were used to make the triangle? ☐ Number of corners = ☐

b How many nails were used to make the square? ☐ Number of corners = ☐

c How many sides on a triangle? ☐

d How many sides on a square? ☐

**2** Every corner of each shape on the geoboard must be at a nail.

a On the geoboard above, draw 3 triangles of different shapes and sizes.
How many sides does each triangle have? ☐

b What shapes can you draw using 4 nails as corners? ☐
Draw 3 of them on the geoboard above. ☐

c A pentagon has 5 sides. How many corners would you need to draw a pentagon? ☐
Draw a pentagon on the geoboard.

d Draw a shape on the geoboard that has 6 corners.
What is this shape called? ☐

ACTIVITY

Olena cut a square of paper into 5 pieces. She asked her friend to put the square back together again. Then she asked her friend to make a picture using the pieces.

- Cut your own square of paper into 5 pieces.
- Discuss the shapes you cut out.
- Mix up your 5 pieces and ask a partner to put the square back together again.
- Use your pieces of paper to make pictures.

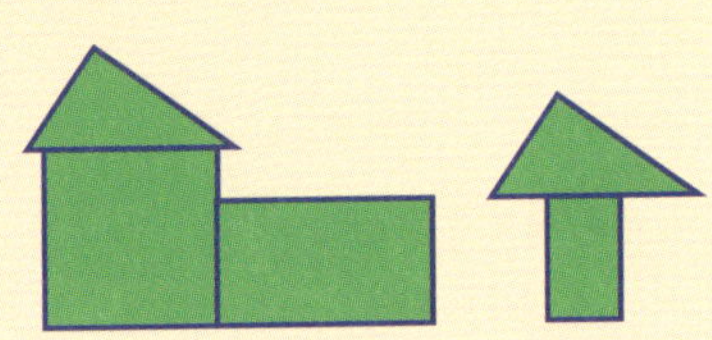

 • *AUSTRALIAN SIGNPOST MATHS 3* • ISBN 978 1 4886 2184 0

# 4:11 Investigating Angles

CONCEPT

- Take 2 straight strips of cardboard and join them at one end with a paper fastener.
- By turning the arms, you can make the movable angle small or large.

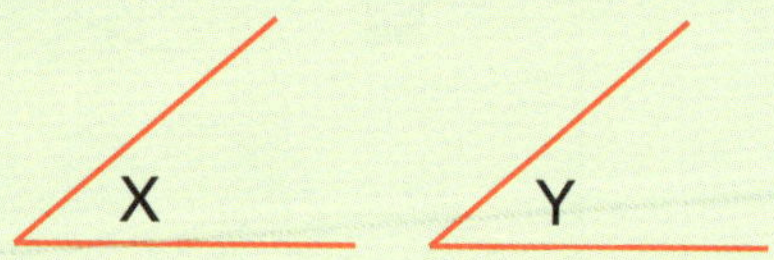


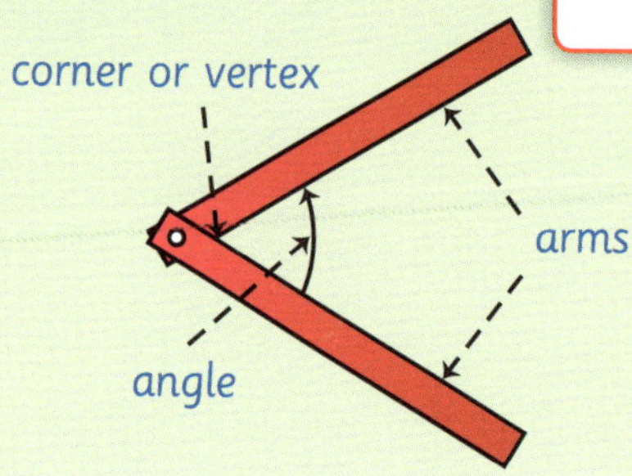

- The size of an angle is the amount of turning between the two arms.

1. Which angle is the largest? ☐
2. Which angle is the smallest? ☐

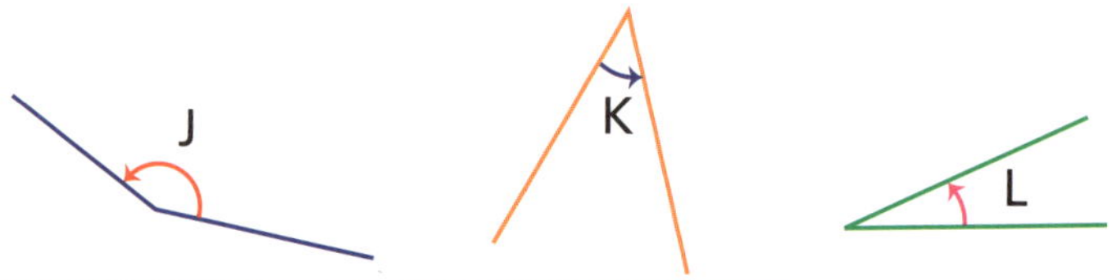


3. Use a movable angle to answer these questions.
   - **a** Which of these angles are the same size as angle **A**? ☐
   - **b** Which angles are the same size as angle **B**? ☐
   - **c** Which angle is smaller (less turn) than angle **B**? ☐
   - **d** Which angles is bigger (more turn) than angle **A**? ☐
   - **e** Is **F** the same size angle as the corner of a square? ☐

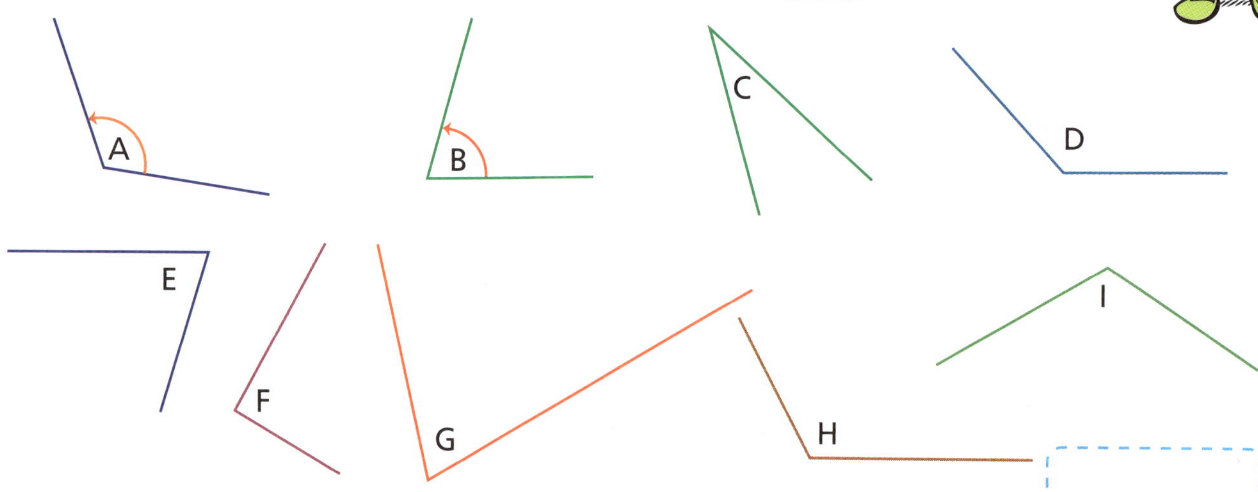


A square corner is called a right angle.

INVESTIGATION

- Compare the angles you can see in your classroom.

# 4:12 Angles

CONCEPT

- The size of an angle is the amount of turn between the arms.

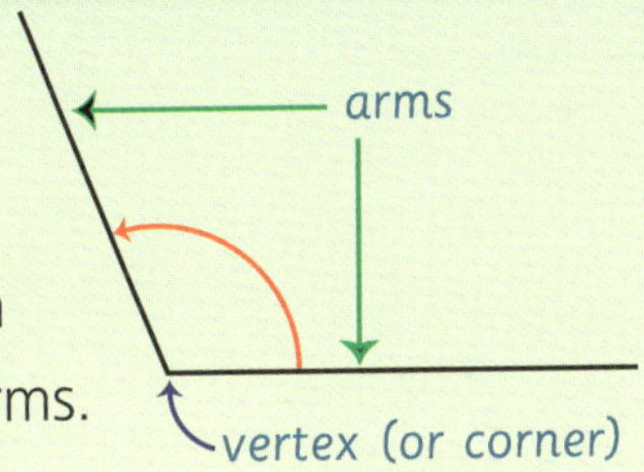


- The stick has been moved through an angle.

- These angles are the same size even though the arms are different lengths.

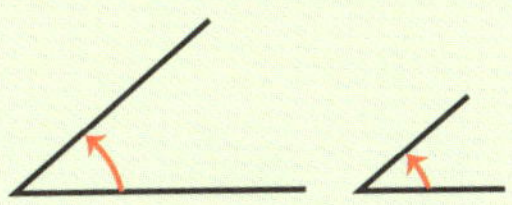

- Sometimes we have to imagine where one of the arms should be.
- The top of the door shows both arms of the angle. The bottom shows one arm, but we can imagine the other.

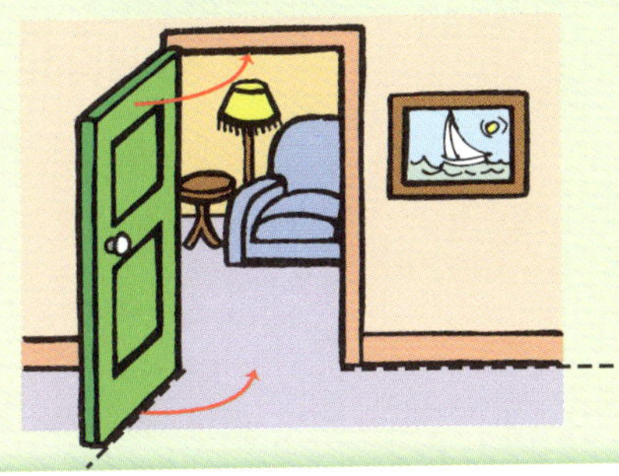

**1** In each part, number the angles from the smallest turn (1) to the largest turn (4).

**a**

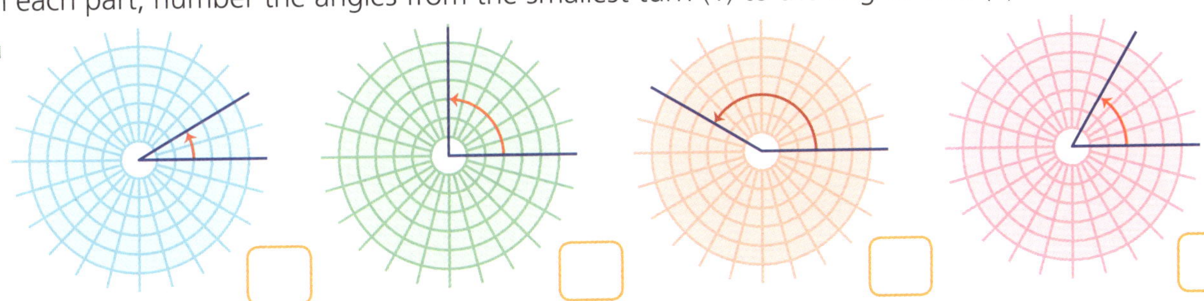

**b**

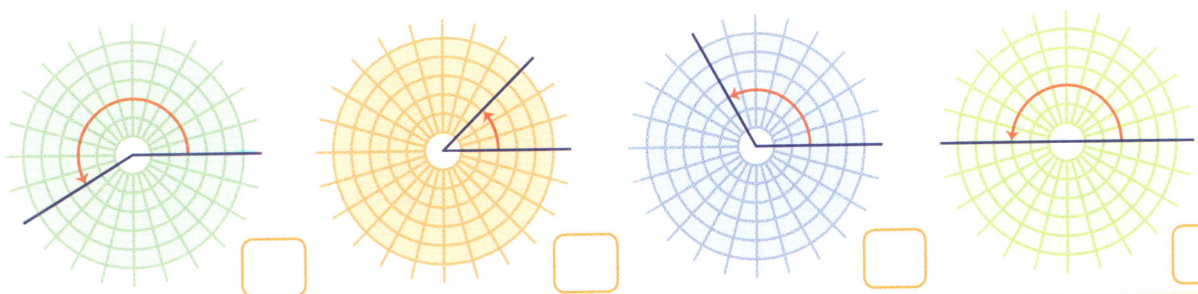

**2** Use a ruler to draw angles on the dot grids.

**a** Draw an angle the same size as **A**.

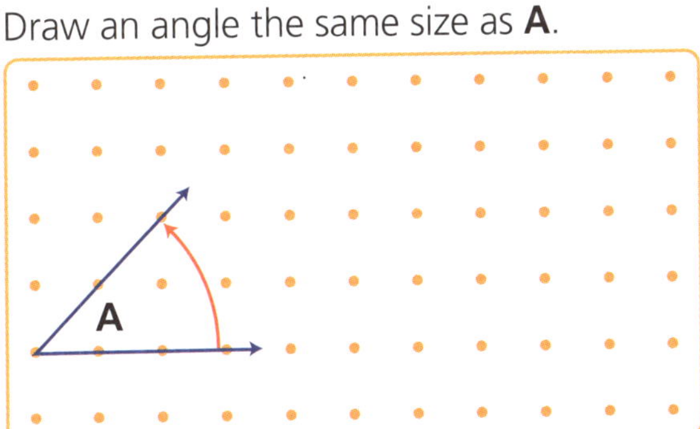


**b** Draw an angle the same size as **B**.

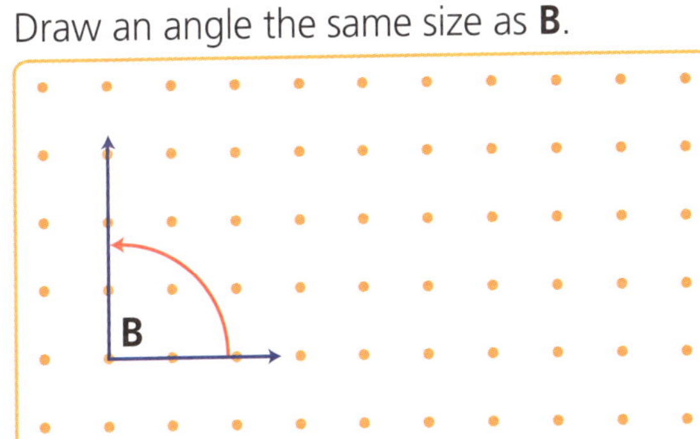


**c** Why is **B** a right angle?

# 4:13 The Trapezium and Parallelogram

CONCEPT

- Arrows are used to show parallel sides. 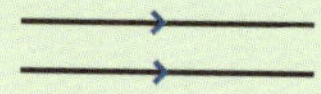

- A **trapezium** has one pair of parallel sides.
- A **parallelogram** has two pairs of parallel sides.
- Both shapes have 4 straight sides.
- A parallelogram is a special trapezium.
- Trapeziums and parallelograms are both quadrilaterals.

**1**

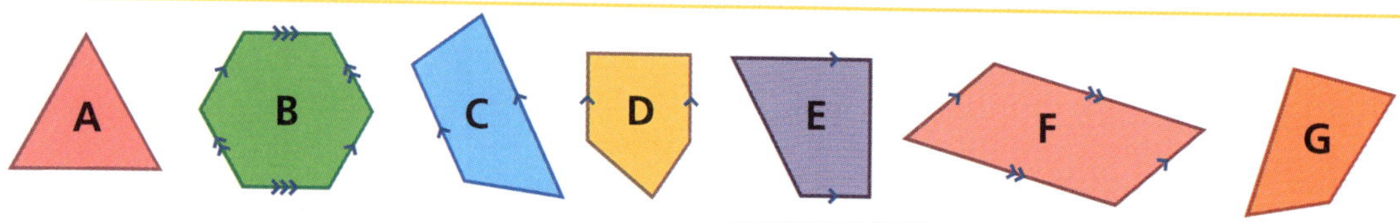


**a** Which of these shapes are trapeziums? ____________

**b** Which shape is a parallelogram? ____________

**2** Match each shape name with one shape picture.

| | | |
|---|---|---|
| oval | | rectangle |
| square | | parallelogram |
| triangle | | pentagon |
| rhombus (diamond) | | trapezium |
| octagon | | hexagon |

**3** **a** On this dot grid, draw three trapeziums.

**b** On this dot grid, draw three parallelograms.

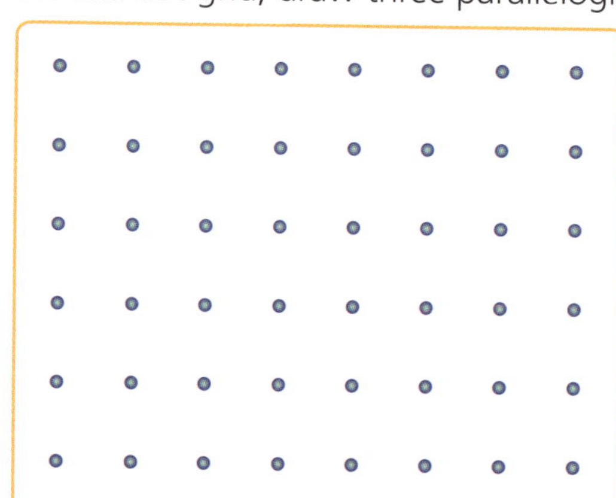

# 4:14 Features of 2D Shapes

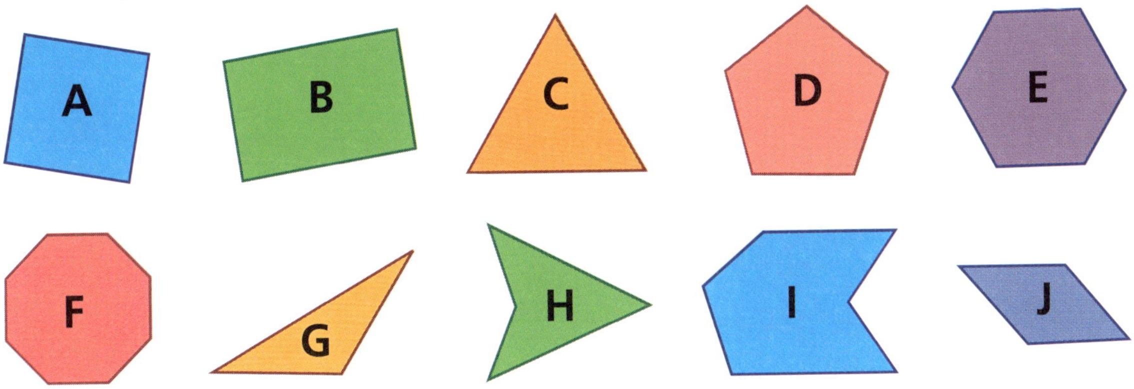


1 Write which of the shapes above have:

a 3 sides

b 4 sides

c 6 sides

d all sides equal

e sides that jut in

CONCEPT

- A **quadrilateral** has 4 sides.
- A **pentagon** has 5 sides.
- A **hexagon** has 6 sides.
- An **octagon** has 8 sides.
- A **rhombus** is a diamond.
- A **kite** has symmetry.

2 a Which of the quadrilaterals above have all sides equal?

b Which of the shapes above is an octagon?

3 What shape is a closed shape with the least possible number of straight sides and no curved sides?

4

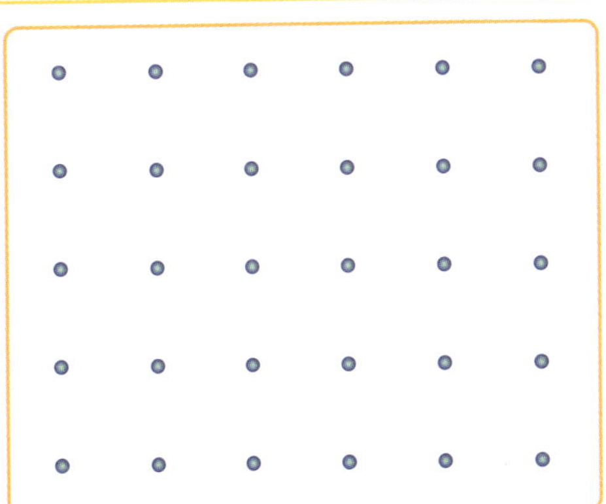

a On this dot grid, draw 5 different quadrilaterals.

b How many different sizes of squares could be drawn on this dot grid?

c How many dots would be inside the largest square you could draw?

d How many dots would be inside the largest hexagon you could draw on this dot grid?

5 How are these shapes different from the shapes at the top of the page?

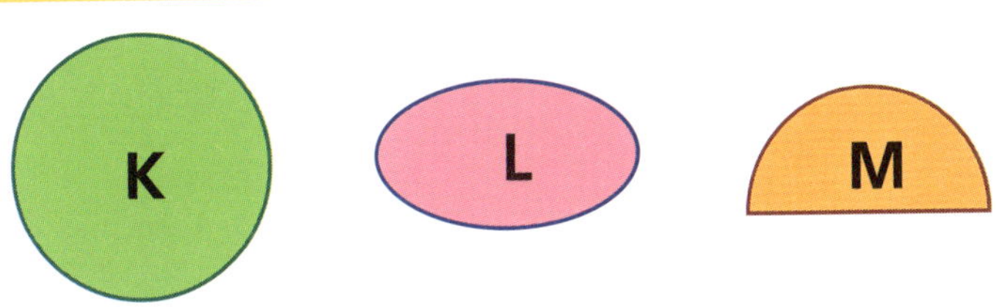


 • *AUSTRALIAN SIGNPOST MATHS 3* • ISBN 978 1 4886 2184 0

# 4:15 Right Angles

CONCEPT

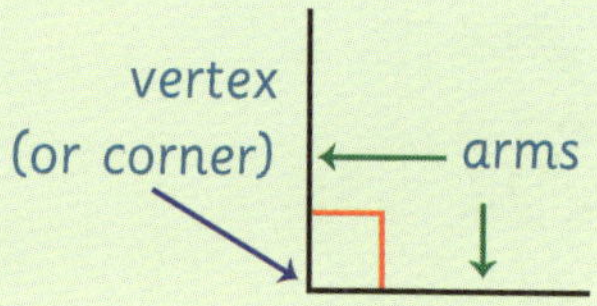


Right Angle

The angle at the corner of a square is called a **right angle**.
Two right angles make a half turn.

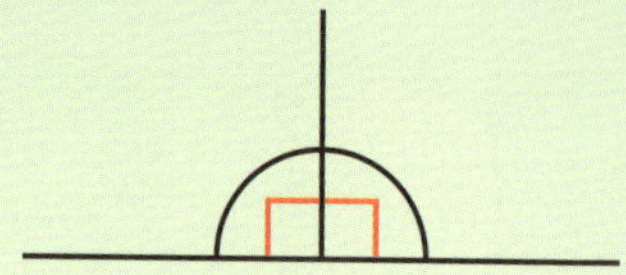

To make a square corner, follow these steps:

1 Fold a piece of paper once to form a half turn.

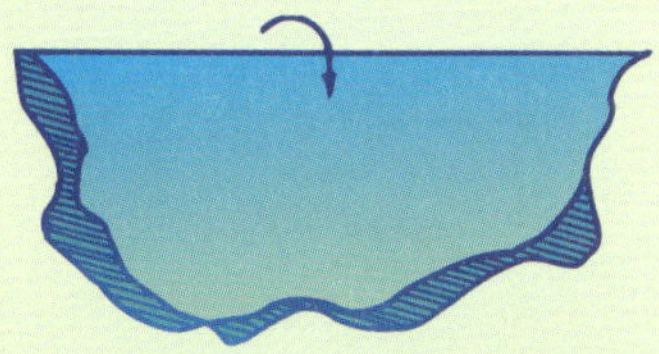

2 Fold the paper again so that the second fold lies exactly on top of the first.

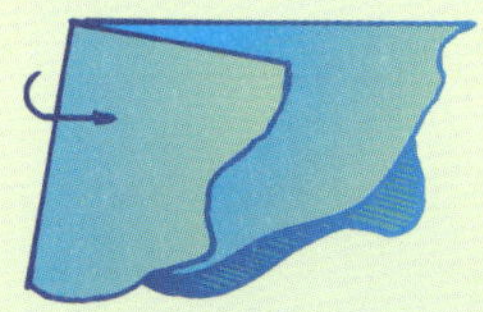

3 Press all the folds down firmly. Where the folds meet, you have a square corner.

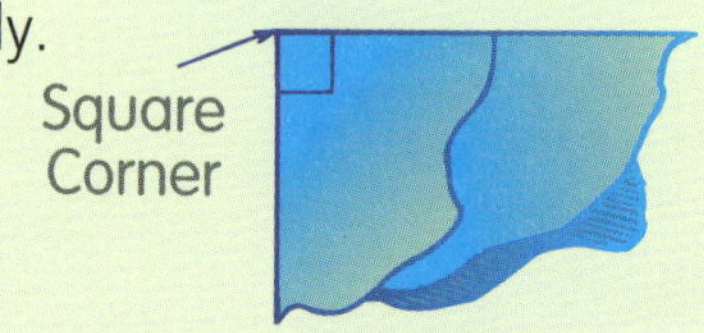


Practise making your own square corners.

1 Use a square corner to test these angles. Which are right angles?

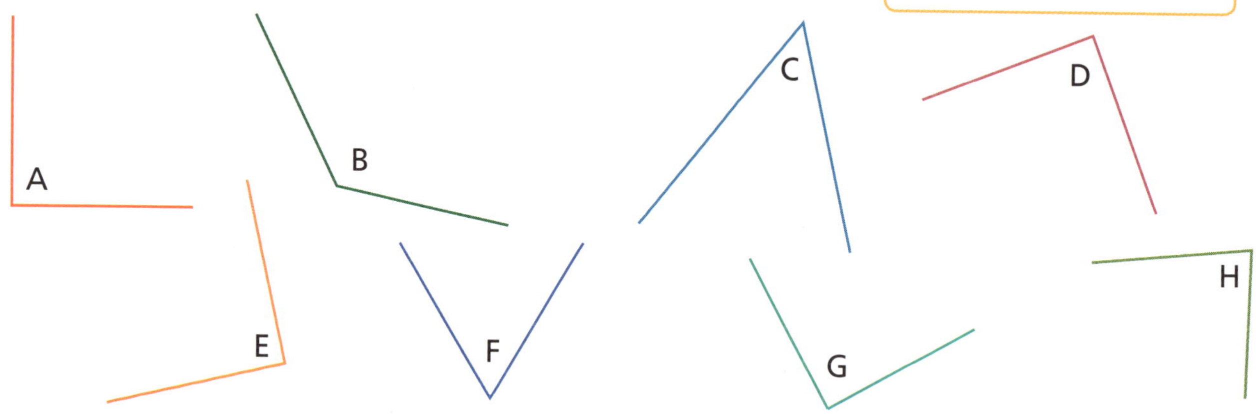


2 Which angle is larger than a right angle?

3 Use a square corner to test the corners of this page. Are they right angles?

INVESTIGATION

Make a list of all the right angles that you can see right now.

# 4:16 Angles

CONCEPT

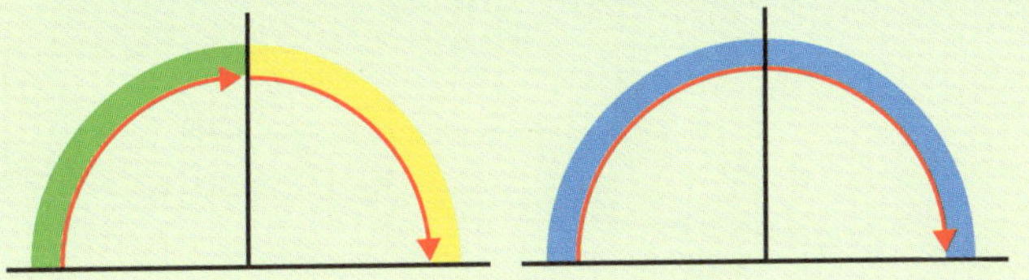

- 2 quarter turns (right angles) are the same as 1 half turn (straight angle).

- Angles are formed where surfaces meet.

**1** Describe each turn as **clockwise** or **anticlockwise**, and as a **quarter**, **half** or **three-quarter turn**.

**a**

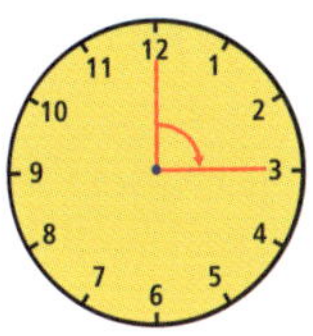

**b**

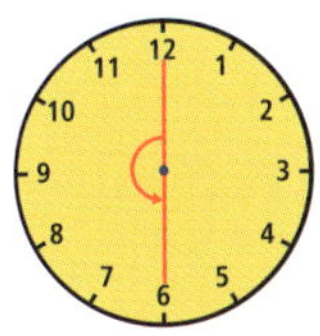

**c**

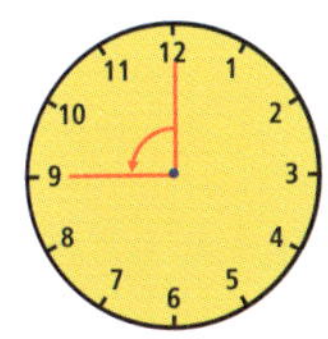

**d**

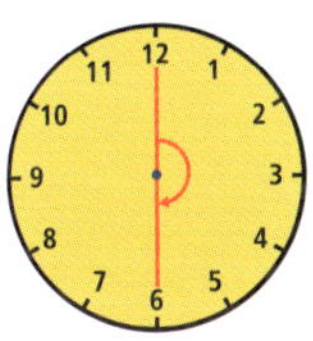

**e**

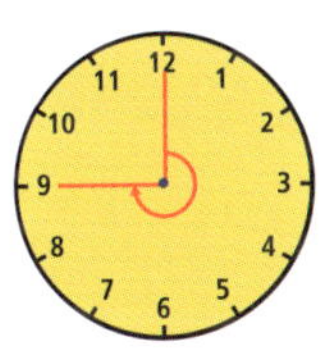

**f**

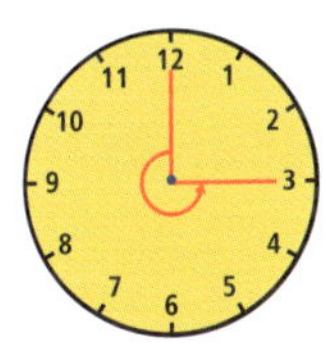

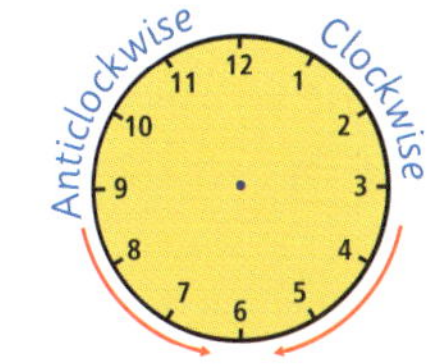


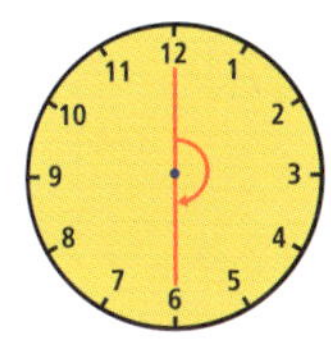

The minute hand turns a half turn in half an hour.

**2** **a** Draw a quarter turn anticlockwise.

**b** Draw a three-quarter turn clockwise.

**c** Is a quarter turn clockwise the same as a three-quarter turn anticlockwise?

INVESTIGATION

List examples of angles made where surfaces meet.

 • *AUSTRALIAN SIGNPOST MATHS 3* • ISBN 978 1 4886 2184 0

# 4:17 Describing Position

1 Which items are in these positions?

a middle of the top shelf

b right side of the top shelf

c left side of the middle shelf

d right side of the bottom shelf

In which positions are these items?

e duck

f book

g bottle

2 Which items are kept in these cupboards?

a left cupboard on the bottom row

b second from the left on the top row

c third from the left on the top row

d second from the left on the bottom row

In which cupboards are these items?

e cups

f bottles

g packets

h jars

3 Each student has their own shelf. Whose shelves are in these positions?

a fourth from the left on the top

b at the far right on the bottom

c fifth from the left on the top

d first on the left on the bottom

e What is the position of Ben's shelf?

f What is the position of Emma's shelf?

# 4:18 Pathways Between Places

1 Move the counter along the path. Write each move in the table.

a

| Moves |
| --- |
| 3 right |
| 4 up |
| 1 |
| 2 |
| |
| |
| |
| |

b

| Moves |
| --- |
| |
| |
| |
| |
| |
| |
| |
| |

2 Follow the directions to colour the path of the counter.

a

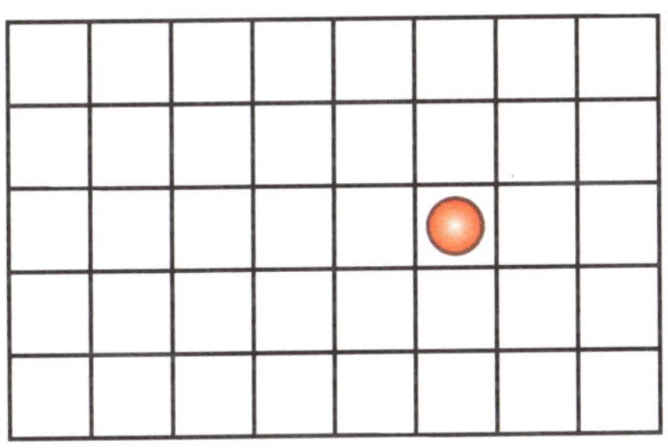

2 right,
2 down,
3 left,
4 up,
1 left,
1 down.

b

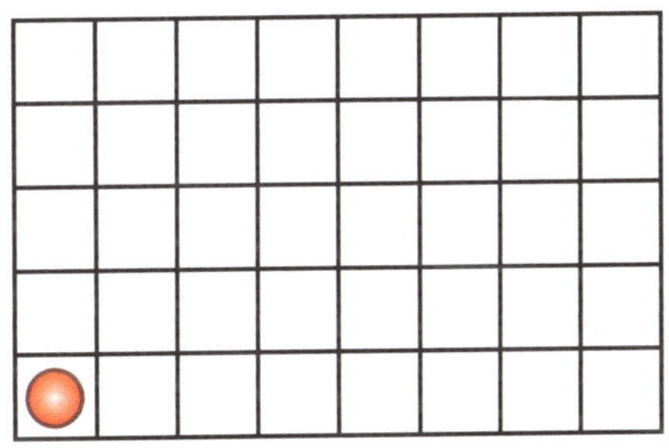

7 right,
4 up,
6 left,
2 down,
4 right,
1 up.

3 Carla has to go to every classroom and then to the office. Starting with room **H**, show the quickest route by writing the letters of the rooms in order.

Carla's shortest route is:

**H C** ______________ **G O**

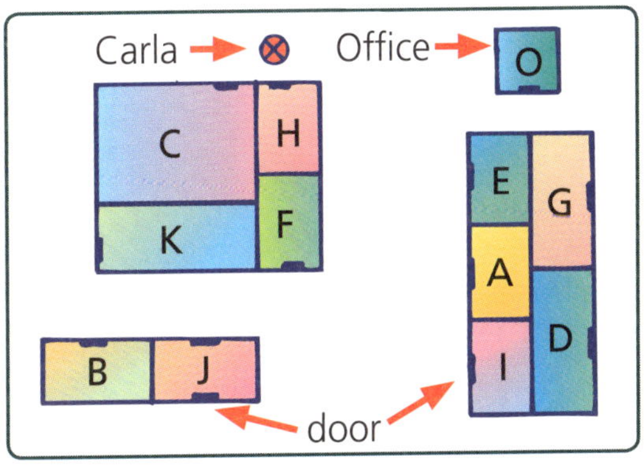


ACTIVITY

- In your school playground, mark a starting point.
- Students work in pairs to follow directions from the starting point to a secret destination. Everyone should measure distances using metre rulers or paces.
- Example directions: Move 3 m forwards, then 5 m to the right, then 7 m forwards, then 10 m to the left.

 • *AUSTRALIAN SIGNPOST MATHS 3* • ISBN 978 1 4886 2184 0

# 4:19 Investigating Prisms and Cylinders

CONCEPT

**Prisms** and **cylinders** are 3D objects.

- Their two ends are the same shape and size.
- Their **cross-section** is the same shape and size as their ends.
- The prism's other **faces** are rectangles.
- The cylinder's other **surface** is curved.

1. Write **prism**, **cylinder** or **neither** under each picture.

a 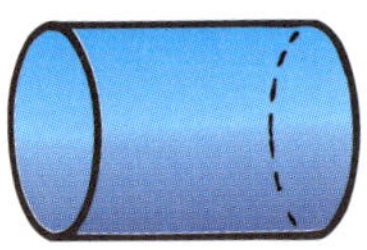

b 

c 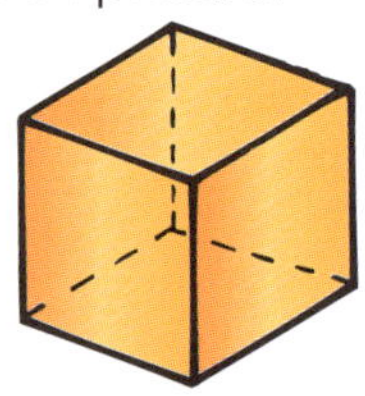

d 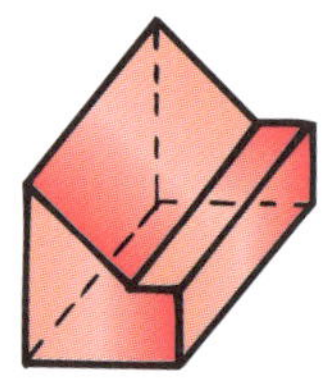

e 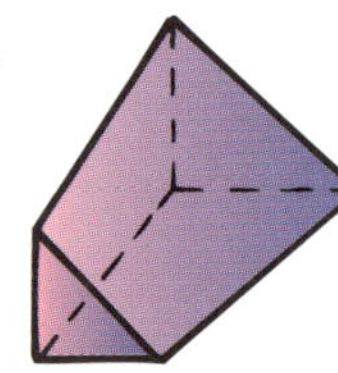

f 

g 

h 

i 

j 

2. Match the object label **X**, **Y** or **Z** with the description.

a I have a curved surface.

b All my faces are rectangles.

c My cross-section is a hexagon.

d My base is a circle.

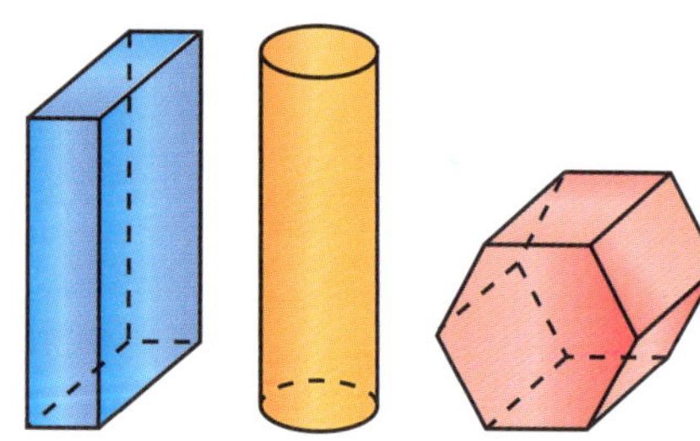

X Y Z

3. a How many faces does object **Z** have?

b How many edges does object **Z** have?

c How many vertices does object **Z** have?

Corners are also called vertices.

4. Draw each face of the prism.

| Prism | Shape of Each Face | | | | |
|---|---|---|---|---|---|
| 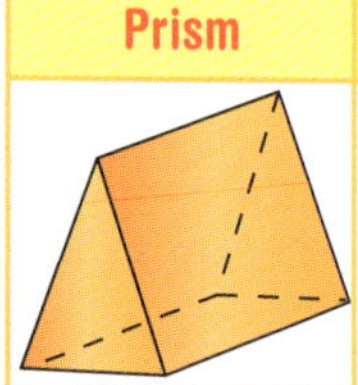 | 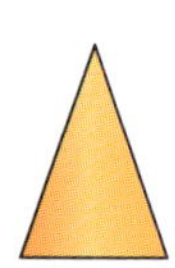 | | | | |



 ISBN 978 1 4886 2184 0

# 4:20 Investigating Pyramids

CONCEPT

**Pyramids** are 3D objects.

- It has one **base**. All the other faces are triangles.
- Its **cross-sections** are the same shape as the base but not the same size.
- One end tapers to a point.

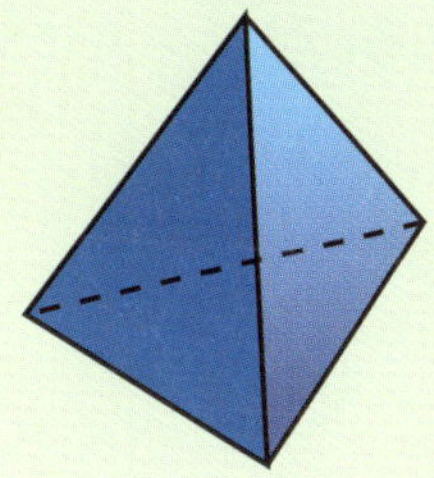

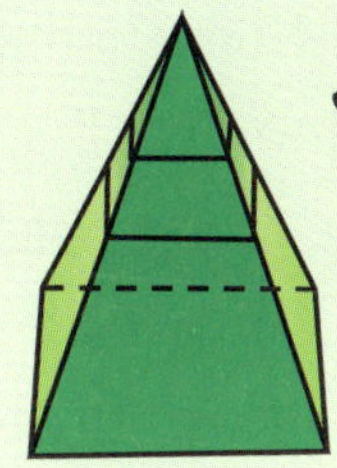

**1** Which of these pictures are pyramids?

a 

b 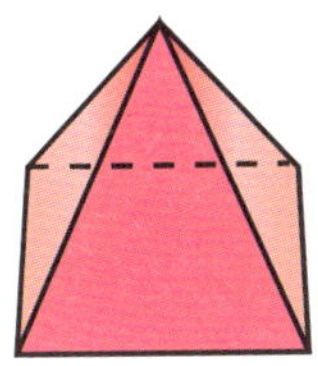

c 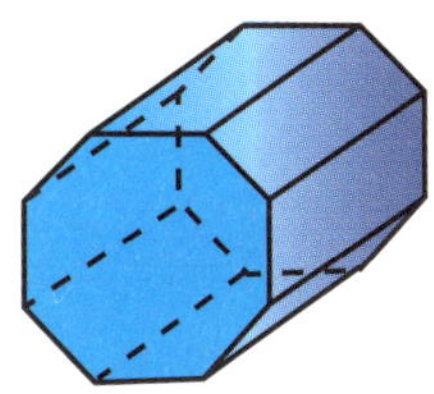

d 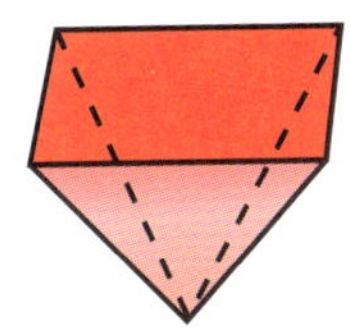

e 

f 

g 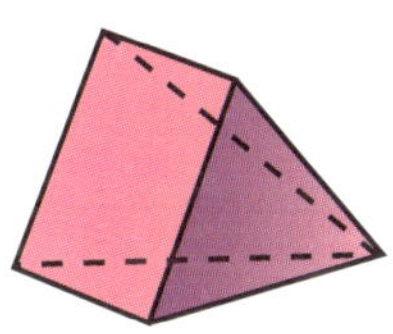

h 

i 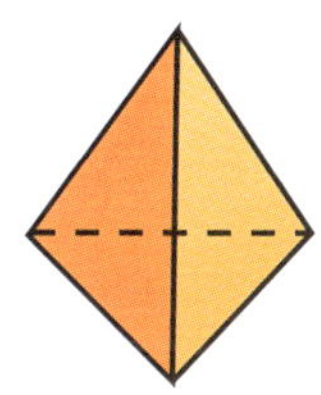

j 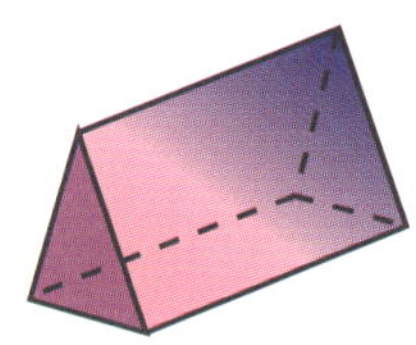

**2** Match the object label **A**, **B** or **C** with the description.

a All my faces are triangles.

b My base is a hexagon.

c I have 4 vertices.

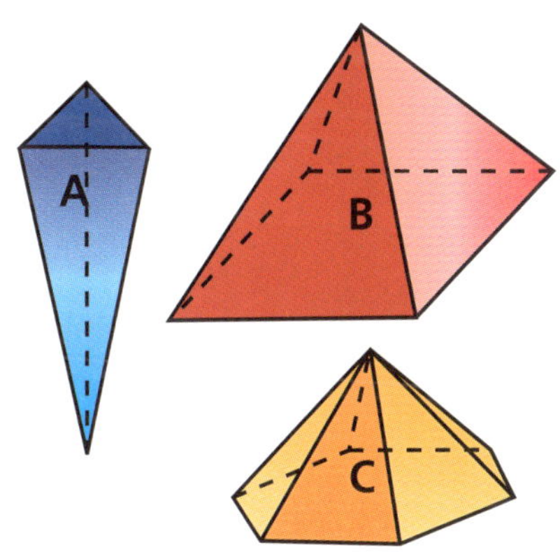


**3** a How many faces does object **B** have?

b How many edges does object **B** have?

c How many vertices does object **B** have?

d How many faces does object **A** have?

**4** Draw each face of the pyramid.

| Pyramid | Shape of Each Face | | | | |
|---|---|---|---|---|---|
| 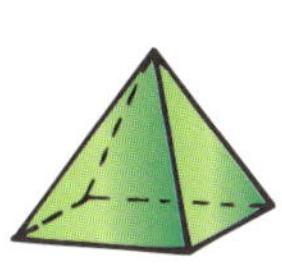 | | | | | |

# 4:21 Creating Maps

CONCEPT

- **Maps** show where things are, as if you were looking down on them from above.
- This is a map of a classroom.
- Below the map is a **key**. It describes the symbols on the map.

**Key**: B = board C = cupboard
D = desk E = entry
N = noticeboard S = sinks
W = window

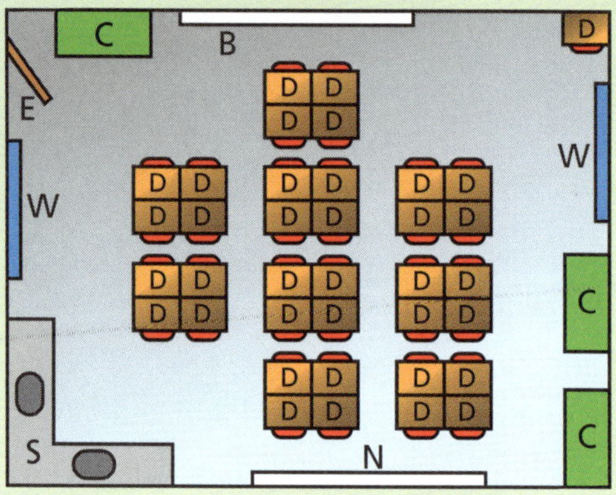


1 Use the classroom map above to answer the questions.

a How many desks are in the classroom?

b How many cupboards are there?

c What is to the left of the board?

d Where is the noticeboard?

2 Draw a map of your own classroom or a room at home.

Key:

INVESTIGATION

Copy a map of your local area. Make a key and use it to show places of interest on the map.

 • *AUSTRALIAN SIGNPOST MATHS 3* • ISBN 978 1 4886 2184 0

# 4:22 Mazes

Which animal reaches the food? Colour their path.

Wallaby

(W)

Echidna

(E)

Koala

(K)

Platypus

(P)

1

2

3

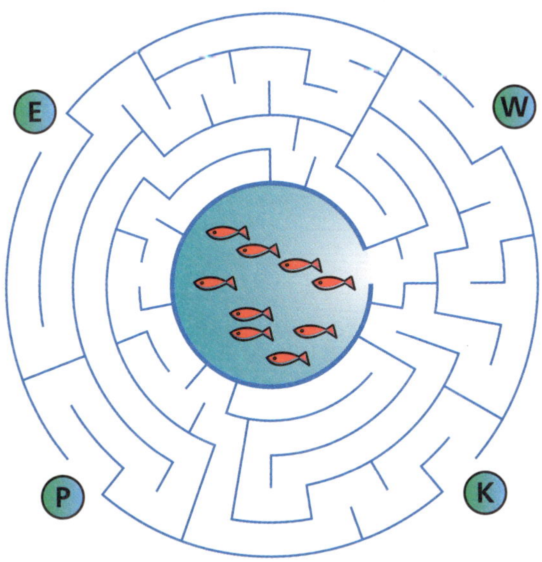


4

Use grid paper to make your own maze.

FUN SPOT

# 4:23 Investigating Spheres

**Spheres** are 3D objects.

- It is like a ball.
- It has one **curved surface**.
- All of the points on the surface are the same distance from the centre.
- Its **cross-section** is always a circle. The circles can be different sizes.

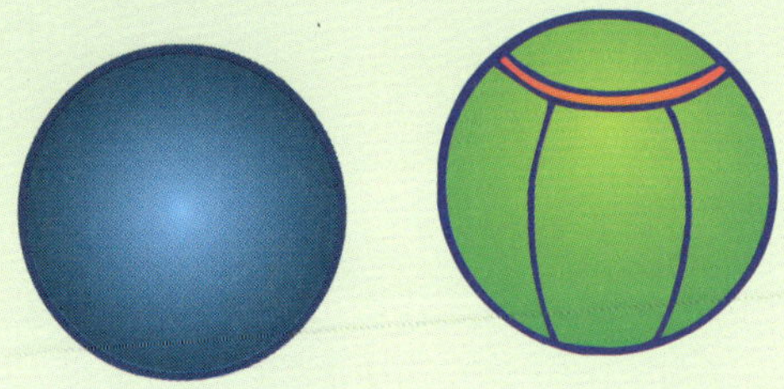

CONCEPT

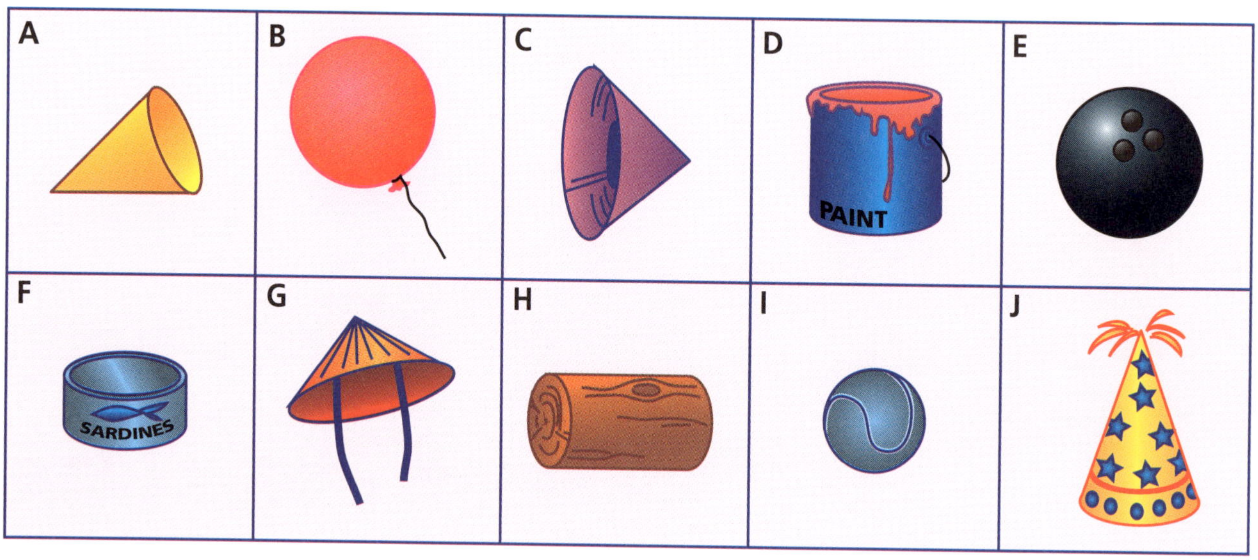


1 a Which pictures show a sphere?

b Which have 2 flat surfaces?

c Which have no vertices (corner points)?

2 Write **true** or **false**. A sphere:

a can stack

b can roll

c can slide

d has a flat surface

e looks like a marble

f looks like a can

Draw objects that have the shape of a sphere or part of a sphere.

ACTIVITY

# 4:24 3D Objects

CONCEPT

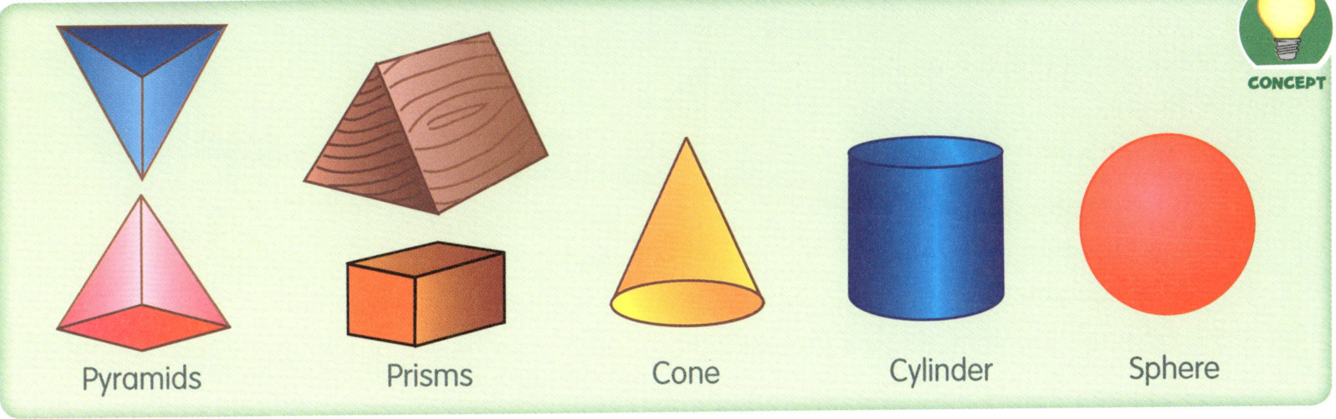


**1** Under each object write **pyramid**, **prism**, **cone**, **cylinder** or **sphere**.

a 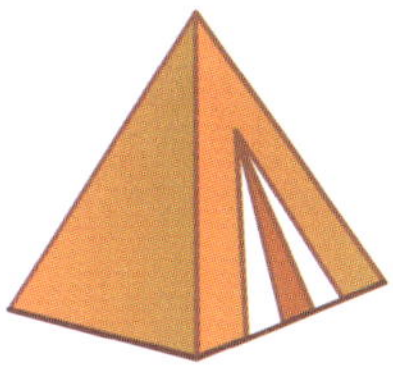

b 

c 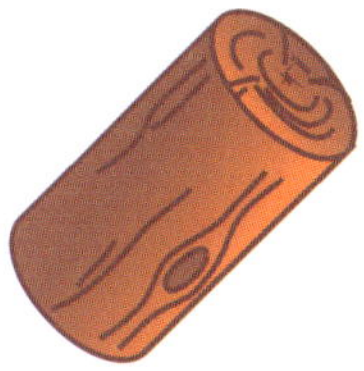

d 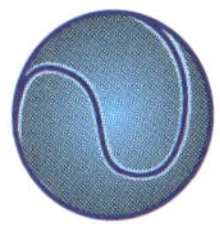

e 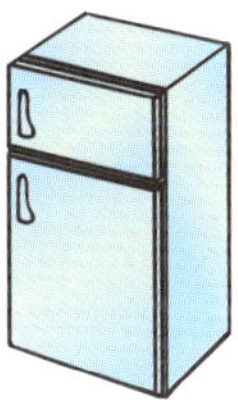

f 

g 

h 

**2** Complete the drawings.

a prism

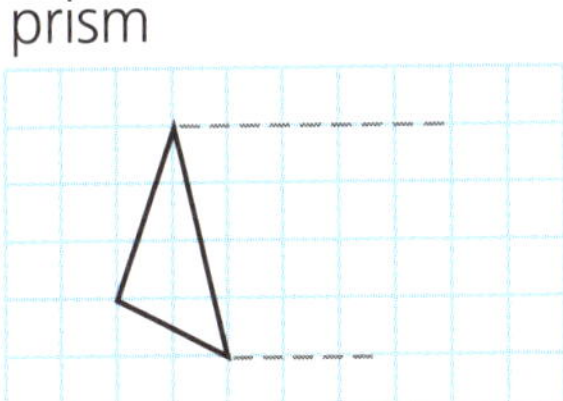

b cone

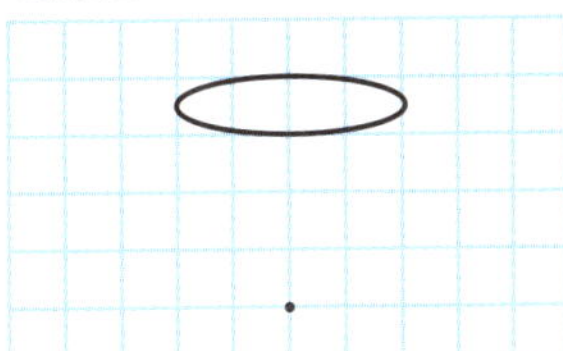

c pyramid

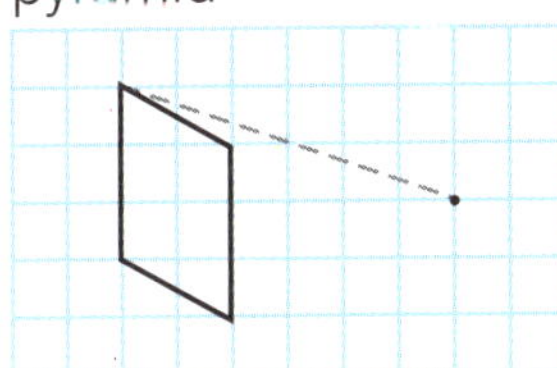

d cylinder

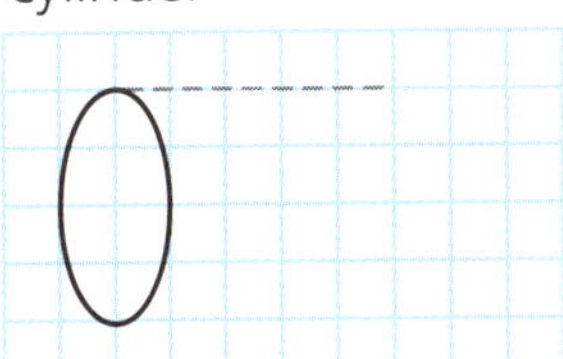

e sphere

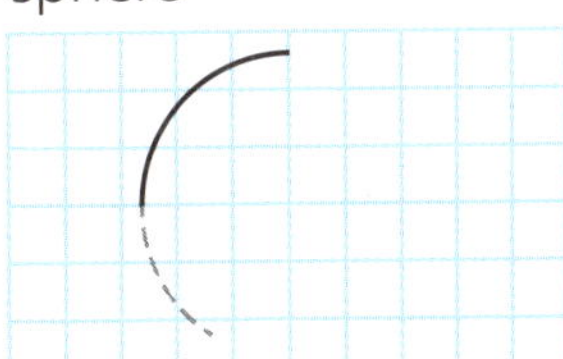

f cube

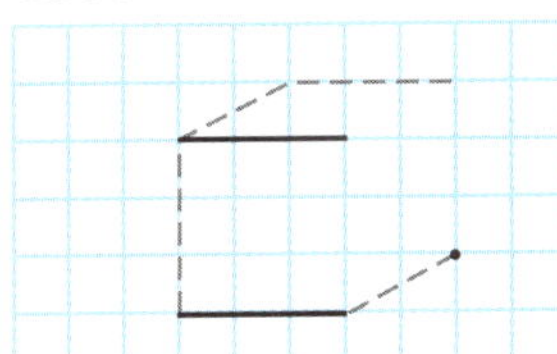

# 4:25 Properties of 2D Shapes

CONCEPT

There are 3 ways to repeat a shape:

**A** flip the shape
**B** slide the shape
**C** turn the shape.

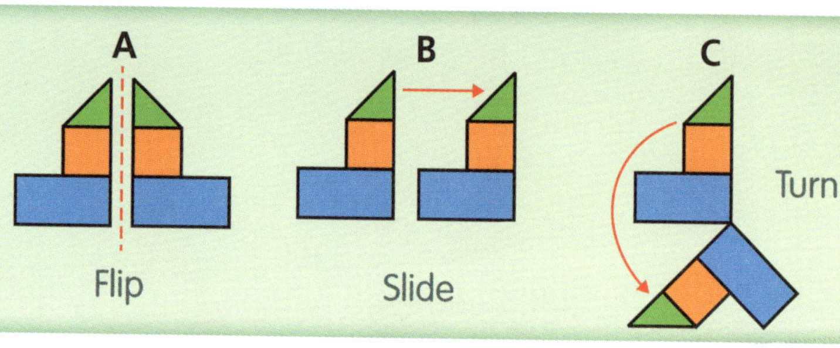


1 Is the coloured shape made by a **flip** (reflection), **slide** (translation) or **turn** (rotation)?

a
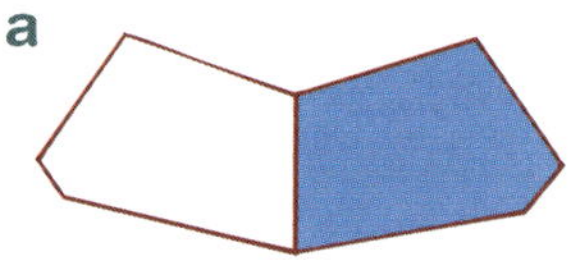

b
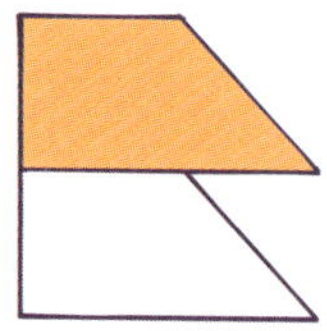

c

d

e
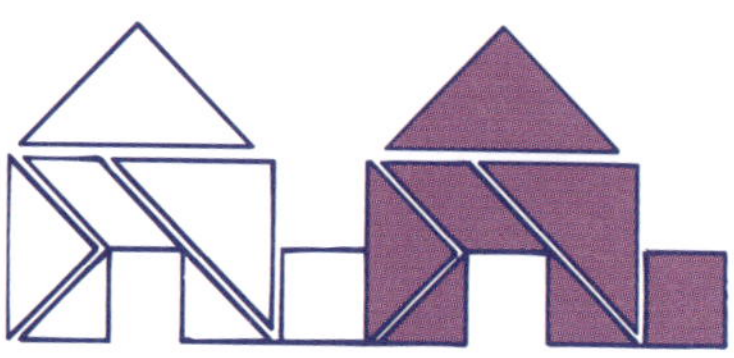

f
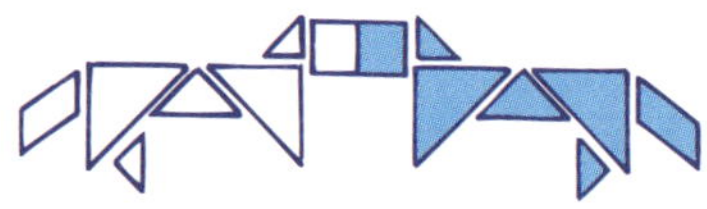

g

h
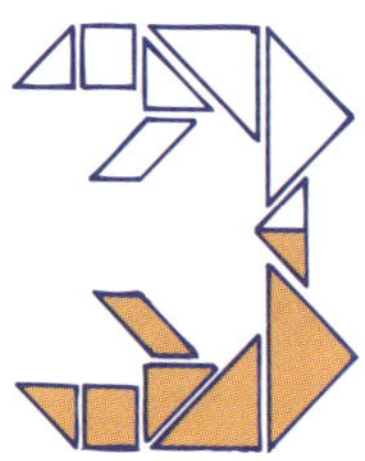

i

j

k Discuss which of these shapes are symmetrical.

## A Chinese Puzzle: Tangrams

FUN SPOT

- Trace or draw this square onto a piece of paper.
- Cut out the square and cut along all the lines.
- Use the seven pieces to make these pictures.
- Try to make the pictures shown in Question 1.
- Use the pieces to make some pictures of your own.

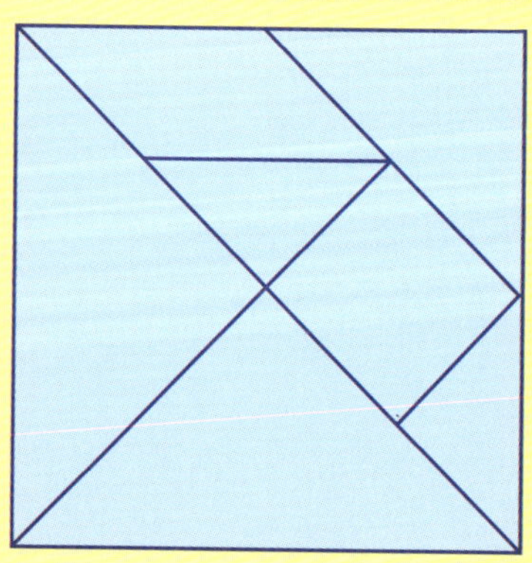

# 4:26 Properties of 2D Shapes

**1** Draw all lines of symmetry. Write if the coloured shape is made by a **flip** (reflection), **slide** (translation) or **turn** (rotation).

a

b
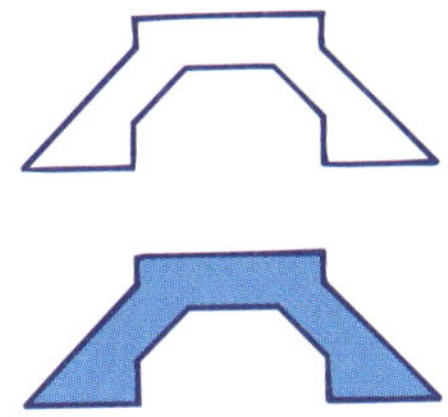

c
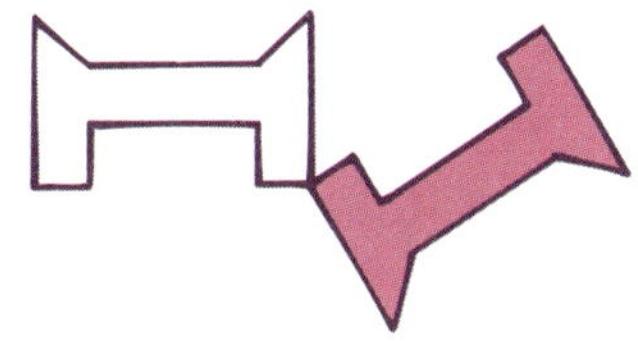

**2** Complete the patterns by flipping to the right and down.

a
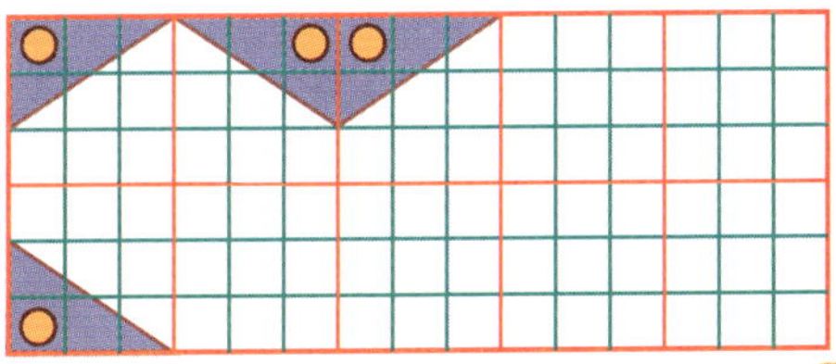

b
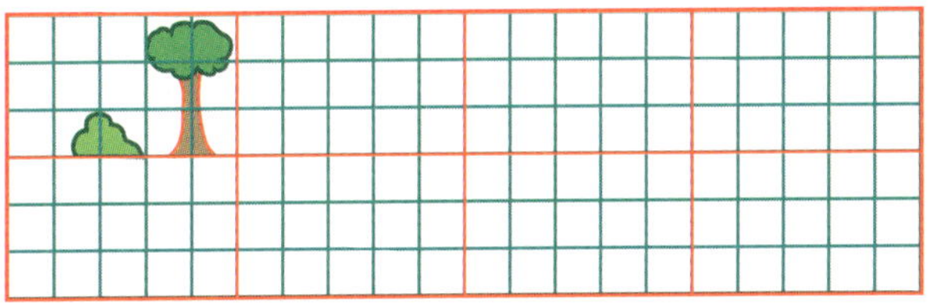

c Are the patterns symmetrical?

**3** Complete the patterns by sliding to the right and down.

a
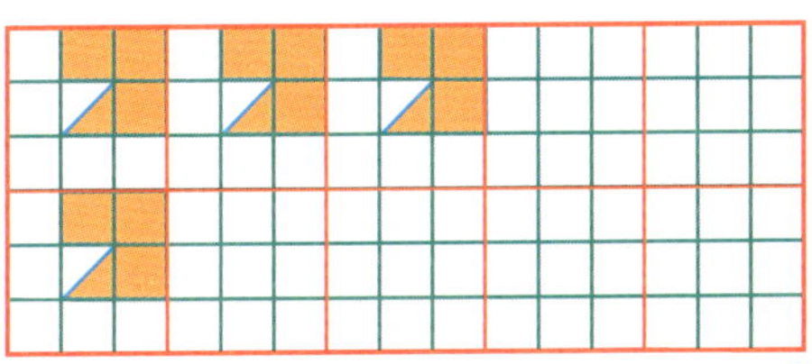

b
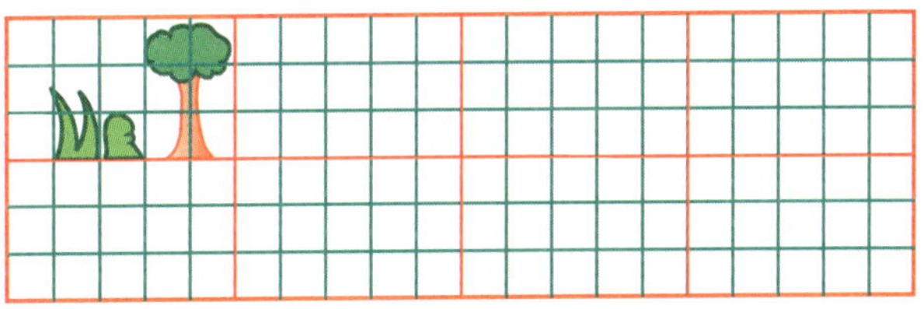

c Are the patterns symmetrical?

> A **tessellation** is made by repeating a shape so that the shapes fit together without any gaps or overlaps.
>
> CONCEPT

**4** Complete the patterns by turning them.

a
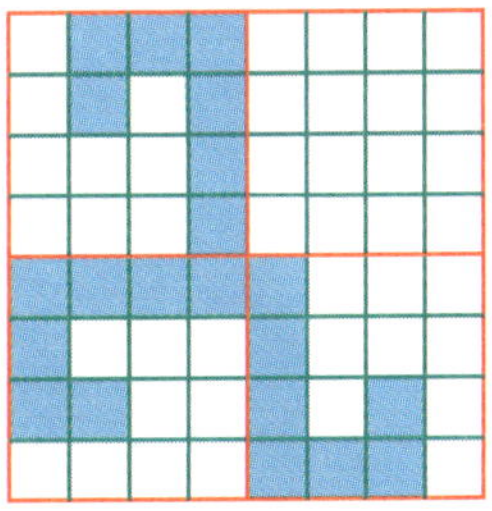

b
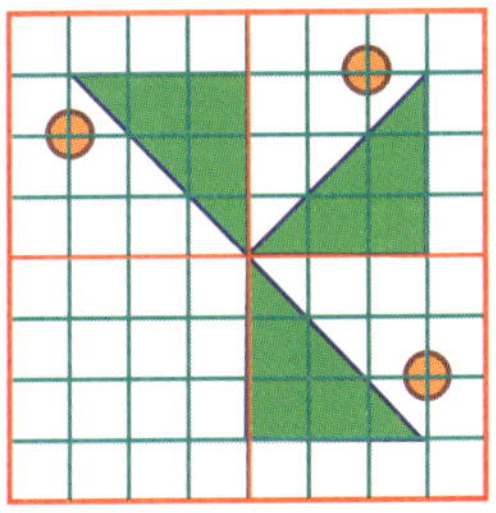

c Are the patterns symmetrical?

**5** Is a jigsaw puzzle a tessellation?

**6** Do these shapes make tessellations?

a
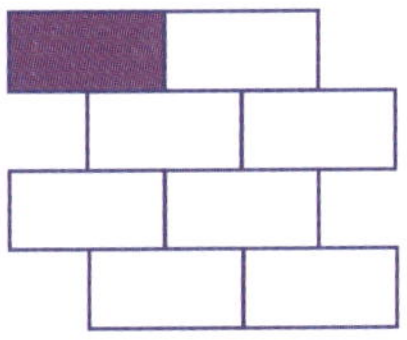

b
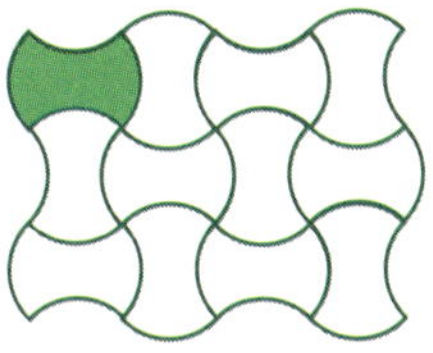

c
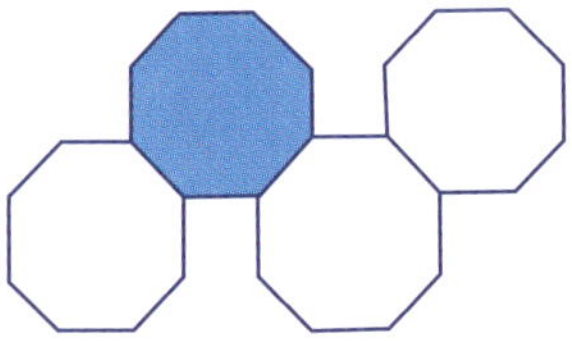

d
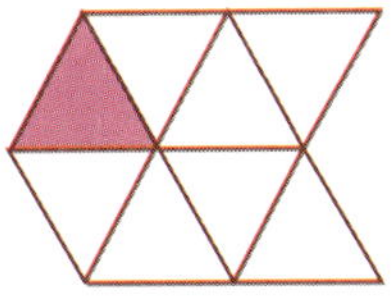

# 4:27 The Net of a Cube

CONCEPT

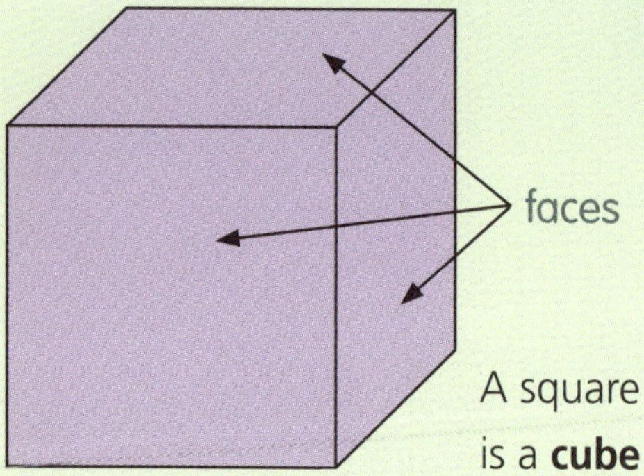


A square box is a **cube**.

ACTIVITY

Trace this net. Cut out and fold your tracing to make your own cube.

1 How many faces does a cube have?

2 Circle the cubes.

3 Which of these nets would **not** fold to make a cube?

**A**

**B**

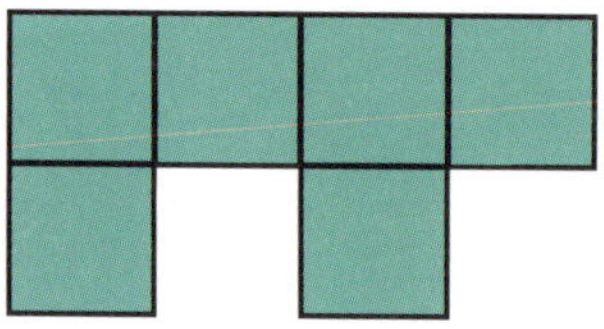

**C**

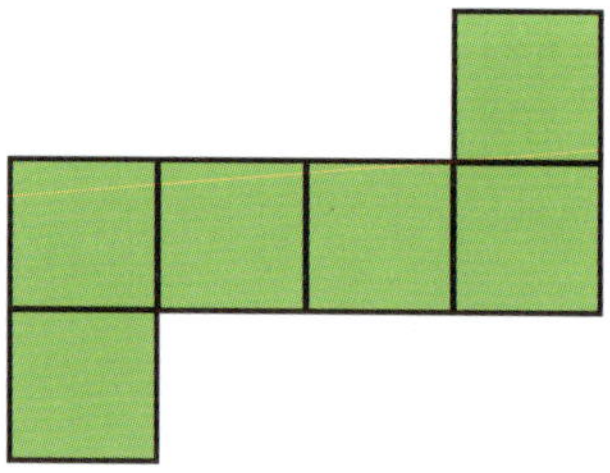



 • *AUSTRALIAN SIGNPOST MATHS 3* • ISBN 978 1 4886 2184 0

# 5:01 Using Blocks in Graphs

1 This graph shows where people were born.

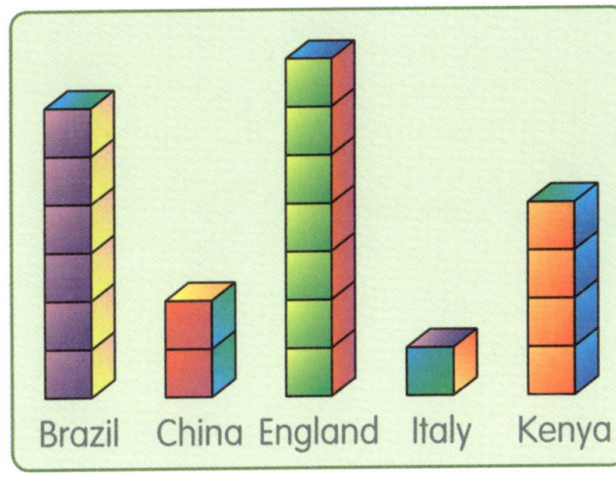


a How many people were born in China? ☐

b How many people were born in England? ☐

c How many more were born in Brazil than in China? ☐

d How many people were there altogether? ☐

e Predict the country that most of your classmates were born in. ☐

f Find out where your classmates were born. Where were most of your class born? ☐

Build a graph with blocks to show what you discovered.

2 Sophie and Kwan made this graph. It shows the colours of all the cars parked at the shops.

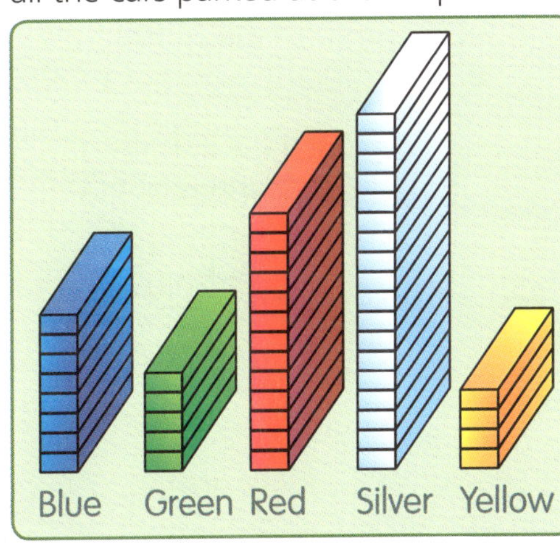


a How many cars were red? ☐

b Which colour was most popular? ☐

c How many cars were not silver? ☐

d How many different colours were seen? ☐

e How many cars were there altogether? ☐

f Predict the most popular car colour in an area nearby. ☐

g Investigate to find the answer. Which colour was most popular? ☐

Build a graph with blocks to show what you discovered.

3 In this picture, how many of each animal can you find?

ACTIVITY

- Fill in this table.

| | |
|---|---|
| Birds | |
| Lions | |
| Monkeys | |

- Make a graph with blocks to show how many animals there are.

 • *AUSTRALIAN SIGNPOST MATHS 3* • ISBN 978 1 4886 2184 0

# 5:02 Tables and Graphs

**1** Each student in the class chose a colour. These are the results.

| Colours Chosen | | | |
|---|---|---|---|
| **Blue** | **Green** | **Orange** | **Red** |
| 12 | 8 | 4 | 6 |

**a** The most popular colour was ______.

**b** The least popular colour was ______.

**c** How many students chose green? ______

**d** How many students are in the class? ______

**2** Use the **Collections** table to answer these questions.

How many buttons were collected by each person?

**a** Darcy ______ **b** Rhonda ______

**c** Rajiv ______ **d** Nasha ______

How many stickers were collected by each person?

**e** Rachel ______ **f** Nasha ______

**g** Rhonda ______ **h** Rajiv ______

**i** How many buttons and stickers altogether were collected by Rachel? ______

| Collections | | |
|---|---|---|
| **Person** | **Number of Buttons** | **Number of Stickers** |
| Darcy | 825 | 173 |
| Nasha | 72 | 201 |
| Rachel | 300 | 88 |
| Rajiv | 166 | 193 |
| Rhonda | 46 | 405 |

**3** How old is each person?

**a** Tom ______

**b** Alana ______

**c** Rachel ______

**d** Deklyn ______

**e** Make up your own question that could be answered using this table.

______

______

| Name | Age | Name | Age |
|---|---|---|---|
| Alana | 27 | Alan | 57 |
| Ally | 38 | Luke | 20 |
| Deklyn | 0 | Rachel | 23 |
| Greg | 10 | Tom | 87 |
| Heather | 17 | Naomi | 22 |

**INVESTIGATION**

- Do you think there are more left-handed or right-handed students in your class?
- Investigate the question by completing this tally.
- By how many is one group bigger than the other? ______

| | | |
|---|---|---|
| **Right-handed** | | |
| **Left-handed** | | |

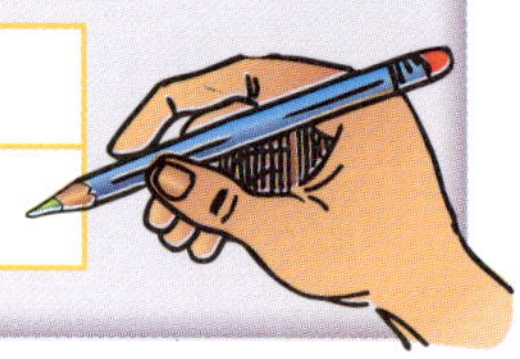



 • *AUSTRALIAN SIGNPOST MATHS 3* • ISBN 978 1 4886 2184 0

# 5:03 Tables and Graphs

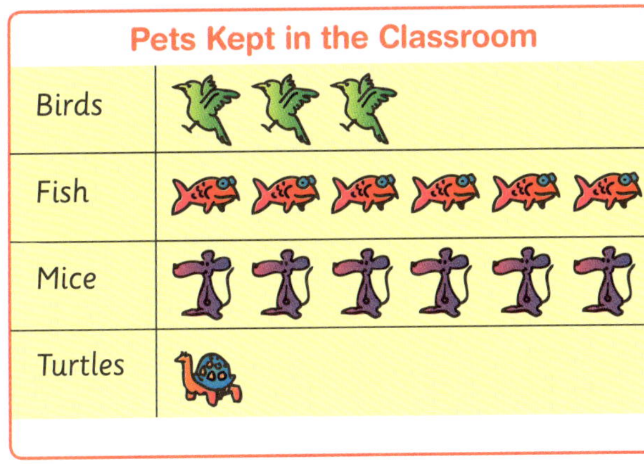


**1** a How many turtles are kept?

b How many birds are kept?

c Which columns are longest?

d Three of which pet are kept?

e How many pets altogether?

**2** Use the graph to complete these tables.

**A**

| Pets Kept in the Classroom | | | |
|---|---|---|---|
| Birds | Fish | Mice | Turtles |
| | | | |

**B**

| Pets Kept in the Classroom | |
|---|---|
| Birds | |
| Fish | |
| Mice | |
| Turtles | |

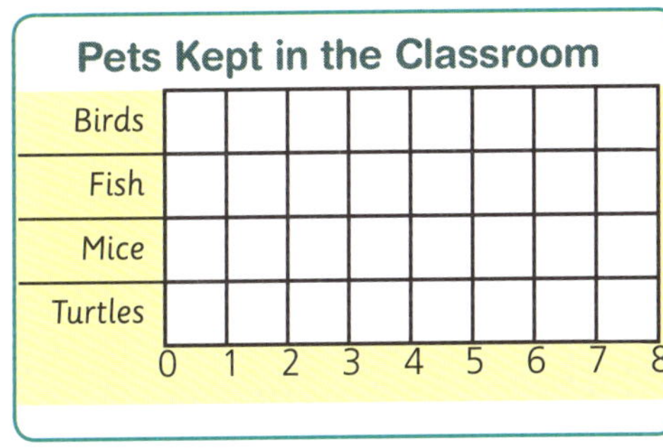


a How many classroom pets can fly?

b How many classroom pets are there altogether?

c Which table do you think looks better, **A** or **B**?

d Complete the graph.

**3** a Use this table to complete the graph.

| Youth Group Attendance | | | | |
|---|---|---|---|---|
| Carmen | Erika | Holly | Paul | Yong |
| 4 | 5 | 3 | 7 | 6 |

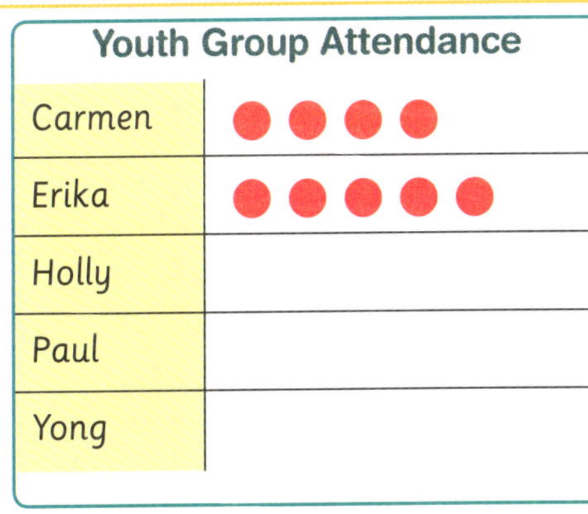


b Who has attended the least?

c Did Yong attend more often than Erika?

d Erika has been away three times. How many times has the group met?

INVESTIGATION

**4** Five flowers are in a row. How many spaces are there between the flowers?

**5** Draw seven trees in a row. How many spaces are there between trees?

**6** How many spaces are there if 40 trees are in a row?

# 5:04 Chance

**1** If you toss a coin, it can come up heads or tails.

**a** Is heads more likely than tails?

**b** Is there an even chance of tossing tails?

**2** If you spin the arrow on a spinner, it can land on any colour.

Even chance means 1 chance out of 2.

**a** Is there 1 chance in 3 of spinning red?

**b** Do all colours have an equal chance of being spun?

**c** Is there an even chance it will land on blue?

**3** Without looking, Jai took a coloured ball from the bowl.

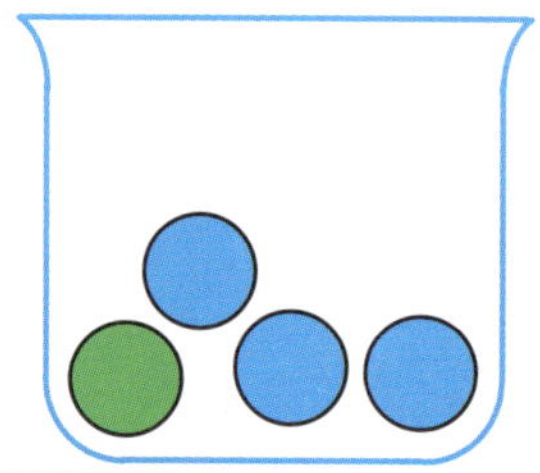

**a** Which colour is more likely to be taken?

**b** Is it possible to take a yellow ball?

**c** Is there an even chance of taking green?

**4** These spinners can stop on red or yellow.

A 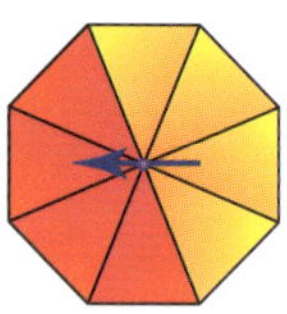
B 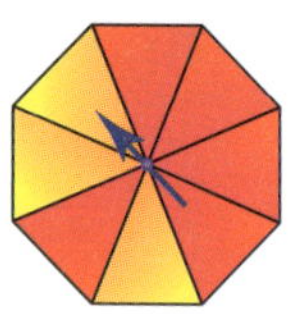
C 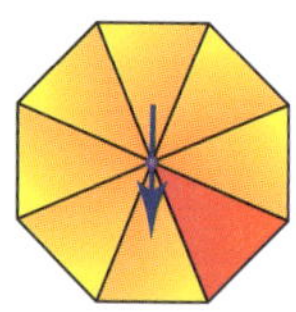
D 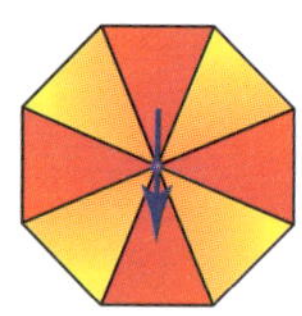
E 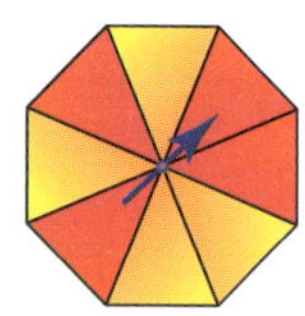

**a** Which spinner is most likely to stop on yellow?

**b** Which spinner is most likely to stop on red?

**c** Which spinners are equally likely to stop on red or yellow?

**d** Is there an even chance that spinner D will stop on red?

**e** Is there an even chance that spinner B will stop on yellow?

**f** On spinner B, is it fair if Ben wins on red and Michelle wins on yellow?

**g** On spinner D, is it fair if Ben wins on red and Michelle wins on yellow?

**INVESTIGATION**

- Make a spinner like this.
- Spin the arrow 50 times. Keep a tally of the colours that you spin.
- Is your spinner fair? How can you tell? Discuss.

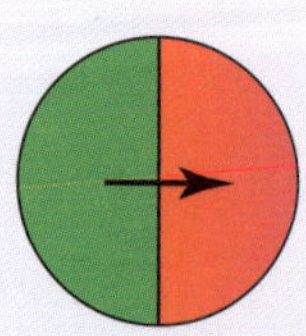

| | |
|---|---|
| Red | 卌 \| |
| Green | \|\|\|\| |



 • *AUSTRALIAN SIGNPOST MATHS 3* • ISBN 978 1 4886 2184 0

# 5:05 Predicting Outcomes

**1** Without looking, Meg took a coloured ball from the bag.

a Which colour is least likely to be taken?

b Which colour is most likely to be taken?

c Is there an even chance of taking a yellow ball?

d Could Meg take the red ball first?

e If Meg took the red ball first, what colours could she take next?

f If Meg took a blue ball first, what colours could she take next?

**2** Put 1 yellow, 3 blue and 5 red counters into a bag.
Pick one counter at random.

a Which colour is most likely to be picked?

b Which colour is least likely to be picked?

**3** This spinner is spun 10 times.

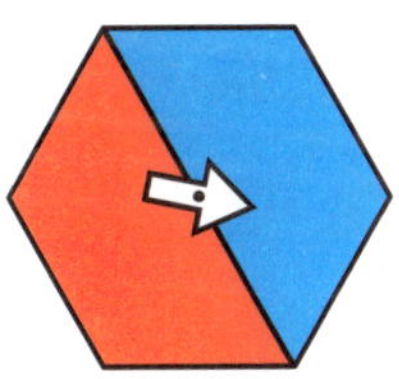

a Would you be surprised if blue was spun every time?

Why or why not?

b How many blue spins would you expect out of 10 spins?

**4** a Are all three colours equally likely to be spun?

b How many red spins would you expect out of 6 spins?

c Is there one chance in three of spinning blue?

INVESTIGATION

- Put 1 yellow, 3 blue and 5 red counters into a bag.
- Randomly pick one counter from the bag.
- Record a tally mark for that colour.
- Put the counter back in the bag.
- Repeat this experiment 50 times.
- What do the tallies show? Discuss your results.

| Taking Counters from a Bag | | |
|---|---|---|
| Blue | | |
| Red | | |
| Yellow | | |

# 5:06 Picture Graphs

**1** a How many trains were sold on Monday?

b On which day were the most trains sold?

c How many trains were sold altogether?

d If there were 30 trains for sale, how many trains were not sold?

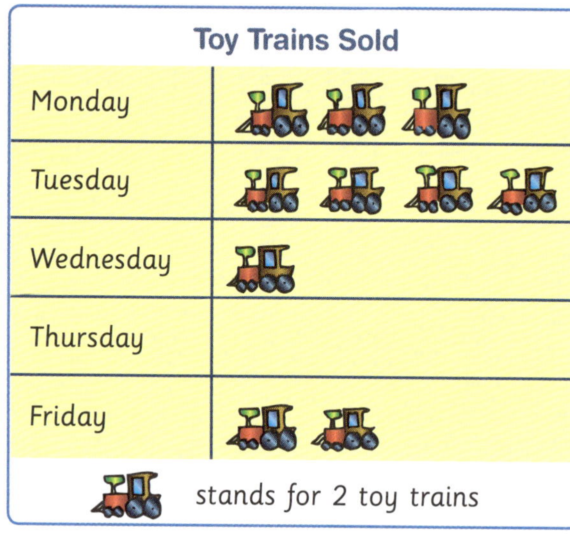


**2** Students cut out shapes to show some of the languages spoken at home. They made this graph using one shape for each student.

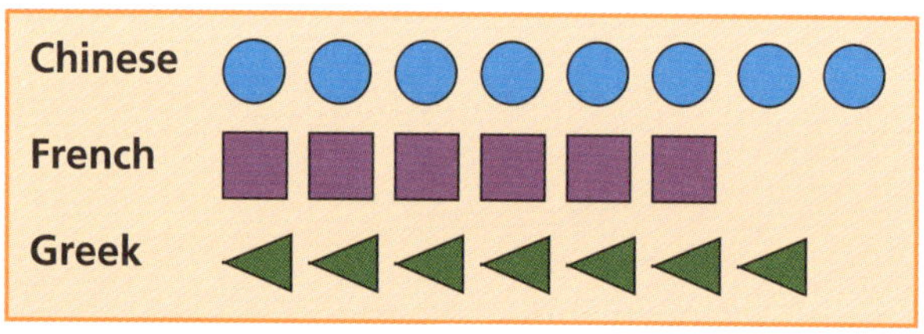


a How many students speak Greek?

b How many students speak French?

c How many students speak Chinese?

d Which row is the longest?

**3** Liz made this graph using stones, blocks and counters. She used a stone for each car she saw, a block for each truck and a counter for each van.

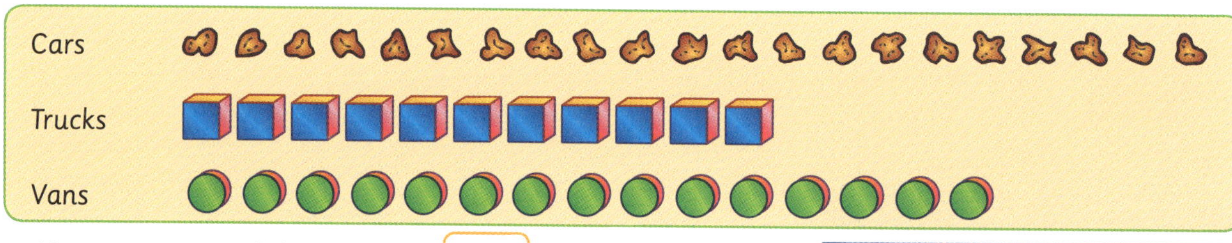


a How many cars did Liz see?

b How many vans did she see?

c How many more cars than trucks did she see?

d How many cars, trucks and vans were seen altogether?

## Car Colour Tallies

INVESTIGATION

- Work in groups. Each group records the colours of cars that drive by in 10 minutes.
- Use tally marks to record the colours of cars.
- Groups discuss and compare their results.
- Draw a graph showing the popular colours.

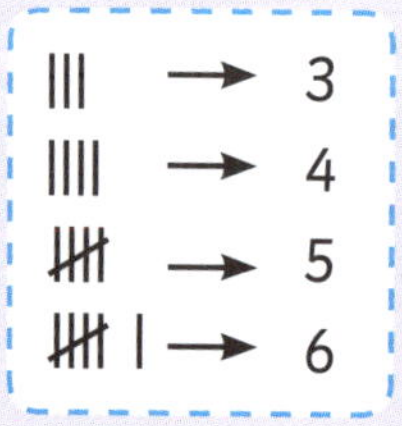

 • *AUSTRALIAN SIGNPOST MATHS 3* • ISBN 978 1 4886 2184 0

# 5:07 Making Graphs

**1** Gino made this table using tally marks. It shows the favourite game of each of his friends. Use this information to complete both graphs.

| Game | Tally | Number |
|---|---|---|
| Handball | 𝍸 III | 8 |
| Hopscotch | 𝍸 II | 7 |
| Marbles | 𝍸 𝍸 | 10 |
| Skipping | IIII | 4 |

Columns in a graph can go up or across.

Key
■ stands for 1 friend

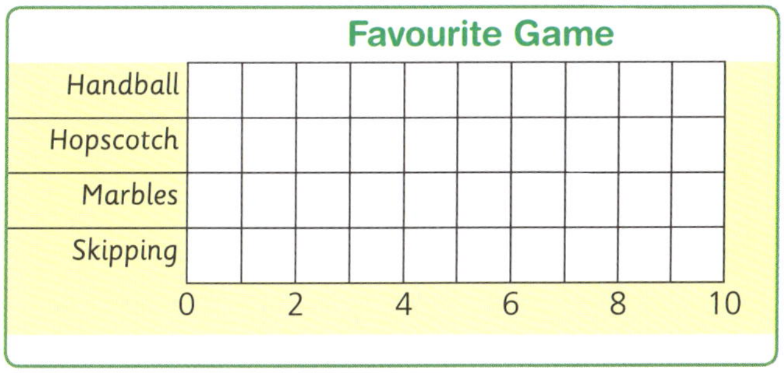


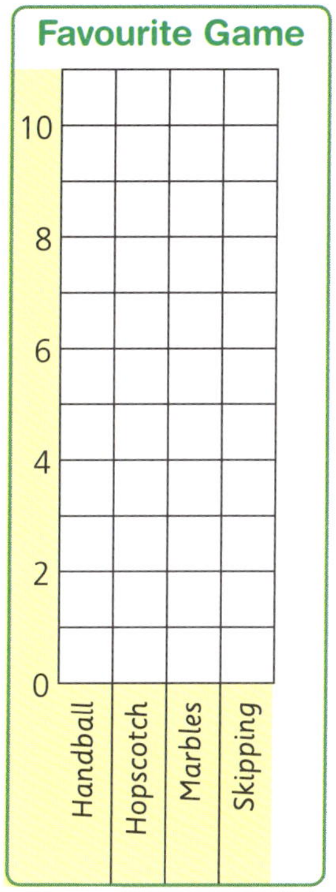


**2** **a** Write a question that could be answered by the investigation above.

**b** Ask people in your class to choose the game they like best. Keep a tally of their answers and draw your own graph.

| Game | Tally | Number |
|---|---|---|
| Handball | | |
| Hopscotch | | |
| Marbles | | |
| Skipping | | |

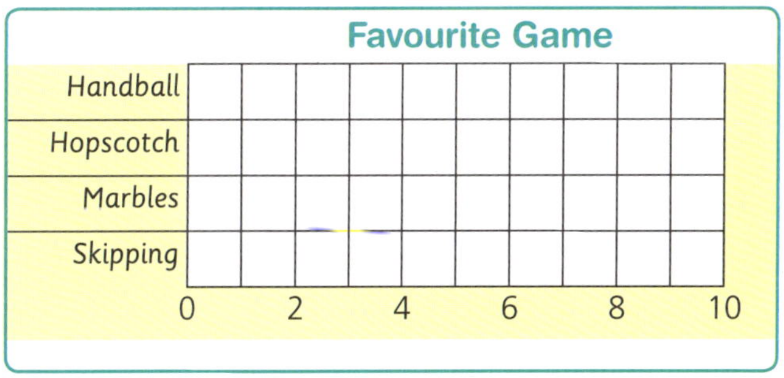


**3** Tina rolled these dice.
Finish the graph to show how many of each number she rolled.

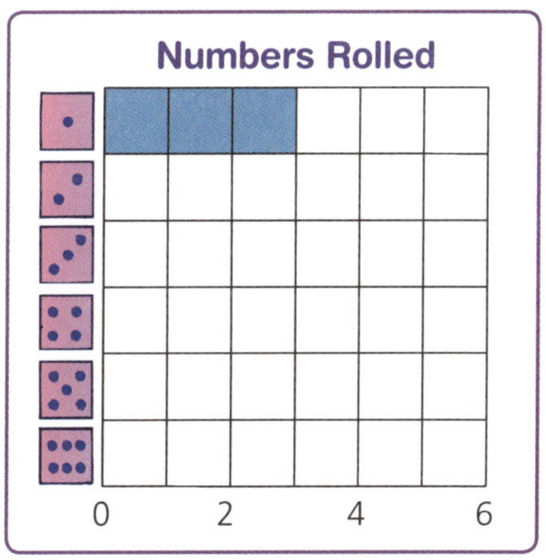


 • AUSTRALIAN SIGNPOST MATHS 3 • ISBN 978 1 4886 2184 0

# 5:08 Reading Tables and Graphs

Use the tables and graphs to answer the questions.

| People at the School Play | | | |
|---|---|---|---|
| | Parents | Students | Teachers |
| Female | 70 | 120 | 6 |
| Male | 80 | 100 | 4 |

**1** How many of these people were at the play?

**a** female teachers

**b** male students

**c** female parents

**d** teachers

**e** males

**f** altogether

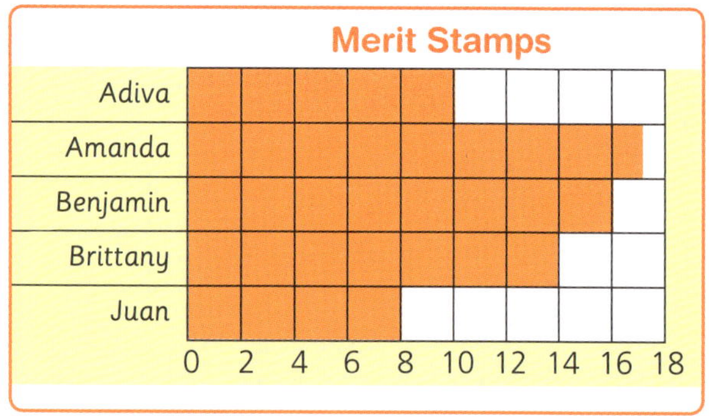


**2** **a** Who has the most merit stamps?

**b** Who has the least stamps?

**c** How many stamps does Adiva have?

**d** How many stamps does Juan have?

**e** How many stamps do Benjamin and Amanda have altogether?

**f** What is the total number of stamps that Adiva and Brittany have?

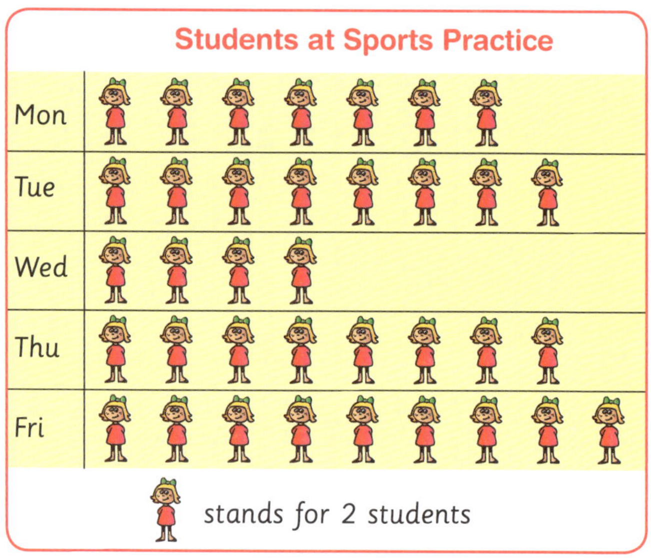


**3** **a** How many students practised on Monday?

**b** How many students practised on Friday?

**c** On which two days did the same number of students practise?

**d** What was the total number of students who practised on Wednesday and Friday?

**e** How many more students practised on Thursday than on Wednesday?

**f** Can we tell how many different students practised during the week?

## Graph Display

ACTIVITY

- Collect different types of graphs.
- Use them to make a display in your classroom.

# 5:09 Reading for Number

Thousands of years before Europeans came to Australia, the continent was home to Indigenous peoples. Groups lived by hunting, fishing and gathering from plants, living closely with the natural environment.

Indigenous Dreaming stories tell of how things began and why they happened. You can still find many old paintings of Dreaming stories on rock walls, some many centuries old.

From the information above, answer these questions.

1. Who were living in Australia before Europeans arrived?
2. Which three words describe how Indigenous peoples got their food?
3. What are the traditional stories of Indigenous peoples called?
4. How many people are shown in the picture?
5. How many people are holding spears?
6. How many spears are in the picture?
7. How many boomerangs are in the picture?
8. How many stone axes are in the picture?
9. On the graph, colour one box in the correct row for each tool shown.

**Tools in the Picture**

| | | | | | | | | |
|---|---|---|---|---|---|---|---|---|
| Boomerangs | | | | | | | | |
| Spears | | | | | | | | |
| Stone axes | | | | | | | | |
| | 0 | 1 | 2 | 3 | 4 | 5 | 6 | 7 | 8 |

© PEARSON AUSTRALIA 2018 • *AUSTRALIAN SIGNPOST MATHS 3* • ISBN 978 1 4886 2184 0

# 5:10 Reading Picture Graphs

1 a How many pears did Peter grow in 2009?

b How many more pears were grown in 2009 than in 2008?

c In which year was the least number of pears grown?

d How many pears were grown in these four years?

e If one pear stands for 100 pears, what would half of a pear stand for?

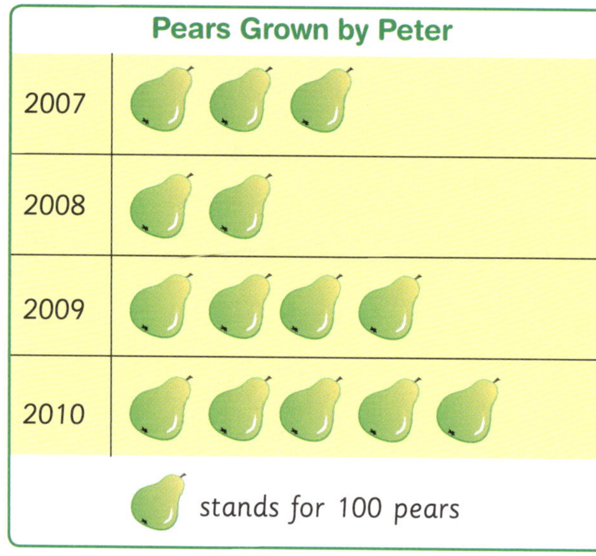

2 a How many toy trains were sold on Tuesday?

b How many toy trains were sold on Saturday?

c On which day were the most trains sold?

d On what two days were the same number of trains sold?

e How many more trains were sold on Thursday than on Friday?

f What was the total number of trains sold?

g Why do you think no sales were made on Sunday?

3 a How many cars crossed the bridge from 1 pm to 2 pm?

b During which hour did the least number of cars cross the bridge?

c During which hour did 30 cars cross?

d How many cars crossed altogether?

e Why do you think so many cars crossed between 4 pm and 6 pm?

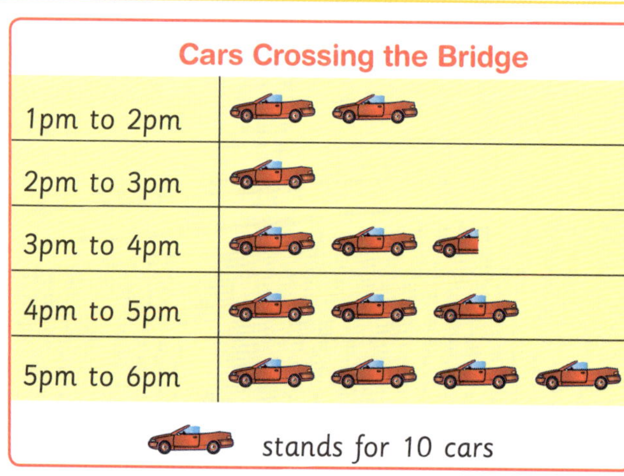

# 5:11 Drawing Graphs

**1** Kai-Lin kept a tally of the hair colour of students in her class. Use the table to finish the graph.

| Hair Colour | Tally | Number |
|---|---|---|
| Black | 卌 卌 \|\| | 12 |
| Blonde | 卌 卌 | 10 |
| Brown | \|\|\|\| | 4 |
| Red | \|\| | 2 |

**a** Which colour was the most common?

**b** How many students had grey hair?

**c** How many more students had blonde hair than brown hair?

**d** How many students were there altogether?

**e** What is the key on the graph used for?

**f** Why do you think people use picture graphs?

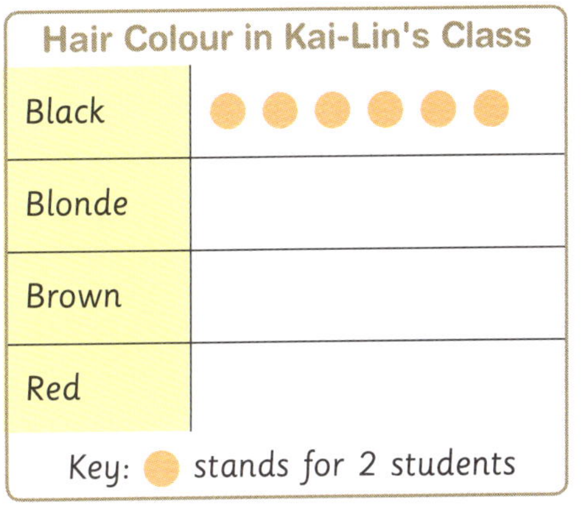


**2** What are the advantages of showing information in these different ways?

**a** tally

**b** picture graph

**c** other graph

| Cars Washed | | |
|---|---|---|
| Monday | Tuesday | Friday |
| 6 | 8 | 4 |

| Cars Washed | |
|---|---|
| Monday | 卌 \| |
| Tuesday | 卌 \|\|\| |
| Friday | \|\|\|\| |

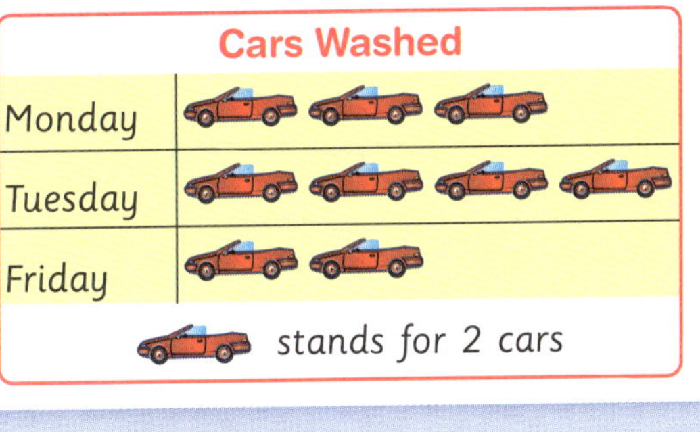


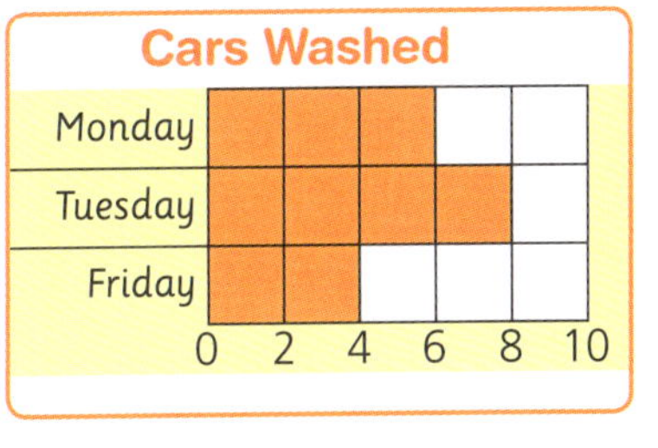


## Using Computers

- Using a computer spreadsheet program, create a table to organise collected data.
- Use the spreadsheet program to make graphs from your table of data.
- Find an interesting graph to share with the class.

ICT

# 5:12 Ordering Events

1 Choose a label to match each statement.

impossible | unlikely | even chance | very likely | certain

a I bought a ticket in a raffle and I will win.

b Our next teacher will be six years old.

c The next school captain will be a student.

d If I drop an egg on the road, it will break.

e We will do maths on our next day at school.

f It will be warmer in December than in May.

g I will see a lion on the way home from school.

h This year, half the students in my class will grow tails.

i The next student I see will have been born after 1900.

j If I toss a coin, it will come up heads.

2 Write the letter of each part of Question 1 in the table below to show each statement's chance.

| Will always happen | Will sometimes happen | Will never happen |
|---|---|---|
| | | |

3 Put the following events in order from least likely to most likely.

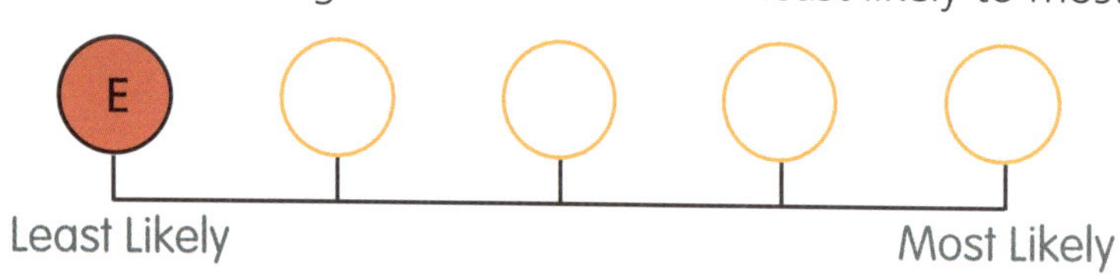


**A** We will have an earthquake today.
**B** I will go to school in November.
**C** I will stay up late tonight.
**D** We will be given homework on our next school day.
**E** A dinosaur will walk into our classroom tomorrow.

- Use this picture to make up four events that could happen. For example, the man could drop the spinning plate.
- Write your four events in order from least likely to most likely.

FUN SPOT



 • *AUSTRALIAN SIGNPOST MATHS 3* • ISBN 978 1 4886 2184 0

# 5:13 Repeating an Experiment

If we toss 10 coins, we are most likely to get 5 heads and 5 tails. However, this doesn't always happen.

CONCEPT

## Coins Experiment

1. Procedure:
   - Toss 10 coins.
   - Count the number of heads.
   - Write the result in the tally table.
   - Repeat this procedure 20 times. What did you discover?

| | Tally | |
|---|---|---|
| 0 heads | | |
| 1 head | | |
| 2 heads | | |
| 3 heads | | |
| 4 heads | | |
| 5 heads | | |
| 6 heads | | |
| 7 heads | | |
| 8 heads | | |
| 9 heads | | |
| 10 heads | | |

## Paper Experiment

INVESTIGATION

- Procedure:
  - Choose 3 sheets of different coloured paper.
  - From each sheet, cut 7 squares.
  - Put the 21 paper squares into a box or bag.
  - Without looking, take out 10 squares, one at a time.
  - Put the 10 squares in three columns, one for each colour. This is a column graph.
  - Compare your graph with Joe's graph.
  - Compare your graph with others in the class.
- Glue your graph into your book.
- Take another 10 squares from your box or bag. Make another graph and compare it with your first graph.

Joe's graph

 • *AUSTRALIAN SIGNPOST MATHS 3* • ISBN 978 1 4886 2184 0

# 3 Answers

## 1:01

1. a 9, 12, 15, 18, 21, 24, 27, 30
   b 12, 16, 20, 24, 28, 32, 36, 40
   c 18, 24, 30, 36, 42, 48, 54, 60
   d 21, 28, 35, 42, 49, 56, 63, 70
   e 24, 32, 40, 48, 56, 64, 72, 80
   f 27, 36, 45, 54, 63, 72, 81, 90
   g 30, 40, 50, 60, 70, 80, 90, 100
   h 300, 400, 500, 600, 700, 800, 900, 1 000

## 1:02

1. a odd 7  b even 12  c even 16
   d odd 13  e odd 29  f even 30
2. 83 100 109 111 118 120 125 127 130
3. They are odd because 1, 3, 5, 7 and 9 can't be drawn in pairs.

## 1:03

1. a

| | | | | | | | | | |
|---|---|---|---|---|---|---|---|---|---|
| 1 | 2 | 3 | 4 | 5 | 6 | 7 | 8 | 9 | 10 |
| 11 | 12 | 13 | 14 | 15 | 16 | 17 | 18 | 19 | 20 |
| 21 | 22 | 23 | 24 | 25 | 26 | 27 | 28 | 29 | 30 |
| 31 | 32 | 33 | 34 | 35 | 36 | 37 | 38 | 39 | 40 |
| 41 | 42 | 43 | 44 | 45 | 46 | 47 | 48 | 49 | 50 |
| 51 | 52 | 53 | 54 | 55 | 56 | 57 | 58 | 59 | 60 |
| 61 | 62 | 63 | 64 | 65 | 66 | 67 | 68 | 69 | 70 |
| 71 | 72 | 73 | 74 | 75 | 76 | 77 | 78 | 79 | 80 |
| 81 | 82 | 83 | 84 | 85 | 86 | 87 | 88 | 89 | 90 |
| 91 | 92 | 93 | 94 | 95 | 96 | 97 | 98 | 99 | 100 |

   b odd numbers  c 78  d 66  e 69
2. They are even because 0, 2, 4, 6 and 8 can be drawn in pairs without one left over.
3. a 4  b 4  c 49  d 20  e 26  f 0
4. a 5  b 3  c 49  d 13  e 26  f 27
5. 38 53 75 14 87 92 66 36 29 41 50 35 74 100
6. a even  b even  c odd  d even
   e even  f odd  g odd  h odd
   i even  j odd  k odd  l even

## 1:04

1. a 316  b 512  c 145  d 224
   e 232  f 129  g 221  h 324
   i 268  j 519  k 674  l 953
2. a 346  b 723  c 962  d 634  e 572  f 914
3. a 137, 446, 653  b 237, 491, 974
   c 106, 567, 819  d 250, 683, 749

## 1:05

1. a 413 

   four hundred and thirteen
   b 324 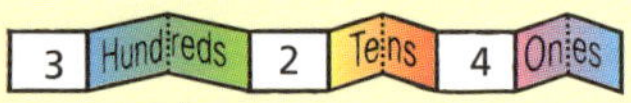

   three hundred and twenty-four
2. a 2  b 3  c 3  d 2  e 3
   f 2  g 1  h 3  i 4  j 3
3. a 260  b 152  c 940  d 718
   e 679  f 534  g 868  h 306

4. a 998, 1 000  b 862, 864  c 658, 660
   d 305, 307  e 498, 500  f 708, 710

## 1:06

1. a–d

| | | | | | | | | | |
|---|---|---|---|---|---|---|---|---|---|
| 1 | 2 | 3 | 4 | 5 | 6 | 7 | 8 | 9 | 10 |
| 11 | 12 | 13 | 14 | 15 | 16 | 17 | 18 | 19 | 20 |
| 21 | 22 | 23 | 24 | 25 | 26 | 27 | 28 | 29 | 30 |
| 31 | 32 | 33 | 34 | 35 | 36 | 37 | 38 | 39 | 40 |
| 41 | 42 | 43 | 44 | 45 | 46 | 47 | 48 | 49 | 50 |
| 51 | 52 | 53 | 54 | 55 | 56 | 57 | 58 | 59 | 60 |
| 61 | 62 | 63 | 64 | 65 | 66 | 67 | 68 | 69 | 70 |
| 71 | 72 | 73 | 74 | 75 | 76 | 77 | 78 | 79 | 80 |
| 81 | 82 | 83 | 84 | 85 | 86 | 87 | 88 | 89 | 90 |
| 91 | 92 | 93 | 94 | 95 | 96 | 97 | 98 | 99 | 100 |

   c (Answers may vary): Every second even number is the pattern for counting by 4s.
   d 8, 16, 24, 32, 40, 48, 56, 64, 72, 80
2. 2, 4, 6, 8 or 0
3. 5 or 0
4. 0
5. a 253, 263, 273, 283, 293  b 800, 795, 790, 785, 780
   c 120, 118, 116, 114, 112  d 700, 600, 500, 400, 300
6. a The rule is start at 223 and add 10 each time.
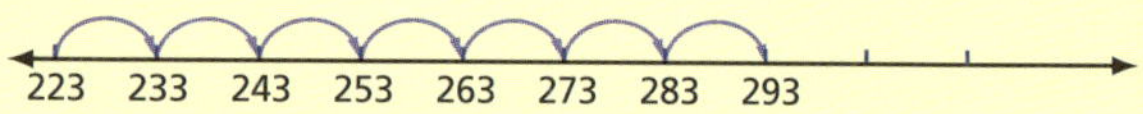

   b The rule is start at 815 and subtract 5 each time.
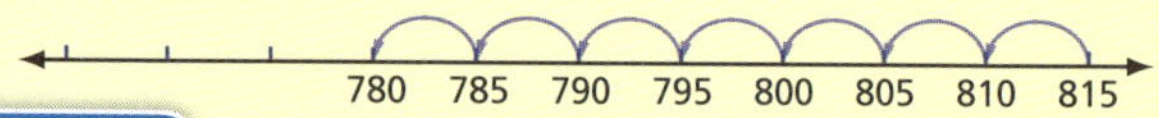


## 1:07

1. a 76, 78, 80, 82, 84, 86, 88, 90, 92, 94, 96, 98, 100
   b 1 000, 900, 800, 700, 600, 500, 400, 300, 200, 100, 0
   c 645, 650, 655, 660, 665, 670, 675, 680, 685, 690
   d 500, 490, 480, 470, 460, 450, 440, 430, 420, 410, 400
2. a 855, 835, 825, 805, 795
   b 615, 610, 600, 595, 590
   c 408, 406, 402, 400, 398
3. 2, 4, 6, 8, 10, 12, 14, 16, 18, 20, 22, 24, 26, 28, 30, 32, 34, 36, 38, 40
   The pattern is counting by 4, or add 4.
4. 5, 10, 15, 20, 25, 30, 35, 40, 45, 50, 55, 60, 65, 70, 75, 80, 85, 90, 95, 100
   The pattern is counting by 10, or add 10.
5. Group the coins into stacks of ten. Group stacks into ten lots of ten to make one hundred. Make sure that you have three groups of ten stacks of ten.
6. a The rule is start at 76 and add 2 each time.
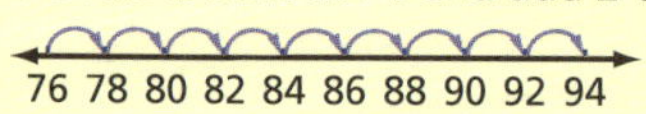

   b The rule is start at 1 000 and subtract 100 each time.
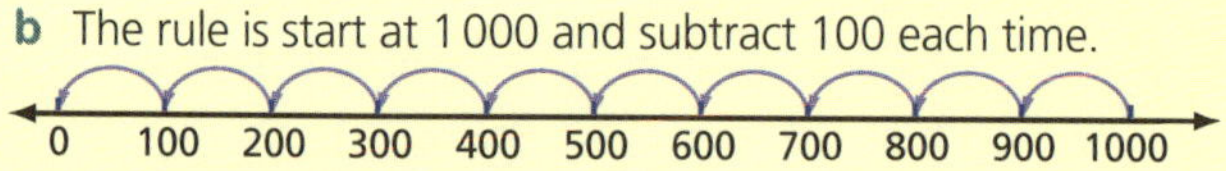

   c The rule is start at 645 and add 5 each time.
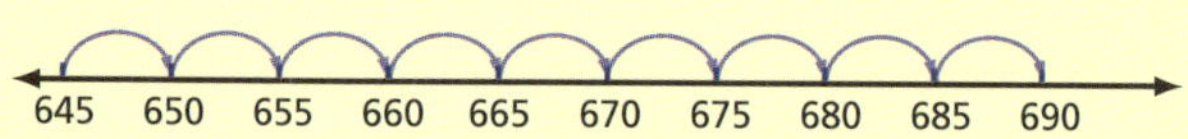



 • *AUSTRALIAN SIGNPOST MATHS 3* • ISBN 978 1 4886 2184 0

## 1:08

1 a

b

c

d

2 a 632 b 817 c 429 d 763
e 238 f 562 g 940 h 351

3 a one hundred and fifty-six b six hundred and seven
c three hundred and nineteen
d eight hundred and forty-one

## 1:09

1 a 746 b 183 c 575 d 397

2 a

b
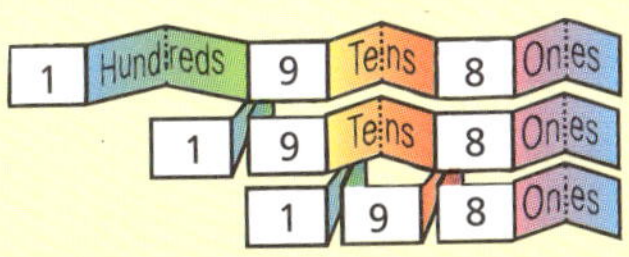


3 a 400 b 800 c 300
d 400 e 700 f 300
g 900 h 500 i 800

4 a 249 < 497 b 963 > 575 c 237 < 999
d 672 < 907 e 364 > 259 f 743 < 816
g 562 < 564 h 419 > 418

## 1:10

1 a 3 of 4 b 4 of 8 c 5 of 6

2 a $\frac{1}{2}$ b $\frac{3}{4}$ c $\frac{7}{8}$ d $\frac{5}{6}$
e $\frac{3}{8}$ f $\frac{7}{10}$ g $\frac{2}{4}$ h $\frac{2}{3}$

3 a
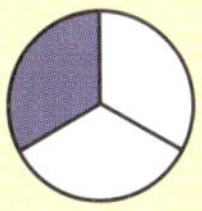
b
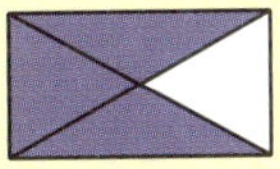
c

d
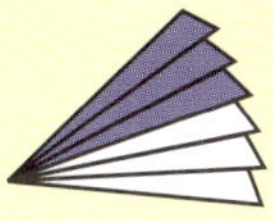
e
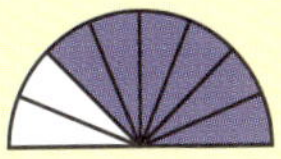
f
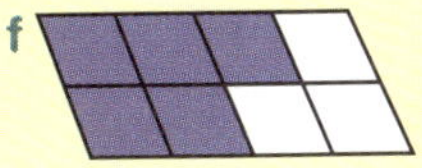
g

h
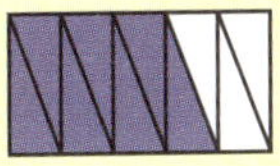
i
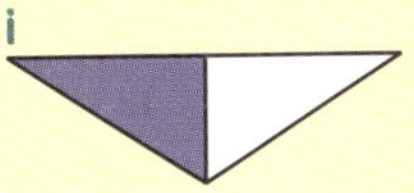

## 1:11

1

| Group | Number Coloured | Total in Group | Fraction Coloured |
|---|---|---|---|
|  | 4 | 5 | $\frac{4}{5}$ |
|  | 3 | 4 | $\frac{3}{4}$ |
|  | 5 | 8 | $\frac{5}{8}$ |
|  | 2 | 4 | $\frac{2}{4}$ |

2 a $\frac{5}{8}$ b $\frac{3}{4}$ c $\frac{5}{6}$ d $\frac{2}{3}$
e $\frac{3}{7}$ f $\frac{5}{9}$ g $\frac{10}{12}$ h $\frac{1}{2}$

3 a 4 fish are coloured b 7 beetles are coloured
c 11 apples are coloured d 3 bats are coloured
e 5 mushrooms are coloured f 3 pencils are coloured

## 1:12

1 a 1518

one thousand five hundred and eighteen

b 1332

one thousand three hundred and thirty-two

2 a 2 b 4 c 1 d 2
e 4 f 2 g 4 h 4

3 a 545 b 27 c 908
d 1000 e 611 f 200

4 a 2547 b 4263 c 5409
d 2157 e 4312 f 10000

## 1:13

1 a 1519 b 1354 c 1395 d 4649
e 4296 f 8125 g 5240 h 8405
i 4092 j 5903 k 6516 l 2051
m 4970

## 1:14

1 a $\frac{1}{4}$ b 1 c $\frac{3}{8}$ d 1
e $\frac{5}{8}$ f 1 g $\frac{3}{4}$ h $\frac{7}{8}$

2 a
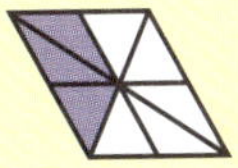
b
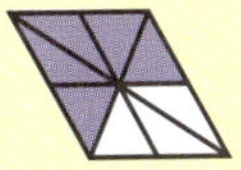
c
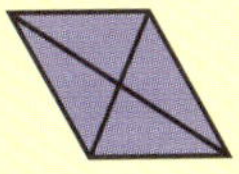
d
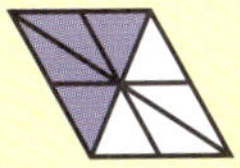
e
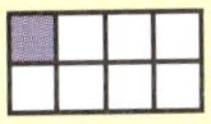
f
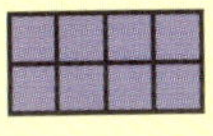
g
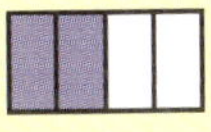
h
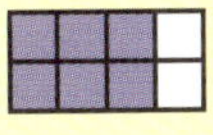

## 1:15

1 a $\frac{4}{8}$ b $\frac{3}{4}$ c $\frac{7}{8}$
d $\frac{3}{4}$ e $\frac{6}{8}$ f $\frac{5}{8}$

2 a
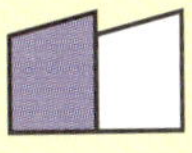
b
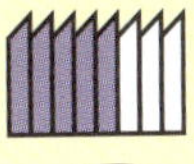
c

d
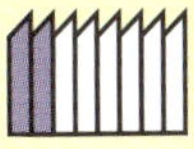
e
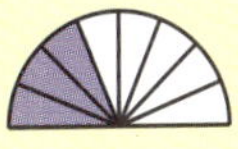
f
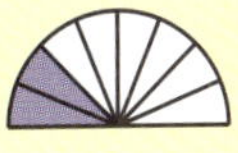
g
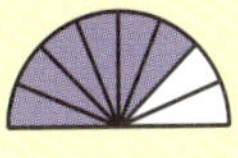
h
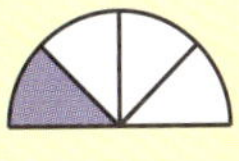

❸ a $\frac{3}{6}$ or $\frac{1}{2}$ b $\frac{4}{6}$ or $\frac{2}{3}$ c $\frac{5}{6}$
d $\frac{3}{8}$ e $\frac{5}{8}$ f $\frac{7}{8}$

## 1:16

❶ a 40, 50, 60 b 10, 12, 14
c 25, 30, 35 d 20, 24, 28

❷ a Start at 4 and double the last number each time.
b Start at 48 and halve the last number each time.

❸ a

| | | | | | | | | | |
|---|---|---|---|---|---|---|---|---|---|
| 1 | 2 | 3 | 4 | 5 | 6 | 7 | 8 | 9 | 10 |
| 11 | 12 | 13 | 14 | 15 | 16 | 17 | 18 | 19 | 20 |
| 21 | 22 | 23 | 24 | 25 | 26 | 27 | 28 | 29 | 30 |

The numbers in the tens pattern are also numbers in the fives pattern. Every fifth number ends in 5 or 0, and every tenth number ends in 0.

b

| | | | | | | | | | |
|---|---|---|---|---|---|---|---|---|---|
| 1 | 2 | 3 | 4 | 5 | 6 | 7 | 8 | 9 | 10 |
| 11 | 12 | 13 | 14 | 15 | 16 | 17 | 18 | 19 | 20 |
| 21 | 22 | 23 | 24 | 25 | 26 | 27 | 28 | 29 | 30 |
| 31 | 32 | 33 | 34 | 35 | 36 | 37 | 38 | 39 | 40 |

The numbers in the eights pattern are also in the twos and fours patterns. The numbers in the fours pattern are also numbers in the twos pattern.

❹ 9, 14, 19, 24, 29, 34, 39, 44. I noticed that the ones digit alternated between 4 and 9.

## 1:17

❶ a 1227
1 Thousands 2 Hundreds 2 Tens 7 Ones
one thousand two hundred and twenty-seven

b 1258
1 Thousands 2 Hundreds 5 Tens 8 Ones
one thousand two hundred and fifty-eight

c 1513
1 Thousands 5 Hundreds 1 Tens 3 Ones
one thousand five hundred and thirteen

d 1342
1 Thousands 3 Hundreds 4 Tens 2 Ones
one thousand three hundred and forty-two

❷ a 3 b 4 c 4 d 3
e 4 f 2 g 4 h 4

❸ a 440 b 10 000 c 1060 d 1002
e 1543 f 3785 g 6959

## 1:18

❶ a 39, 67, 86, 94, 106, 116
b 109, 130, 134, 153, 167, 186
c 192, 289, 373, 430, 546, 751
d 352, 436, 581, 872, 964, 978

❷ a 642 b 971 c 640 d 953
e 871 f 530 g 986 h 752

❸ a 234 b 105 c 269 d 378
e 608 f 149 g 507 h 268
(Note: 3-digit numbers cannot begin with zero.)

❹ a 94 > 86 b 134 > 126 c 335 < 465
d 672 < 756 e 549 < 594 f 806 < 926
g 964 < 999 h 483 > 438 i 1935 > 1724

❺ a true b true c true
d false e false f false

❻ Answers will vary. Examples:
a 4 < 38, 84 > 3, 8 – 4 > 3 b 47 > 9, 97 > 4, 9 – 7 < 4
c 16 > 4, 1 × 4 < 6, 6 ÷ 1 > 4

## 1:19

❶ a 200 b 400 c 500 d 600
e 400 f 300 g 800 h 300

❷ a 100 b 400 c 500 d 500
e 200 f 800 g 900 h 600
i 200 j 500 k 400 l 900
m 400 n 200 o 700 p 600

❸ a 163, 220, 186, 217, 237, 205 will be circled.
b 467, 456, 483, 532, 521 will be circled.

❹ a 300 b 900 c 200 d 800 e 700 f 400

❺ a false b false c true d true

## 1:20

❶ a b c d e f

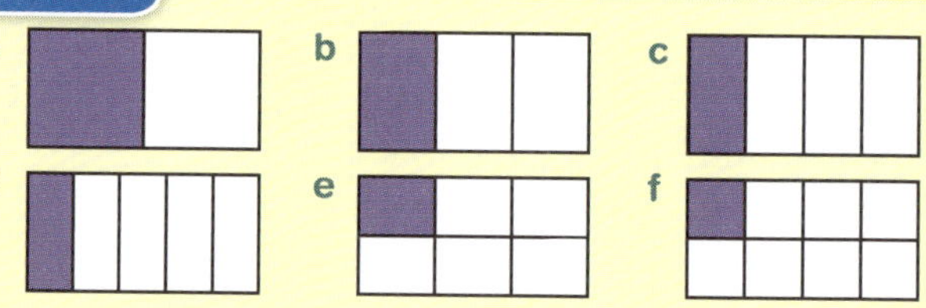

❷ a $\frac{1}{2}$ b $\frac{1}{3}$ c $\frac{1}{4}$
d $\frac{1}{3}$ e $\frac{1}{5}$ f $\frac{1}{2}$

❸ a b c d e f g h

❹ a b

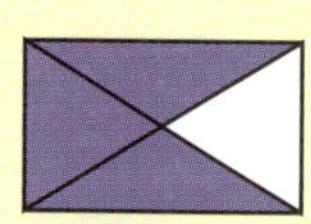

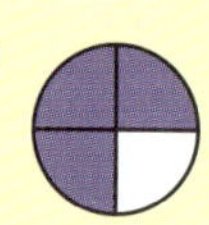

## 1:21

❶ a b c d

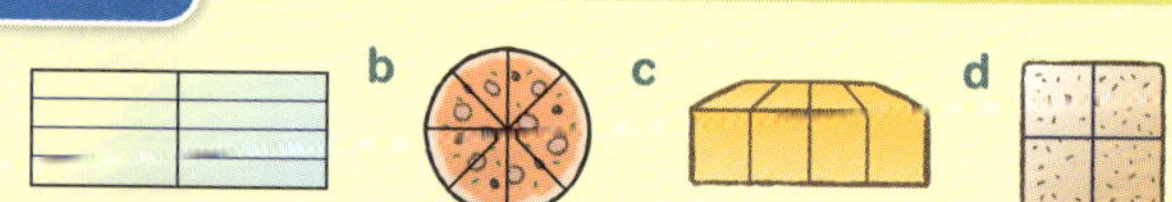

❷ a b c d

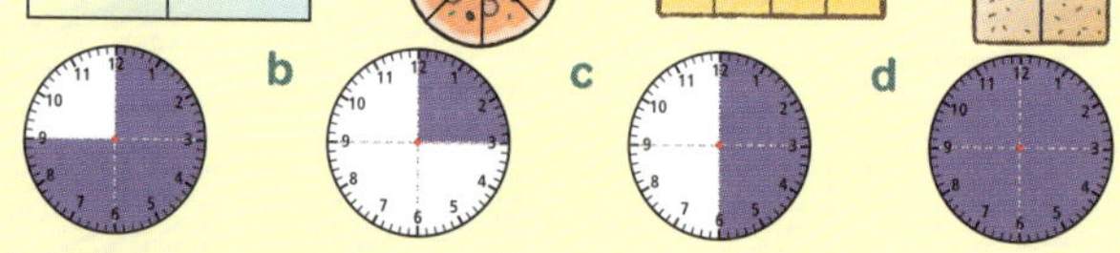

❸ a

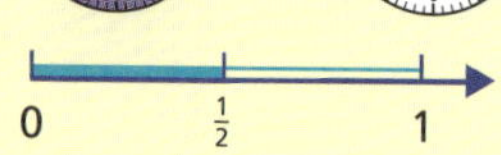


b

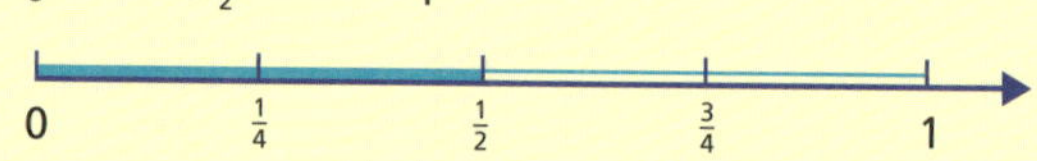

c
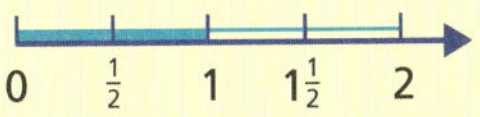


d
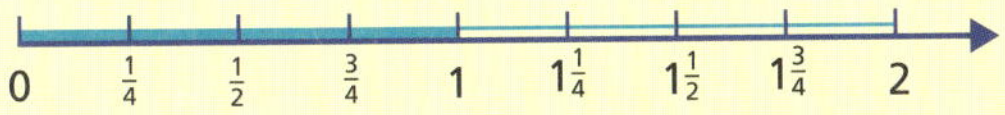


4 a 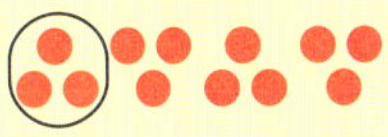
b 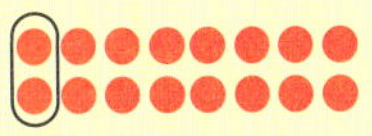
c 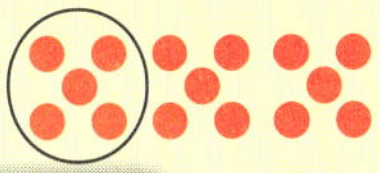
d 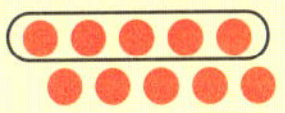

## 1:22

1 a 1594 b 1625 c 1465
d 1613 e 1079 f 1802

2 1514

1 Thousands 5 Hundreds 1 Tens 4 Ones

one thousand five hundred and fourteen

3 a 3 b 2 c 4 d 3
e 2 f 4 g 4 h 1

4 a 460, 470, 480 b 690, 700, 710 c 849, 859, 869
d 786, 796, 806 e 720, 820, 920 f 639, 739, 839

5 a 347 b 680 c 1000 d 1317
e 4877 f 3208 g 2650 h 3641
i 2200 j 5011 k 10000 l 1002

## 1:23

1 a 1 Thousands 4 Hundreds 7 Tens 9 Ones
b 3 Thousands 5 Hundreds 6 Tens 8 Ones
c 4 Thousands 2 Hundreds 3 Tens 1 Ones
d 2 Thousands 3 Hundreds 0 Tens 7 Ones

2 a 5763 b 8956 c 6275 d 4649 e 9582

3 a six thousand eight hundred and twenty-four
b two thousand three hundred and nineteen

4 a 5000 b 3000 c 10000

## 1:24

1 a 16, 20, 24 b 20, 25, 30
c 12, 15, 18 d 24, 30, 36

2 a 21, 31, 41, 51, 61 b 40, 45, 50, 55, 60
c 11, 13, 15, 17, 19 d 22, 24, 26, 28, 30

3

| | | | | | | | | | |
|---|---|---|---|---|---|---|---|---|---|
| 1 | 2 | 3 | 4 | 5 | 6 | 7 | 8 | 9 | 10 |
| 11 | 12 | 13 | 14 | 15 | 16 | 17 | 18 | 19 | 20 |
| 21 | 22 | 23 | 24 | 25 | 26 | 27 | 28 | 29 | 30 |
| 31 | 32 | 33 | 34 | 35 | 36 | 37 | 38 | 39 | 40 |
| 41 | 42 | 43 | 44 | 45 | 46 | 47 | 48 | 49 | 50 |
| 51 | 52 | 53 | 54 | 55 | 56 | 57 | 58 | 59 | 60 |

4 a Start at 25 and count by 5s.
b Start at 79 and count back by 10s.

## 1:25

1 a 7 b 16 c 40
d 8 e 30 f 11
g 22 h 60 i 62

2 a Add 2 b Add 3 c Add 10
d Subtract 4 e Subtract 1 f Subtract 11
g Add 4 h Subtract 5 i Add 8

3 a 27, 33, 39 b 40, 39, 38 c 72, 70, 68
d 42, 50, 58 e 43, 63, 83 f 36, 32, 28
g 6, 12, 24 h 20, 35, 50 i 53, 43, 33 j 2, 4, 8

4 a 6, 9, 12, Add 3 b 15, 20, 25, Add 5

## 1:26

1 a 3769 b 4637 c 2581
d 5476 e 3958 f 6843

2 a 4000 + 200 + 40 + 7 b 3000 + 100 + 60 + 9
c 2000 + 600 + 70 + 5 d 5000 + 100 + 90 + 2

3

| | Thousands | Hundreds | Tens | Units |
|---|---|---|---|---|
| a | 4 | 3 | 7 | 8 |
| b | 6 | 2 | 4 | 9 |
| c | 3 | 8 | 2 | 1 |
| d | 5 | 8 | 7 | 3 |
| e | 2 | 6 | 3 | 4 |
| f | 1 | 9 | 8 | 6 |

4 a 900 b 4000 c 9 d 80
e 3000 f 5 g 600 h 30

5 a 7483 b 9638 c 6963 d 9847

## 1:27

1 a 642 b 978 c 1523 d 5951

2 a 665, 675 b 779, 789

3 a 396 b 9999 c 1347
d 2581 e 1435 f 6723

4 a 723 b 965 c 631
d 1264 e 2753 f 4194

5 a true b false c false d true
e true f true g false h true

6 a +

| | | |
|---|---|---|
| 3 | 9 | 12 |
| 4 | 8 | 12 |
| 7 | 17 | 24 |

b +

| | | |
|---|---|---|
| 8 | 6 | 14 |
| 4 | 5 | 9 |
| 12 | 11 | 23 |

c +

| | | |
|---|---|---|
| 6 | 4 | 10 |
| 6 | 7 | 13 |
| 12 | 11 | 23 |

d +

| | | |
|---|---|---|
| 9 | 7 | 16 |
| 8 | 5 | 13 |
| 17 | 12 | 29 |

## 1:28

1 a Start at 50 and add 50. Every second number increases by 100.
b Start at 100 and add 100. The number in the hundreds column increases by 1 and the two zeroes do not change.
c Start at 47 and subtract 10. The digit in the ones column does not change.
d Start at 166 and add 10. The number in the tens column increases by 1 and the ones column does not change.

2 a Start at 1 and add $\frac{1}{2}$.
b Start at $\frac{1}{4}$ and add $\frac{1}{4}$.
c Start at $\frac{1}{3}$ and add $\frac{1}{3}$.

3 a 6, 10, 12 b 60, 50 c 16, 24
d 150, 160, 180 e 16, 48 f 6, 18

4

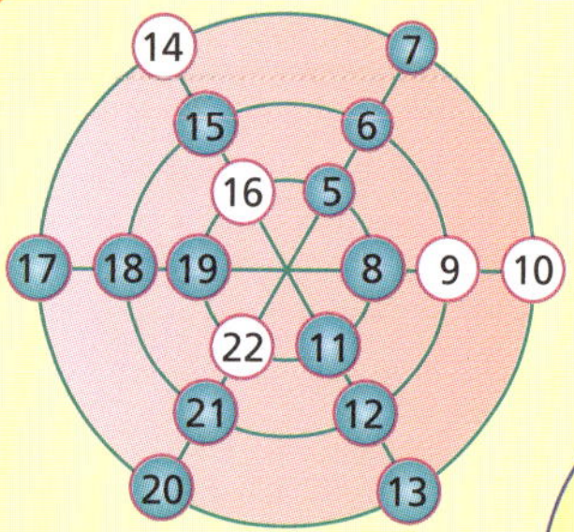


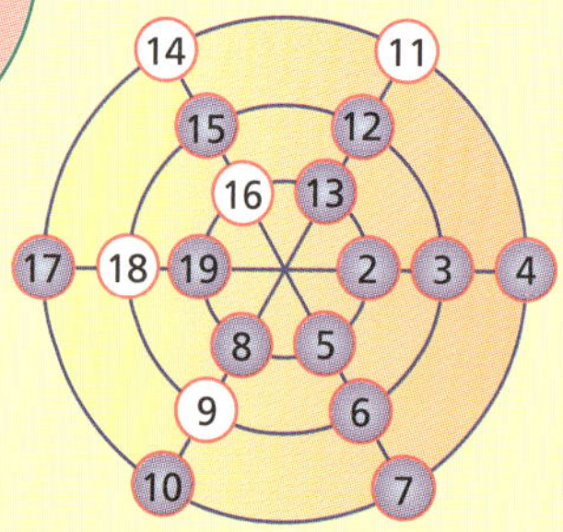


## 1:29

1 a

| | | | | | | |
|---|---|---|---|---|---|---|
| First Number | 1 | 2 | 3 | 4 | 5 | 6 |
| Second Number | 8 | 9 | 10 | 11 | 12 | 13 |

Rule: Add 7

b

| | | | | | | |
|---|---|---|---|---|---|---|
| First Number | 10 | 20 | 30 | 40 | 50 | 60 |
| Second Number | 4 | 14 | 24 | 34 | 44 | 54 |

Rule: Subtract 6

c

| | | | | | | |
|---|---|---|---|---|---|---|
| First Number | 2 | 4 | 6 | 8 | 10 | 12 |
| Second Number | 1 | 2 | 3 | 4 | 5 | 6 |

Rule: Divide by 2

d

| | | | | | | |
|---|---|---|---|---|---|---|
| First Number | 29 | 30 | 31 | 32 | 33 | 34 |
| Second Number | 31 | 32 | 33 | 34 | 35 | 36 |

Rule: Add 2

e

| | | | | | | |
|---|---|---|---|---|---|---|
| First Number | 1 | 2 | 3 | 4 | 5 | 6 |
| Second Number | 2 | 4 | 6 | 8 | 10 | 12 |

Rule: Multiply by 2

f

| | | | | | | |
|---|---|---|---|---|---|---|
| First Number | 20 | 18 | 16 | 14 | 12 | 10 |
| Second Number | 10 | 9 | 8 | 7 | 6 | 5 |

Rule: Divide by 2

g

| | | | | | | |
|---|---|---|---|---|---|---|
| First Number | 5 | 6 | 7 | 8 | 9 | 10 |
| Second Number | 25 | 30 | 35 | 40 | 45 | 50 |

Rule: Multiply by 5

h

| | | | | | | |
|---|---|---|---|---|---|---|
| First Number | 30 | 25 | 20 | 15 | 10 | 5 |
| Second Number | 6 | 5 | 4 | 3 | 2 | 1 |

Rule: Divide by 5

i

| | | | | | | |
|---|---|---|---|---|---|---|
| First Number | 3 | 4 | 5 | 6 | 7 | 8 |
| Second Number | 30 | 40 | 50 | 60 | 70 | 80 |

Rule: Multiply by 10

j

| | | | | | | |
|---|---|---|---|---|---|---|
| First Number | 100 | 80 | 60 | 40 | 20 | 10 |
| Second Number | 10 | 8 | 6 | 4 | 2 | 1 |

Rule: Divide by 10

2 Answers will vary.

## 1:30

1 a 1352 b 1234 c 1392 d 9646 e 1255
f 9708 g 2055 h 4249 i 9030 j 4016
k 1515 l 6540 m 8409

2 a 600 or 6 hundreds b 6000 or 6 thousands
c 60 or 6 tens

## 1:31

1 a 1639 b 1741 c 2956
d 4583 e 3810 f 5307

2 a 2000 + 300 + 50 + 6 b 1000 + 900 + 70 + 5
c 6000 + 200 + 80 + 3 d 4000 + 800 + 6

3

| | Thousands | Hundreds | Tens | Ones |
|---|---|---|---|---|
| a | 1 | 9 | 6 | 3 |
| b | 2 | 4 | 7 | 9 |
| c | 2 | 3 | 0 | 7 |
| d | 3 | 5 | 8 | 2 |
| e | 5 | 1 | 9 | 0 |
| f | 4 | 0 | 6 | 3 |

4 a 200 or 2 hundreds b 7 or 7 ones
c 40 or 4 tens d 1000 or 1 thousand
e 60 or 6 tens f 4000 or 4 thousands
g 700 or 7 hundreds h 6 or 6 ones

5 a 6835 b 9276 c 3806 d 8795

6 a 966 b 1235 c 2190 d 4629
e 5728 f 6309 g 3399 h 7019

## 1:32

1 a 2584 b 5738 c 2495 d 9734

2

| | Thousands | Hundreds | Tens | Ones |
|---|---|---|---|---|
| a | 2 | 5 | 6 | 0 |
| b | 1 | 2 | 3 | 4 |
| c | 6 | 4 | 5 | 2 |
| d | 5 | 6 | 2 | 9 |
| e | 7 | 2 | 0 | 4 |

3 a 600 or 6 hundreds b 2 or 2 ones
c 90 or 9 tens d 2000 or 2 thousands
e 50 or 5 tens f 3 or 3 ones

## 1:33

1 a 1 Thousands 7 Hundreds 6 Tens 4 Ones
b 3 Thousands 7 Hundreds 9 Tens 5 Ones
c 4 Thousands 0 Hundreds 1 Tens 3 Ones
d 9 Thousands 6 Hundreds 5 Tens 8 Ones

2 a 15 b 12 c 23 d 61 e 74 f 93

3 a 174 b 162 c 490 d 726

4 a 1210 b 5641 c 6301 d 2516

5 a 300 tens = 30 hundreds
b 70 tens = 7 hundreds

## 1:34

1 a 1970 b 2391 c 3489 d 5370 e 1725
f 4960 g 6400 h 7999 i 7406

2 a 86 < 92 b 235 > 230 c 409 > 406
d 763 < 769 e 1511 > 1501 f 3637 > 3256
g 5206 < 5260 h 6431 > 6314 i 9725 < 9752
j 4311 > 4131

3 a true b false c false d true e true f true

4 a 10 or 1 ten b 500 or 5 hundreds
c 9 or 9 ones d 6000 or 6 thousands
e 10 or 1 ten f 900 or 9 hundreds
g 50 or 5 tens h 7000 or 7 thousands
i 80 or 8 tens

5 a three thousand six hundred and ninety-one
b five thousand two hundred and sixteen
c six thousand five hundred and nine

## 1:35

1 28, 37, 46, 55, 64, 73, 82, 91
The pattern is "the tens digit is increasing by 1 each time and the ones digit is decreasing by 1 each time".

2 90, 82, 74, 66, 58, 50, 42, 34
77, 69, 61, 53, 45, 37, 29, 21
The pattern is "the ones digit is increasing by 2 each time".

3 50, 100, 150, 200, 250, 300, 350
63, 113, 163, 213, 263, 313, 363
The pattern is "the last two digits are repeating every two times".

4 25, 50, 75, 100, 125, 150, 175
47, 72, 97, 122, 147, 172, 197
The pattern is "the last two digits are repeating every four times".

5 I now have 97 bottles. (Explanations will vary.)

## 1:36

1 a 10 b 20 c 60 d 80
e 100 f 300 g 400 h 600

2 a 90 b 80 c 90 d 100 e 80 f 70
g 90 h 100 i 90 j 90 k 20 l 60
m 10 n 30 o 30 p 40

3 a 95 b 80 c 93 d 99 e 75 f 73
g 85 h 103 i 89 j 93 k 27 l 57
m 11 n 30 o 29 p 38

4

| | Number sentence | Estimate | Actual answer | Difference |
|---|---|---|---|---|
| a | 76 + 39 | 120 | 115 | 5 |
| b | 52 + 71 | 120 | 123 | 3 |
| c | 38 + 33 | 70 | 71 | 1 |
| d | 66 + 28 | 100 | 94 | 6 |

5

| | Number sentence | Estimate | Actual answer | Difference |
|---|---|---|---|---|
| a | 56 – 18 | 40 | 38 | 2 |
| b | 59 – 22 | 40 | 37 | 3 |
| c | 82 – 46 | 30 | 36 | 6 |
| d | 78 – 27 | 50 | 51 | 1 |

6

| | Number sentence | Estimate | Actual answer | Difference |
|---|---|---|---|---|
| a | 163 + 196 | 400 | 359 | 41 |
| b | 237 + 348 | 500 | 585 | 85 |
| c | 631 – 420 | 200 | 211 | 11 |
| d | 788 – 159 | 600 | 629 | 29 |

## 2:01

1 5 cents, 2 dollars, 10 cents, 1 dollar, 20 cents, 50 cents

2

3 $5, $10, $20, $50, $100

## 2:02

1 a 5, 5, 1, 4 b 6, 6, 2, 4
c 10, 10, 9, 1 d 10, 10, 8, 2

2

| + | 0 | 1 | 2 | 3 | 4 | 5 | 6 | 7 | 8 | 9 | 10 |
|---|---|---|---|---|---|---|---|---|---|---|---|
| 0 | 0 | 1 | 2 | 3 | 4 | 5 | 6 | 7 | 8 | 9 | 10 |
| 1 | 1 | 2 | 3 | 4 | 5 | 6 | 7 | 8 | 9 | 10 | 11 |
| 2 | 2 | 3 | 4 | 5 | 6 | 7 | 8 | 9 | 10 | 11 | 12 |
| 3 | 3 | 4 | 5 | 6 | 7 | 8 | 9 | 10 | 11 | 12 | 13 |
| 4 | 4 | 5 | 6 | 7 | 8 | 9 | 10 | 11 | 12 | 13 | 14 |
| 5 | 5 | 6 | 7 | 8 | 9 | 10 | 11 | 12 | 13 | 14 | 15 |
| 6 | 6 | 7 | 8 | 9 | 10 | 11 | 12 | 13 | 14 | 15 | 16 |
| 7 | 7 | 8 | 9 | 10 | 11 | 12 | 13 | 14 | 15 | 16 | 17 |
| 8 | 8 | 9 | 10 | 11 | 12 | 13 | 14 | 15 | 16 | 17 | 18 |
| 9 | 9 | 10 | 11 | 12 | 13 | 14 | 15 | 16 | 17 | 18 | 19 |
| 10 | 10 | 11 | 12 | 13 | 14 | 15 | 16 | 17 | 18 | 19 | 20 |

Matching pairs of numbers will be coloured.

3 a 10, 10, 10, 10, 10, 10, 10, 10, 10
b 1, 2, 3, 4, 5, 6, 7, 8, 9

## 2:03

1 a 13 b 14 c 15 d 5 e 2
f 5 g 1 h 10 i 7 j 2
k 17 l 13 m 4 n 12

2 a 13, 4 b 16, 9 c 8, 2

3 a true b false

## 2:04

1 a 2 b 4 c 6 d 8 e 10
f 12 g 14 h 16 i 18 j 20
k 4 l 8 m 12 n 16 o 20

2 a 16, 6, 22 b 14, 10, 24 c 8, 18, 26 d 20, 12, 8

3 a 14 b 16 c 22 d 30

4 a

| | 3 | 10 | 2 | 1 | 5 | 4 | 7 | 9 | 8 | 6 |
|---|---|---|---|---|---|---|---|---|---|---|
| ×2 | 6 | 20 | 4 | 2 | 10 | 8 | 14 | 18 | 16 | 12 |

b

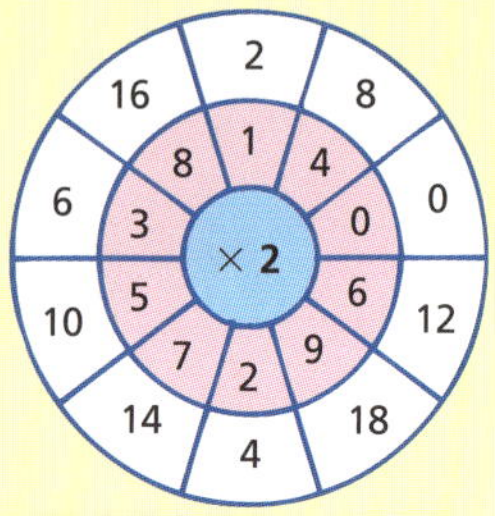


c

| | 6 | 4 | 8 | 2 | 5 | 10 | 1 | 7 | 9 | 3 |
|---|---|---|---|---|---|---|---|---|---|---|
| × 2 | 12 | 8 | 16 | 4 | 10 | 20 | 2 | 14 | 18 | 6 |

## 2:05

1. a 13 b 13 c 11 d 14 e 16 f 15
2. a 6 + 5 + 4 = 15 b 5 + 9 + 5 = 19
   c 3 + 7 + 4 = 14 d 2 + 8 + 6 = 16
3. a 19 b 13 c 20 d 19 e 20
4. a 14 b 20 c 13 d 23 e 21
5. a 27 b 28 c 33 d 24 e 30

## 2:06

1. The tens digit increases by 1.
   The ones decreases by 1.
   So, to add 9 we can add 10 and then subtract 1.
2. The tens digit decreases by 1.
   The ones digit increases by 1.
   So, to add 9 we can subtract 10 and then add 1.
3. a 16 b 25 c 37 d 51
   e 8 f 25 g 72, 63, 54, 45
4. a 14 b 27 c 61 d 83
   e 15 f 59 g 56, 64, 72, 80
5. a 26 b 70 c 84 d 34
   e 64 f 38 g 52 h 84

## 2:07

1. a 27 b 37 c 57 d 47 e 76 f 86
2.

| | | | | | | | | | |
|---|---|---|---|---|---|---|---|---|---|
| 1 | 2 | 3 | 4 | 5 | 6 | 7 | 8 | 9 | 10 |
| 11 | 12 | 13 | 14 | 15 | 16 | 17 | 18 | 19 | 20 |
| 21 | 22 | 23 | 24 | 25 | 26 | 27 | 28 | 29 | 30 |
| 31 | 32 | 33 | 34 | 35 | 36 | 37 | 38 | 39 | 40 |
| 41 | 42 | 43 | 44 | 45 | 46 | 47 | 48 | 49 | 50 |
| 51 | 52 | 53 | 54 | 55 | 56 | 57 | 58 | 59 | 60 |
| 61 | 62 | 63 | 64 | 65 | 66 | 67 | 68 | 69 | 70 |
| 71 | 72 | 73 | 74 | 75 | 76 | 77 | 78 | 79 | 80 |
| 81 | 82 | 83 | 84 | 85 | 86 | 87 | 88 | 89 | 90 |
| 91 | 92 | 93 | 94 | 95 | 96 | 97 | 98 | 99 | 100 |

3. a

| | 13 | 23 | 33 | 43 | 53 | 63 | 73 | 83 | 93 | 103 | 113 |
|---|---|---|---|---|---|---|---|---|---|---|---|
| − 4 | 9 | 19 | 29 | 39 | 49 | 59 | 69 | 79 | 89 | 99 | 109 |
| − 6 | 7 | 17 | 27 | 37 | 47 | 57 | 67 | 77 | 87 | 97 | 107 |

b

| | 12 | 22 | 32 | 42 | 52 | 62 | 72 | 82 | 92 | 102 | 112 |
|---|---|---|---|---|---|---|---|---|---|---|---|
| − 5 | 7 | 17 | 27 | 37 | 47 | 57 | 67 | 77 | 87 | 97 | 107 |
| − 9 | 3 | 13 | 23 | 33 | 43 | 53 | 63 | 73 | 83 | 93 | 103 |

4. a 28 − 23 = 5 b 55 − 40 = 15 c 61 − 56 = 5
5. a 31 − 4 = 27, 41 − 4 = 37, 51 − 4 = 47
   b 34 − 8 = 26, 44 − 8 = 36, 54 − 8 = 46
   c 35 − 6 = 29, 45 − 6 = 39, 55 − 6 = 49

6. Answers will vary.

## 2:08

1. a 5 b 10 c 15 d 20 e 25
   f 30 g 35 h 40 i 45 j 50
   k 10 l 20 m 30 n 40 o 50
2. a 60 b 70 c 80 d 90 e 100
3. a

| | 3 | 10 | 2 | 1 | 5 | 4 | 7 | 9 | 8 | 6 |
|---|---|---|---|---|---|---|---|---|---|---|
| × 10 | 30 | 100 | 20 | 10 | 50 | 40 | 70 | 90 | 80 | 60 |

b

| | 3 | 10 | 2 | 1 | 5 | 4 | 7 | 9 | 8 | 6 |
|---|---|---|---|---|---|---|---|---|---|---|
| × 5 | 15 | 50 | 10 | 5 | 25 | 20 | 35 | 45 | 40 | 30 |

c yes d yes e yes

## 2:09

1. a 6 b 7 c 8 d 0 e 0 f 0
2. a 2 b 20 c 14 d 10 e 4 f 40
   g 30 h 12 i 40 j 70 k 6 l 20
   m 18 n 50 o 9 p 0
3. a 10 b 90 c 35 d 7 e 25 f 60
   g 8 h 45 i 16 j 80 k 15 l 0
4. a

| | 3 | 10 | 2 | 1 | 5 | 4 | 7 | 9 | 8 | 6 |
|---|---|---|---|---|---|---|---|---|---|---|
| × 1 | 3 | 10 | 2 | 1 | 5 | 4 | 7 | 9 | 8 | 6 |

b

| | 3 | 10 | 2 | 1 | 5 | 4 | 7 | 9 | 8 | 6 |
|---|---|---|---|---|---|---|---|---|---|---|
| × 0 | 0 | 0 | 0 | 0 | 0 | 0 | 0 | 0 | 0 | 0 |

## 2:10

1. a 700 b 900 c 700 d 600
2. a 27 b 34 c 78 d 34
3. a 46 b 33 c 43 d 61
4. a 56 b 92 c 67 d 64
5. a 73 b 89 c 39 d 48 e 57 f 80
6. a 89

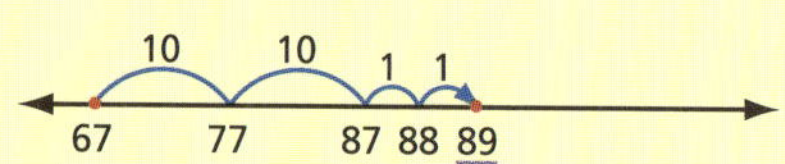


b 78

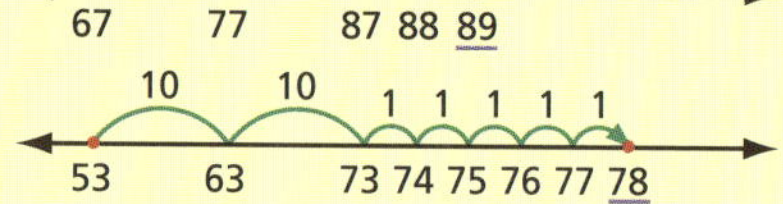


c 61

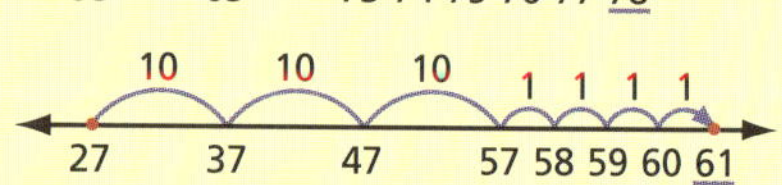


d 62

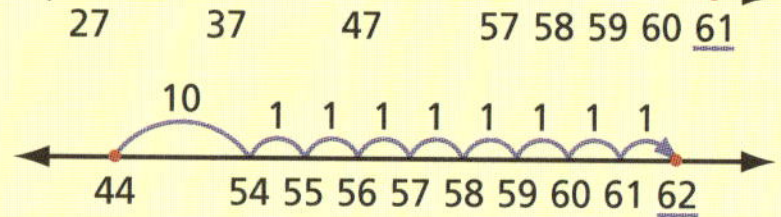


7. Strategies used may vary.
   a 88, E b 43, B c 53, C or D
   d 45, C or D e 800, A f 66, D or F

## 2:11

1. a 17 b 25 c 17 d 28 e 34 f 24 g 32
   h 19 i 23 j 26 k 15 l 27 m 31
2. a 29

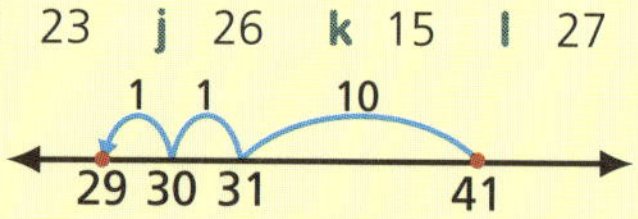


b 28

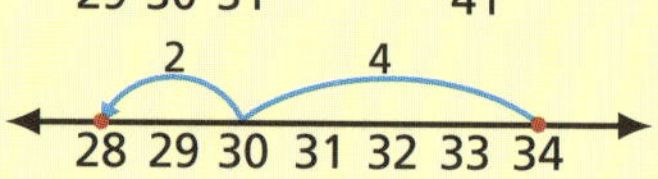


c 9

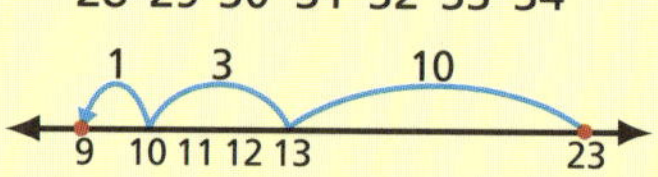




 • *AUSTRALIAN SIGNPOST MATHS 3* • ISBN 978 1 4886 2184 0

d 19

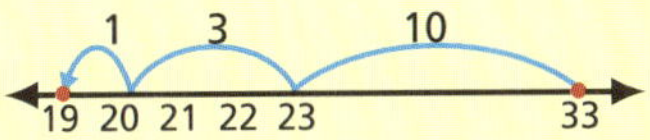


e 48

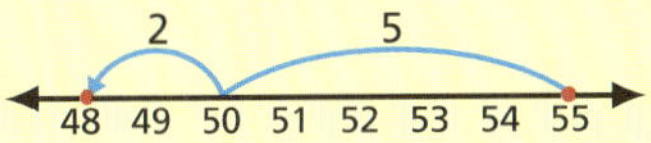


f 19

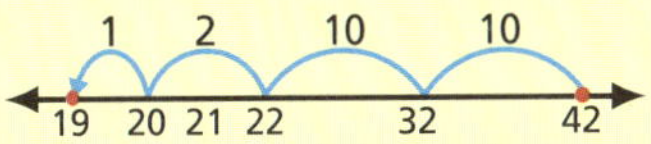


g 77

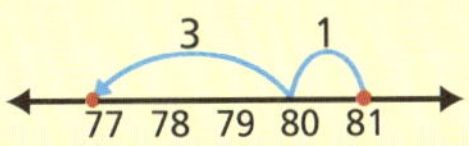


h 58

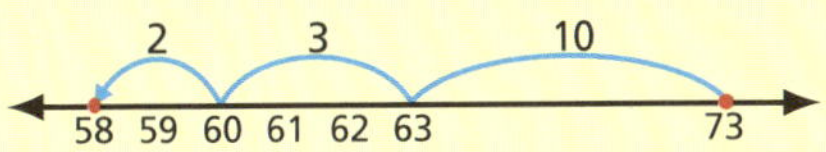


i 37

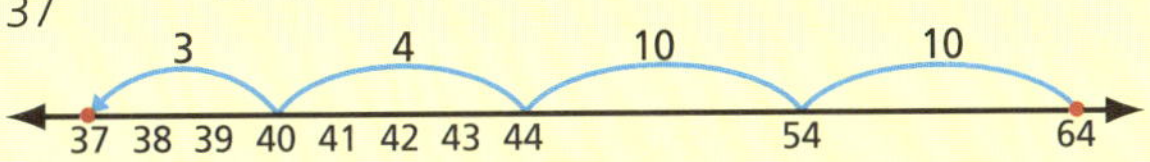


## 2:12

1 a 12 b 35 c 14 d 80 e 70 f 16
g 25 h 50 i 18 j 30 k 60 l 90
m 40 n 22 o 45 p 55

2 a 2, 2 b 10, 10 c 20, 20
d 5, 5 e 0, 0 f 50, 50

3 a 0, 7, 30, 50, 6, 25, 40, 12, 40, 70
b 0, 6, 45, 100, 4, 20, 50, 18, 35, 80
c 0, 9, 40, 20, 10, 15, 20, 14, 45, 60
d 0, 8, 35, 10, 8, 10, 30, 16, 30, 90

## 2:13

1 a 3 b 6 c 9 d 12 e 15
f 18 g 21 h 24 i 27 j 30

2 a 0 b 0 c 30 d 30
e 6 f 6 g 3 h 3
i 15 j 15 k 24 l 12
m 27 n 18 o 9 p 21

3 a 5 b 5 c 10 d 10
e 0 f 0 g 50 h 50
i 20 j 20 k 25 l 100

4 a $3 \times 8 = 24$ b $3 \times 9 = 27$

5 a

| | 4 | 7 | 9 | 5 | 8 | 6 |
|---|---|---|---|---|---|---|
| × 3 | 12 | 21 | 27 | 15 | 24 | 18 |

b

| | 4 | 7 | 9 | 5 | 8 | 6 |
|---|---|---|---|---|---|---|
| × 5 | 20 | 35 | 45 | 25 | 40 | 30 |

c

| | 4 | 7 | 9 | 5 | 8 | 6 |
|---|---|---|---|---|---|---|
| × 2 | 8 | 14 | 18 | 10 | 16 | 12 |

d

| | 4 | 7 | 9 | 5 | 8 | 6 |
|---|---|---|---|---|---|---|
| × 10 | 40 | 70 | 90 | 50 | 80 | 60 |

## 2:14

1 a 21 + 38 = 59 b 31 + 34 = 65 c 26 + 32 = 58
d 25 + 34 = 59 e 47 + 11 = 58 f 33 + 32 = 65

2 a 89 b 78 c 79 d 79 e 97
f 86 g 96 h 65 i 88

3 a 24 + 31 = 55 b 31 + 24 = 55
c The order of addition does not matter. Numbers added together in any order give the same answer.
No, 31 – 24 does not equal 24 – 31.

## 2:15

1 a 5 + 12 = 17 b 13 + 32 = 45

2 a 27 b 43 c 37 d 53
e 48 f 39 g 48 h 29

3 a 4 tens and 7 ones b 4 tens and 9 ones
c 3 tens and 9 ones

## 2:16

1 a 8 × 2 = 16 b 3 × 10 = 30 c 2 × 4 = 8

2 a 15 b $10 \times 9 = 90$

3 a yes b 8 c 8 d yes

4 a 4 b 4 c 50 d 50 e 70
f 20 g 20 h 10 i 10 j 40

5

| | | | | | | | | | |
|---|---|---|---|---|---|---|---|---|---|
| 1 | 2 | 3 | 4 | 5 | 6 | 7 | 8 | 9 | 10 |
| 11 | 12 | 13 | 14 | 15 | 16 | 17 | 18 | 19 | 20 |
| 21 | 22 | 23 | 24 | 25 | 26 | 27 | 28 | 29 | 30 |
| 31 | 32 | 33 | 34 | 35 | 36 | 37 | 38 | 39 | 40 |
| 41 | 42 | 43 | 44 | 45 | 46 | 47 | 48 | 49 | 50 |

## 2:17

1 a 7 × 5 = 35 b 4 × 10 = 40 c 5 × 5 = 25
d 8 × 5 = 40 e 6 × 5 = 30

2 a 18 b 20 c 80 d 14 e 4 f 12 g 30
h 10 i 15 j 3 k 70 l 0

3 a 10 b 35 c 45 d 0 e 6
f 45 g 70 h 40 i 20 j 9

4

| | | | | | | | | | |
|---|---|---|---|---|---|---|---|---|---|
| 1 | 2 | 3 | 4 | 5 | 6 | 7 | 8 | 9 | 10 |
| 11 | 12 | 13 | 14 | 15 | 16 | 17 | 18 | 19 | 20 |
| 21 | 22 | 23 | 24 | 25 | 26 | 27 | 28 | 29 | 30 |
| 31 | 32 | 33 | 34 | 35 | 36 | 37 | 38 | 39 | 40 |
| 41 | 42 | 43 | 44 | 45 | 46 | 47 | 48 | 49 | 50 |
| 51 | 52 | 53 | 54 | 55 | 56 | 57 | 58 | 59 | 60 |
| 61 | 62 | 63 | 64 | 65 | 66 | 67 | 68 | 69 | 70 |
| 71 | 72 | 73 | 74 | 75 | 76 | 77 | 78 | 79 | 80 |
| 81 | 82 | 83 | 84 | 85 | 86 | 87 | 88 | 89 | 90 |
| 91 | 92 | 93 | 94 | 95 | 96 | 97 | 98 | 99 | 100 |

The table of fives could stop at 50.
The table of tens should be both yellow and blue (it may look green).
The table of twos could stop at 20.

5

| × | 0 | 1 | 2 | 3 | 4 | 5 | 6 | 7 | 8 | 9 | 10 |
|---|---|---|---|---|---|---|---|---|---|---|---|
| 0 | 0 | 0 | 0 | 0 | 0 | 0 | 0 | 0 | 0 | 0 | 0 |
| 1 | 0 | 1 | 2 | 3 | 4 | 5 | 6 | 7 | 8 | 9 | 10 |
| 2 | 0 | 2 | 4 | 6 | 8 | 10 | 12 | 14 | 16 | 18 | 20 |
| 3 | 0 | 3 | 6 | | | 15 | | | | | 30 |
| 4 | 0 | 4 | 8 | | | 20 | | | | | 40 |
| 5 | 0 | 5 | 10 | 15 | 20 | 25 | 30 | 35 | 40 | 45 | 50 |
| 6 | 0 | 6 | 12 | | | 30 | | | | | 60 |
| 7 | 0 | 7 | 14 | | | 35 | | | | | 70 |
| 8 | 0 | 8 | 16 | | | 40 | | | | | 80 |
| 9 | 0 | 9 | 18 | | | 45 | | | | | 90 |
| 10 | 0 | 10 | 20 | 30 | 40 | 50 | 60 | 70 | 80 | 90 | 100 |

## 2:18

1 a 6, 6 b 15, 15 c 30, 30
d 3, 3 e 0, 0 f 18, 18

2 a 0, 7, 18, 50, 9, 25, 40, 12, 24, 70
b 0, 6, 9, 100, 6, 20, 50, 27, 21, 80
c 0, 9, 24, 18, 15, 15, 20, 21, 27, 60
d 0, 3, 21, 10, 12, 10, 30, 24, 18, 90

3 a 6 b 21 c 3 d 12 e 15 f 24
g 25 h 30 i 27 j 30 k 9 l 18
m 40 n 15 o 45 p 33

## 2:19

1 a 10, 5, 10, 5 b 18, 6, 18, 6 c 25, 5, 5
d 40, 4, 4 e 35, 7, 7 f 16, 8, 8

2 a 9 b 6 c 4 d 7 e 5 f 8 g 10

3 a 6 b 7 c 9 d 7 e 7 f 10 g 8

## 2:20

A: 65c, B: $2.35, C: $1.15, D: 65c, E: $1.30, F: $2.35, G: 35c, H: $2.60

1 a D b B c G d H

2 a $1.30 b $3.65 c $1.50

3 a C b E c A d G

A: $15, B: $50, C: $20, D: $20, E: $60, F: $30, G: $200, H: $60

## 2:21

1 Coloured coins:
a 50c, 20c, 20c b 50c, 20c, 10c, 5c
c 50c, 20c, 20c, 10c d $1, 20c, 10c, 10c, 10c, 10c
e $1, 50c, 20c, 20c, 5c f $1, 20c

2 Answers will vary. Examples:
a $2, $1, 50c, 20c, 10c b $2, 50c, 20c, 20c, 5c
c $2, $2, 20c, 10c, 5c

## 2:22

1 a 8, 8, 8 b 7, 7, 7 c 9, 9, 9
d 6, 6, 6 e 19, 19 f 28, 28

## 2:23

1 a 5 b 8 c 5 d 10 e 10 f 16

2 a $26 + $57 – $57 = $26, $26 b 75 + 68 – 68 = 75, 75
c 35 – 16 + 16 = 35, 35
d $83 – $37 + $37 = $83, $83

3 a 28 b 45 c 91 d 40

4 a 32 b 92 c 29 d 89 e 7 f 29

## 2:24

1 a 4 b 3 c 2 d 3 e 4
f 7 g 7 h 4 i 5

2 a 3 b 2 c 6 d 2 e 5 f 6

3 a 6 pieces each b 4 pieces each c 3 pieces each

4 a 3 mugs each with 1 left over
b 2 mugs each with 2 left over c 2 mugs each

## 2:25

1 a 7, 7, 7 b 3, 3, 3 c 3, 3, 3 d 3, 3, 3

2 a 5, 5, 4, 4 b 4, 4, 2, 2 c 7, 7, 70 d 6, 6, 5

3 a 3 b 4 c 7 pens

## 2:26

1 Coloured coins:
a 50c, $1, 20c, 5c = $1.75 b $1, 20c, 10c, 5c = $1.35
c 50c, 20c, 10c, 5c = 85c

2 b 5c, 10c = 15c c 5c, 20c, 20c = 45c
d 5c, 20c = 25c

## 2:27

1 a $15 b $30 c $60 d $35

2 a

b

c

3 a $25, $35, $50 b $30, $65, $95 c $45, $70, $85

## 2:28

1 a 58 b 78 c 77 d 89 e 96 f 77
g 89 h 98 i $89 j $97 k $79 l $89

2 a 8 tens and 8 ones b 8 tens and 7 ones
c 7 tens and 9 ones d 8 tens and 9 ones
e 8 tens and 9 ones f 7 tens and 8 ones

3 a $96 b 29 c 57 d $77

## 2:29

1 a 36 – 4 = 32 b 55 – 23 = 32 c 47 – 5 = 42
d 68 – 34 = 34 e 76 – 42 = 34 f 59 – 46 = 13

2 a 36 b 31 c 23 d 27 e 35 f 34
g 42 h 34 i $63 j $40 k $4 l $52

## 2:30

1

| | 2 | 5 | 1 | 10 | 3 | 7 | 4 | 9 | 6 | 8 |
|---|---|---|---|---|---|---|---|---|---|---|
| × 5 | 10 | 25 | 5 | 50 | 15 | 35 | 20 | 45 | 30 | 40 |
| × 10 | 20 | 50 | 10 | 100 | 30 | 70 | 40 | 90 | 60 | 80 |

a yes b yes c yes d no e yes
f yes g yes, because 5 is half of 10

2 a 8 + 2 = 10 b 20 + 20 = 40
c 100 + 30 = 130 d 50 + 5 = 55

3

| | 7 | 9 | 8 | 11 | 14 | 21 | 12 | 6 | 13 | 16 |
|---|---|---|---|---|---|---|---|---|---|---|
| × 2 | 14 | 18 | 16 | 22 | 28 | 42 | 24 | 12 | 26 | 32 |

a 20 + 12 = 32 b 60 + 8 = 68 c 400 + 2 = 402

## 2:31

1 a 20, 20 b 4, 4 c 5, 5

2 a 20, 20 b 5, 5 c 4, 4

3 a 24, 24 b 3, 3 c 8, 8

4 a 18, 18 b 9, 9 c 2, 2

5 a 5, 2 b 10, 3 c 7, 2 d 9, 3

6 a 4, 4 b 7, 7 c 5, 5 d 8, 8


 • AUSTRALIAN SIGNPOST MATHS 3 • ISBN 978 1 4886 2184 0

## 2:32

1. **a** 10, 12, 14, 16, 18, 20, 22, 24
   **b** 20, 25, 30, 35, 40, 45, 50, 55
   **c** 12, 15, 18, 21, 24, 27, 30, 33
2. **a** 2, 10, 12, 16, 20 **b** 5, 10, 15, 20 **c** 10, 20
3. **a** 2, 10 **b** 5, 2 **c** 3, 4
4. 4, 9

## 2:33

1. **a** 4, 4 **b** 10, 10 **c** 5, 5 **d** 4, 4
   **e** 9, 9 **f** 6, 6 **g** 7, 7 **h** 8, 8
2. **a** 2, 10 **b** 6, 2 **c** 5, 3 **d** 8, 5
3. **a** 1 **b** 10 **c** 6 **d** 5
   **e** 1 **f** 10 **g** 2 **h** 8
4. **a** 5 **b** 4 **c** 7 **d** 6 **e** 5 **f** 6
   **g** 8 **h** 9 **i** 2 **j** 8 **k** 9 **l** 7
5. **a** 5, 5 **b** no full groups of ten **c** no

## 2:34

1. **a** 25 + 46 = 71 **b** 27 + 44 = 71 **c** 37 + 45 = 82
   **d** 34 + 28 = 62 **e** 35 + 25 = 60 **f** 28 + 39 = 67
2. **a** 62 **b** 81 **c** 74 **d** 92 **e** 82
   **f** 80 **g** 90 **h** 83 **i** 86
3. **a** 53
   **b** 82

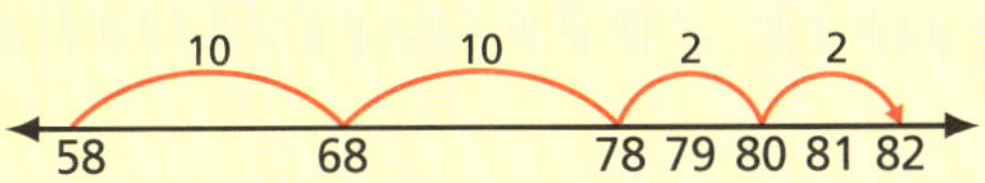


   **c** 67

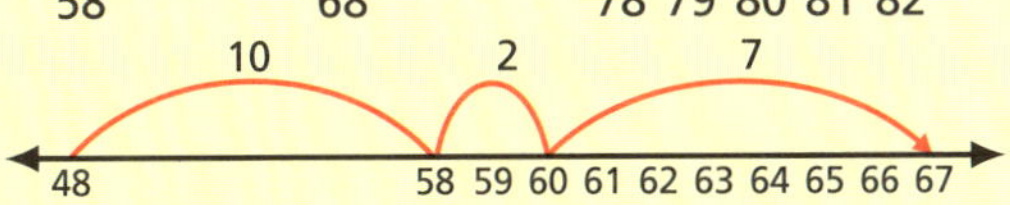


   **d** 91

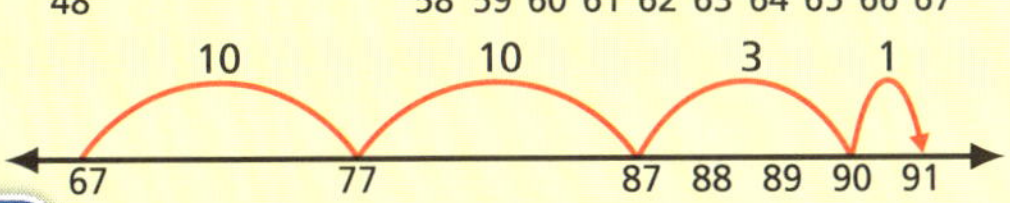


## 2:35

1.

| + | | | | | | | | | |
|---|---|---|---|---|---|---|---|---|---|
| | 24 | 21 | 37 | 43 | 45 | 42 | 30 | 35 | 39 |
| | 32 | 29 | 45 | 51 | 53 | 50 | 38 | 43 | 47 |
| | 39 | 36 | 52 | 58 | 60 | 57 | 45 | 50 | 54 |

2. **a** 12 + 23 = 35 **b** 14 + 36 = 50
   **c** 25 + 17 = 42 **d** 26 + 27 = 53
3. **a** 91 **b** 80 **c** 71 **d** 93
   **e** 59 **f** 77 **g** 72 **h** 84
   **i** 75 **j** 76 **k** 68 **l** 71

## 2:36

1. **a** 31 **b** 23 **c** 41 **d** 55
   **e** 61 **f** 54 **g** 80 **h** 42
   **i** \$94 **j** \$63 **k** \$30 **l** \$77
2. **a** 4 + 29 = 33 **b** 6 + 37 = 43
   **c** 65 + 8 = 73 **d** 57 + 6 = 63

## 2:37

1. **a** 40 **b** 72 **c** 61 **d** 61 **e** 93 **f** 70
   **g** 93 **h** 52 **i** 94 **j** 95 **k** 82 **l** 82
2. **a** \$54 **b** \$96 **c** \$90 **d** \$80

## 2:38

1. 100 − 40 = 60
   60 bones
2. How many pies?
   3 × 10 = 30
   30 pies
3. How many friends?
   24 ÷ 2 = 12
   12 friends
4. How many books?
   24 + 55 = 79
   79 books
5. How many got off?
   15 + 10 + 8 = 33
   33 people got off.
6. How many bottles?
   (3 × 5) + 2 = 17
   17 bottles
7. How many rows?
   45 ÷ 5 = 9
   9 rows
8. **a** 899 **b** 890 **c** 590 **d** 490
   **e** 698 **f** 198 **g** yes

## 2:39

1. How many cards?
   43 + 25 = 68
   68 cards
2. How many books each?
   45 ÷ 5 = 9
   9 books each
3. How many sheep left?
   96 − 28 = 68
   68 sheep left
4. How many cans?
   54 + 37 = 91
   91 cans
5. How many seeds?
   (6 × 5) + 4 = 34
   34 seeds
6. **a** 10 **b** 5 **c** 3

## 2:40

1. **a** 72 + 34 = 106, 106 **b** 92 + 45 = 137, 137
   **c** 74 + 43 = 117, 117 **d** 82 + 54 = 136, 136
2. **a** 107 **b** 136 **c** 127 **d** 119 **e** 188
   **f** 125 **g** 119 **h** 118 **i** 137

## 2:41

1. **a**

133 balls

**b**

| | Tens | Ones |
|---|---|---|
| | 3 | 2 |
| + | 7 | 5 |
| 1 | 0 | 7 |

107 trees

**c**

| | Tens | Ones |
|---|---|---|
| | 6 | 4 |
| + | 8 | 2 |
| 1 | 4 | 6 |

146 boats

**d**

| | Tens | Ones |
|---|---|---|
| | 9 | 6 |
| + | 8 | 3 |
| 1 | 7 | 9 |

179 cards

**2** **a** 14 tens and 7 ones **b** 12 tens and 9 ones
**c** 17 tens and 8 ones **d** 12 tens and 4 ones
**e** 12 tens and 8 ones **f** 11 tens and 7 ones

**3** **a** 125 **b** 138 **c** 167
**d** 154 **e** 127 **f** 108
**g** 133 **h** 119 **i** 119

## 2:42

**1** **a**

| | Tens | Ones |
|---|---|---|
| 1 | 1 | 6 |
| − | 5 | 1 |
| | 6 | 5 |

65 left

**b**

| | Tens | Ones |
|---|---|---|
| 2 | 2 | 5 |
| − | 3 | 4 |
| 1 | 9 | 1 |

$191 left

**c**

| | Tens | Ones |
|---|---|---|
| 3 | 4 | 2 |
| − | 6 | 1 |
| 2 | 8 | 1 |

281 left

**d**

| | Tens | Ones |
|---|---|---|
| 4 | 0 | 8 |
| − | 6 | 4 |
| 3 | 4 | 4 |

$344 left

**2** **a** 7 tens and 6 ones **b** 16 tens and 2 ones
**c** 25 tens and 1 one **d** 49 tens and 3 ones
**e** 59 tens and 2 ones **f** 39 tens and 1 one

**3** **a** 91 **b** 281 **c** 126
**d** 581 **e** 380 **f** 673
**g** 455 **h** 780 **i** 886

## 2:43

**1** **a** 162 + 41 = 203, 203 **b** 171 + 28 = 199, 199

**2** **a** 287 **b** 437 **c** 226 **d** 317 **e** 524 **f** 507
**g** 615 **h** 745 **i** 724 **j** 927 **k** 628 **l** 529

**3** **a** 822 **b** 415 **c** 334 **d** 617 **e** 842 **f** 607

## 2:44

**1** **a** 22 **b** $45 **c** 38 **d** 16

**2** **a** 19 km **b** 34 km **c** 28 km

**3** **a** 12 km **b** 74 km **c** 62 km
**d** 35 km **e** 39 km **f** 27 km

## 2:45

**1** **a** 20 **b** 10 **c** 3 **d** 30 **e** 8 **f** 60

**2** **a** 7, 14, 14 **b** 3, 15, 15 **c** 40, 4, 4
**d** 20, 4, 4 **e** 10, 5, 5 **f** 10, 20, 20

**3** **a** 6 **b** 10 **c** 30 **d** 4 **e** 8
**f** 50 **g** 9 **h** 7 **i** 45 **j** 6

## 2:46

**1** **a** 20c, 20c, $1, $20, $10, $31.40
**b** 5c, 20c, 10c, 50c, $2, $10, $12.85
**c** 5c, 10c, 50c, $2, $1, $5, $8.65
**d** 50c, $5, $20, $10, $50, $85.50
**e** 20c, 50c, $2, $1, $5, $8.70
**f** 5c, 20c, 20c, $2, $5, $10, $17.45

**2** **a** 10c, 50c, $5, $5.60
**b** 10c, $10, $10.10
**c** 5c, 20c, 50c, $2, $2, $10, $14.75

**3** Answers will vary. Examples:
5c, $1, $1, $2, $2, $2
5c, 50c, 50c, $1, $1, $5
5c, 10c, 20c, 20c, 50c, $2, $5

## 2:47

**1** **a** $1.40 **b** $2.55

**2** **a** $6.20 **b** $5.35 **c** $7.85 **d** $4.75 **e** $9.10
**f** $4.00 **g** $8.70 **h** $0.55 **i** $10.00

**3** **a** $3.75 **b** $1.85 **c** $2.10 **d** $6.75 **e** $4.00
**f** $8.25 **g** $1.60 **h** $3.55 **i** $3.00

**4** **a** $11.95 **b** $30.90 **c** $4.65

## 2:48

**1** **a** $10 **b** $20 **c** $25 **d** $40 **e** $30 **f** $35

**2** **a** 30 **b** 50 **c** 90 **d** 100

**3** 20

**4** **a** 500c or $5 **b** 400c or $4 **c** 600c or $6

**5** **a** 10 + 3 = 13 **b** 10 + 4 = 14
**c** 30 − 6 = 24 **d** 600 − 6 = 594

**6** 2 × 10, 10 × 2, 1 × 20, 20 × 1, 4 × 5, 5 × 4

**7** 1 × 30, 30 × 1, 2 × 15, 15 × 2, 3 × 10, 10 × 3, 5 × 6, 6 × 5

## 2:49

**1** **a**

| Tens | Ones |
|---|---|
| 2 | 14 |
| ~~3~~ | ~~4~~ |

**b**

| Tens | Ones |
|---|---|
| 4 | 17 |
| ~~5~~ | ~~7~~ |

**c**

| Tens | Ones |
|---|---|
| 3 | 16 |
| ~~4~~ | ~~6~~ |

**d**

| Tens | Ones |
|---|---|
| 6 | 12 |
| ~~7~~ | ~~2~~ |

**e**

| Tens | Ones |
|---|---|
| 5 | 13 |
| ~~6~~ | ~~3~~ |

**f**

| Tens | Ones |
|---|---|
| 7 | 15 |
| ~~8~~ | ~~5~~ |


 • *AUSTRALIAN SIGNPOST MATHS 3* • ISBN 978 1 4886 2184 0

2 a 18 b 16 c 29 d 49
e 68 f 57 g 28 h 37

## 2:50

1 a

| Tens | Ones |
| --- | --- |
| 3 | 15 |
| ~~4~~ | ~~5~~ |

b

| Tens | Ones |
| --- | --- |
| 1 | 17 |
| ~~2~~ | ~~7~~ |

c

| Tens | Ones |
| --- | --- |
| 2 | 11 |
| ~~3~~ | ~~1~~ |

d

| Tens | Ones |
| --- | --- |
| 6 | 12 |
| ~~7~~ | ~~2~~ |

2 a 38 b 54 c 28 d 19
e 14 f 29 g 25 h 57

## 3:01

1 a C b C c A
2 a A, B, C b B, A, C c C, B, A
3 a G b B c G
d G e Y f Y, R, B, G
4 a B b C c yes

## 3:02

1 a a quarter past 6 or 15 past 6
b half past 9 or 9 thirty
c 1 o'clock
d a quarter to 2 or 1 forty-five
e a quarter past 2 or 2 fifteen
f half past 5 or 5 thirty
g 6 o'clock
h a quarter to 9 or 8 forty-five

2 a

b

c

d

e 8:15 f 2:30 g 12:45 h 7:00

## 3:03

1 a 5 past 4 b 20 past 1 c 10 past 11 d 25 past 7
e 20 past 5 f 5 past 9 g 10 past 6 h 25 past 9
2 a 50 past 7, 10 to 8 b 40 past 3, 20 to 4
c 55 past 1, 5 to 2 d 35 past 4, 25 to 5

## 3:04

1 a 4:10, 10 past 4 b 7:25, 25 past 7
c 12:20, 20 past 12 d 4:05, 5 past 4
e 10:25, 25 past 10 f 2:35, 35 past 2
g 11:45, 45 past 11 h 8:50, 50 past 8
i 3:55, 55 past 3 j 4:40, 40 past 4
k 35 past 4 l 40 past 2
m 25 past 8 n 45 past 3
o 5 past 5 p 10 past 6
q 35 past 12 r 50 past 11
s 5 past 7 t 55 past 9

## 3:05

1 a 10:25, 25 past 10
b 11:15, 15 past 11 (quarter past 11)
c 2:20, 20 past 2 d 9:10, 10 past 9
e 4:05, 5 past 4
f 5:45, 45 past 5 (quarter to 6)
g 2:35, 35 past 2 h 4:40, 40 past 4
i 7:50, 50 past 7 j 1:55, 55 past 1
k 30 past 9 (half past 9) l 25 past 2
m 40 past 5 n 45 past 8 (quarter to 9)
o 55 past 6

2 a
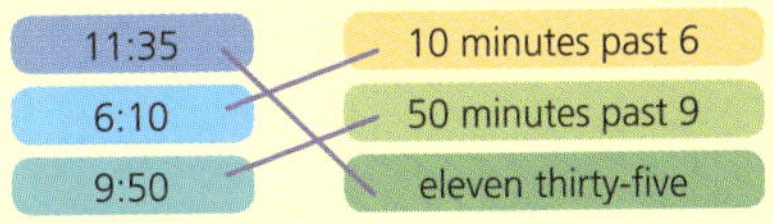

b
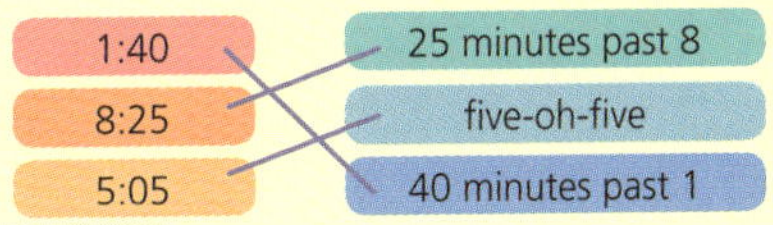


## 3:06

1 Answers will vary.
2 Answers will vary.
3 Some answers may vary.
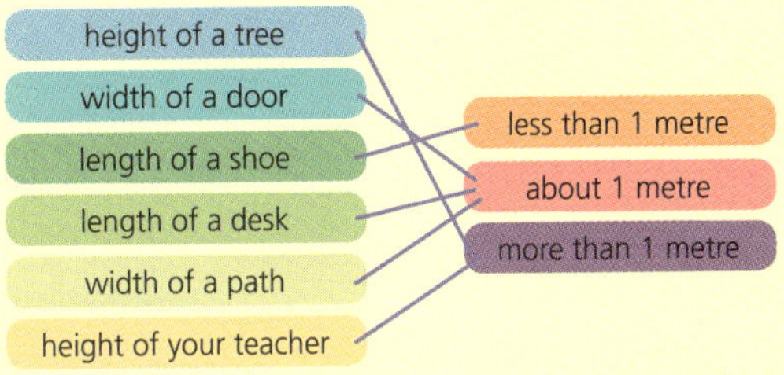


## 3:07

1 Answers will vary.
2
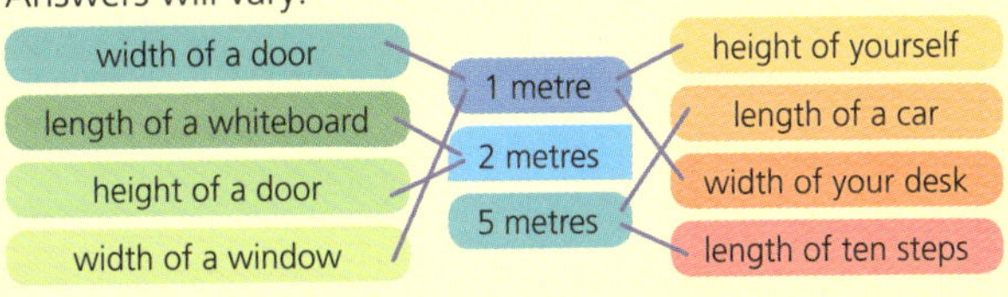

3 Answers will vary.

## 3:08

1 a 9 cm b 6 cm c 10 cm
2 a 5 cm b 9 cm c 4 cm d 7 cm
3 a 7 cm b 6 cm c 3 cm d 5 cm e 4 cm

## 3:09

1 a 7 cm b 4 cm c 8 cm
d 3 cm e 2 cm f 9 cm
2 a
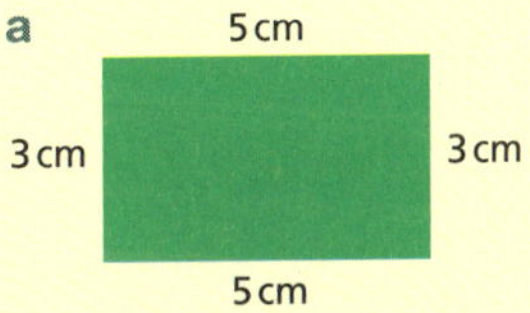

b
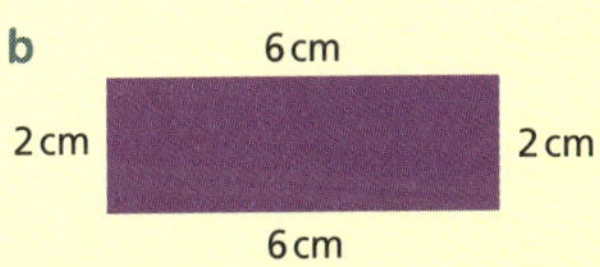

c
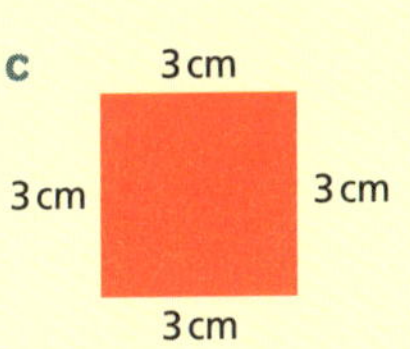

d
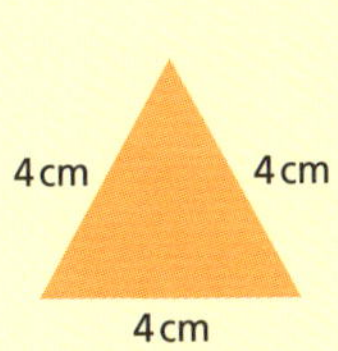

e
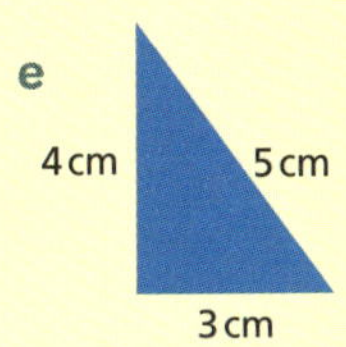

f The shapes in **a** and **b** are the longest all the way around.

## 3:10

1 a 
b 
c 

2 a 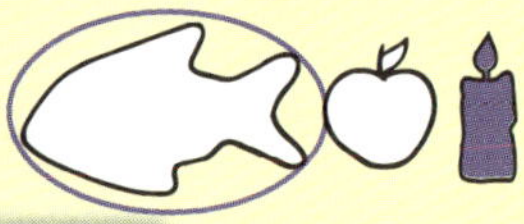
b 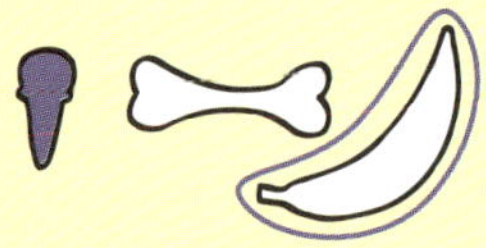

## 3:11

1 a 9 b 12 c 12 d 6
2 Answers will vary. d B e C
3 a 7, 5, 8 b E c I d I e E

## 3:12

1 

2 Answers will vary.
3 Answers will vary.

## 3:13

Answers will vary.

## 3:14

1 Answers will vary.
2 Answers will vary.
3 **a**, **c**, **e**, **f**, **g** and **j** should be circled.
4 **a**, **b** and **e** should be circled.

## 3:15

1 a 3 kg b 10 kg c 8 kg d 5 kg e 7 kg
f 11 kg g 12 kg h 20 kg i 16 kg
2 a yes b yes c no d yes e yes f no
g yes h yes i no j yes k yes l no
3 a 4 kg b 10 kg c 29 kg d 27 kg
e 23 kg f 8 kg g 45 kg h 21 kg
4 a yes b no c yes d yes e no
f yes g yes h yes i no j yes

## 3:16

1 a 7 to 2 b 16 to 1 c 18 to 10 d 3 to 9
e 18 to 12 f 27 to 5 g 5 to 2 h 12 to 6
2 a 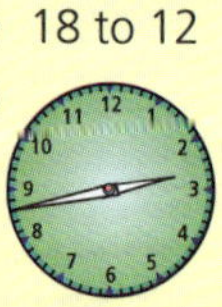 b  c  d 

3 a 16 to 6 b 11 to 9 c 21 to 3
d 16 to 10 e 21 to 11 f 6 to 7
4 a 8 to 6 b 18 to 11 c 23 to 9
d 13 to 12 e 23 to 8 f 3 to 12

## 3:17

1 a 22 past 7 b 3 to 7 c 18 past 12 d 6 past 10
e 12 to 1 f 8 past 12 g 22 to 4 h 7 to 8
2 a 27 past 6 b 18 past 7 c 10 past 8 d 4 past 12
e 26 past 4 f 27 to 6 g 19 to 2 h 6 to 5
i 18 to 12 j 1 to 7

## 3:18

1 a 1 m 70 cm b 3 m 36 cm c 4 m 26 cm
d 7 m 13 cm e 6 m 83 cm f 5 m 42 cm
g 9 m 59 cm h 8 m 77 cm
2 a m b cm c cm d m e cm f cm g m h cm

## 3:19

1 a 126 cm b 173 cm c 251 cm
d 297 cm e 316 cm f 342 cm
g 168 cm h 206 cm i 330 cm
2 a 2 m 38 cm b 2 m 13 cm c 3 m 58 cm
d 3 m 7 cm e 1 m 49 cm f 2 m 27 cm
3 a 18 cm b 12 cm c 10 cm
d 14 cm e 16 cm f 6 cm

## 3:20

1 31
2 a Tuesday b Saturday c Wednesday
3 a 7/5/12 b 25/5/12 c 30/5/12
4 a 23 b 8
5 a 26th and 27th May b Monday 30th April
c Friday 1st June

## 3:21

1 a 55 minutes past 2 is the same as 5 to 3, so it is nearly 3 o'clock.
b 52 minutes past 7 is the same as 8 to 8, so it is nearly 8 o'clock.
c 48 minutes past 11 is the same as 12 to 12, so it is nearly 12 o'clock.
2 a cook a sausage b sing a song
c watch a movie d make your bed
3 Answers will vary. Example:
a 4 b 2 c 1 d 3
4 a 3:30, 5:30 b 6:00, 6:30
5 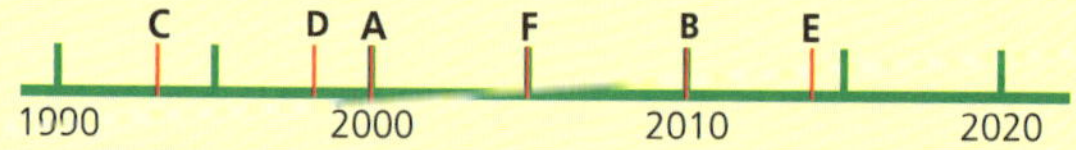

Greg's Life

## 3:22

1 a 24 mm b 16 mm c 31 mm d 62 mm e 47 mm
2 a 16 mm b 27 mm c 12 mm d 45 mm
e 39 mm f 43 mm g 36 mm h 24 mm
3 a 1 cm 8 mm b 4 cm 7 mm c 2 cm 7 mm
d 1 cm 3 mm e 3 cm 4 mm f 5 cm 3 mm
g 2 cm 1 mm h 3 cm 2 mm i 4 cm 5 mm

## 3:23

1. a 28 mm b 58 mm c 66 mm d 37 mm e 18 mm f 74 mm
2. a 29 mm b 14 mm c 23 mm d 54 mm e 43 mm
3. a 34 mm b 16 mm c 65 mm d 88 mm e 47 mm f 92 mm g 73 mm h 29 mm
4. a 1 cm 9 mm b 6 cm 2 mm c 3 cm 7 mm d 7 cm 1 mm e 4 cm 5 mm f 8 cm 6 mm g 5 cm 8 mm h 2 cm 3 mm i 9 cm 4 mm j 9 cm 9 mm k 6 cm 9 mm l 8 cm 1 mm

## 3:24

1. a 24, 12, 32 b Yes (even though the number of informal units is different in each case)
2. a 19, 16, 18 b M, N, L
3. They each have the same area, 20 squares.

## 3:25

1. a 12 b 12 c 6 d 7
2. a b c d

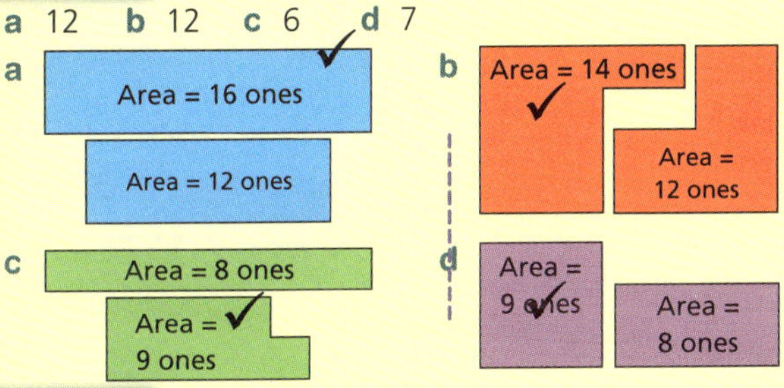


## 3:26

1. a 40 g b 100 g c 600 g d 300 g e 80 g f 50 g g 20 g h 70 g i 400 g
2. a 300 g b 800 g c 700 g d 600 g e 400 g f 700 g

## 3:27

1. a 60 g b 100 g c 40 kg d 30 g e 9 kg f 70 g g 80 g h 7 kg i 200 g j 10 kg k 60 kg l 50 kg
2. a g b g c kg d kg e g

## 3:28

1. a 1 cm, 10 mm b 10 cm, 100 mm
2. Answers will vary. Example: Because centimetres are always the same but hand spans vary from person to person.

## 3:29

1. a yes b no c no d yes e no f no g yes h no i no j no k yes l yes
2. a no b yes c yes d no e yes f yes g no h no i yes

## 3:30

1. a no b yes c no d yes e yes f no g yes h no i yes
2. a 3 L b 5 L c 1 L d 6 L e 9 L f 4 L g 250 mL h 400 mL i 750 mL j 300 mL k 900 mL l 500 mL
3. a 1 000 mL b 2 000 mL c 5 000 mL d 3 000 mL e 8 000 mL f 6 000 mL
4. a 1 L b 4 L c 7 L

## 3:31

1. a 2 L b 4 L c 1 L d 10 L e 7 L f 5 L g 19 L h 20 L i 14 L
2. a C b D c C d D e A f 16 L

## 3:32

1. a 30 b Thursday c Saturday d Thursday e 2nd November f 6th November g 27th November h 24th November i 5th November, 12th November, 19th November, 26th November j 4 k Wednesday
2. a 7 b 14 c 31 d 30 e 31 f 31 g 30 h 31 i 31 j 30 k 30 l 31 m 31 n 28

## 3:33

1. 

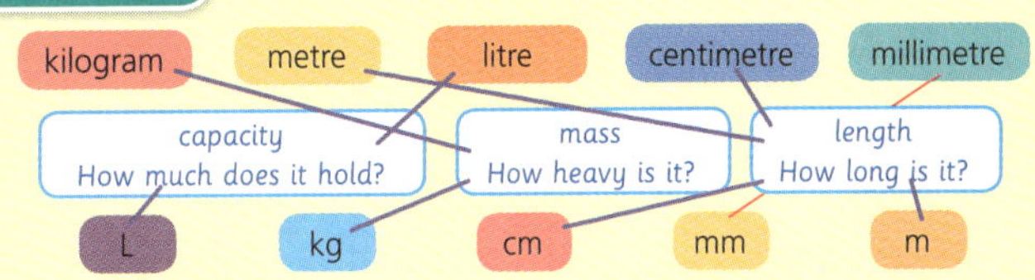


2. a capacity b mass c length
3. Answers will vary.

## 4:01

1. a C, E, H b F c B, G d D
2. E, H
3. a C b H c A d D
4. a D, F b A, F c C, H d 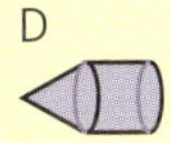
5. a 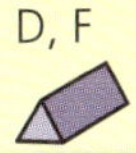 b 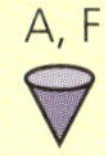

## 4:02

1. 

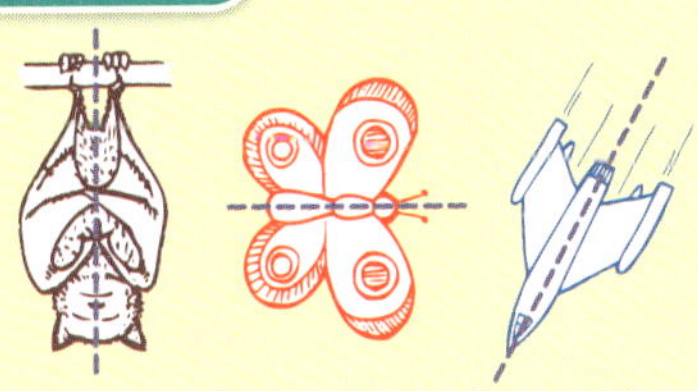

2. 

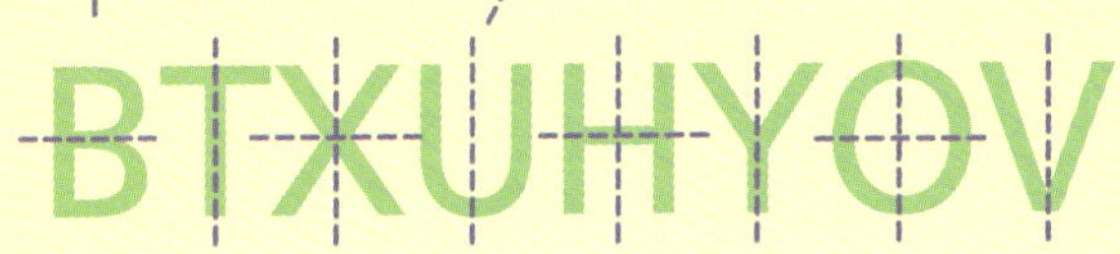


3. 

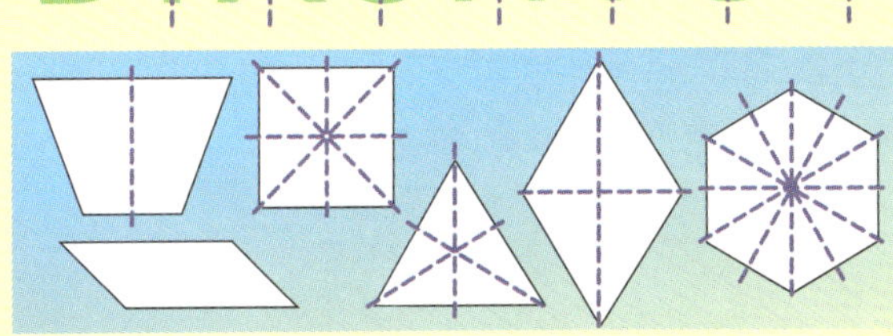

## 4:03

1. a 6, cube b 2, cylinder c 1, cone d 0, sphere e 5, (triangular) prism
2. a square b circle c circle d circle e triangle

3 a e b a c a d c e e
f e g d h b i b, c, d

4 a 2 b 1 c 3 d 4

## 4:04

1 

2 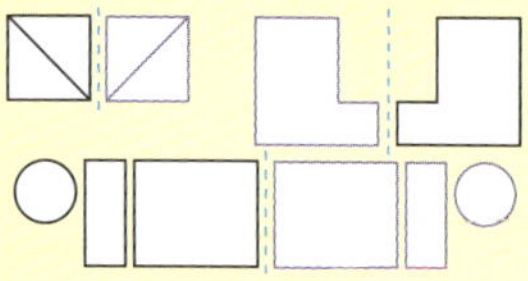

3 flip 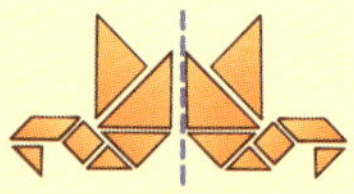

## 4:05

1 a 0 b 1 c 3 d 0 e 3 f 0 g 1 h 3

2 a, d, f. The lines are going in the same direction and will not meet.

3 Where the lines meet, they form a right angle (square corner).

4 square, rectangle

5 a Chen, Alan b Chen, Alan c Chen, Alan

6 Answers will vary.

## 4:06

1 a triangle 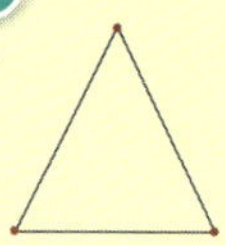 b square 

c pentagon 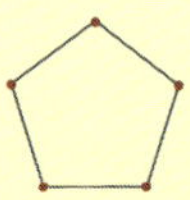 d hexagon 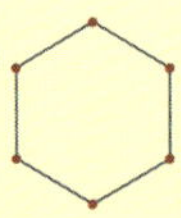

2 b (square), d (hexagon)

3 a triangle 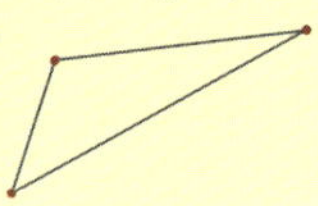 b quadrilateral 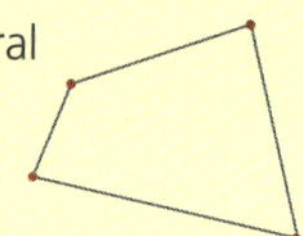

c pentagon 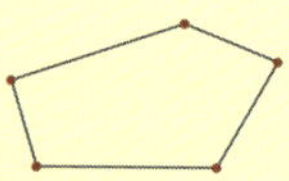 d hexagon 

4 a 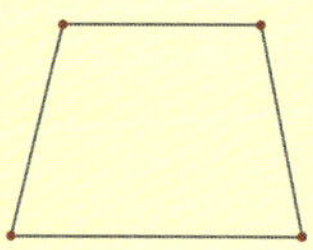 b 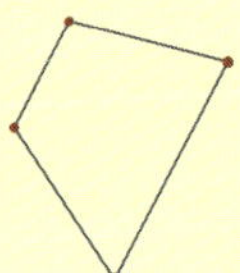 c 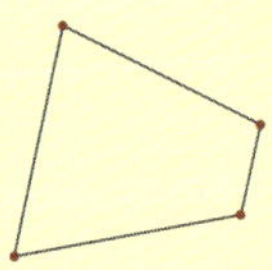

5 a 3 b yes
c E F H M N W Z (and perhaps I)

6 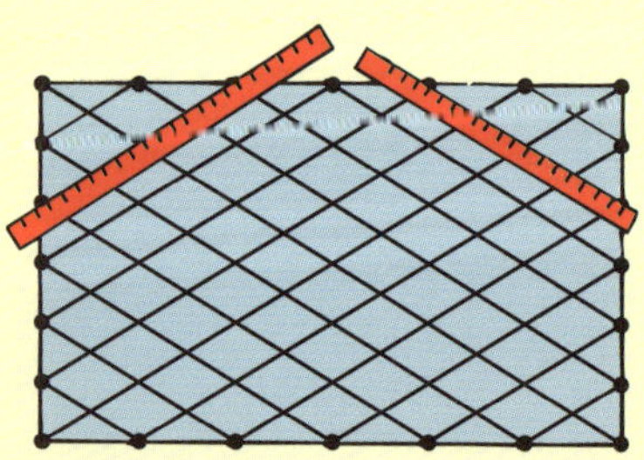

parallelogram or rhombus or diamond

## 4:07

1 a turn left, then first left
b turn right, then first left, then first right
c turn right, then first right, then first left
d turn right, then first right, then second right

2 a G (Post office) b H (Hospital)

3 a H b E c N d E

## 4:08

1 a D b A c A d C e C f F
g Answers will vary. Example: 3 right, 7 up.

2 a 3 up, 2 right, 4 up b 2 up, 1 left, 5 up
c 6 up, 1 left, 3 down, 1 left, 4 up
d 5 down, 1 right, 2 down e 4 down, 2 left, 3 down

3 a 3 forwards
b 1 forwards, turn right, 2 forwards, turn left, 4 forwards
c 1 forwards, turn right, 6 forwards, turn right, 2 forwards
d 2 forwards, turn left, 2 forwards, turn right, 2 forwards
e 2 forwards, turn left, 2 forwards, turn left, 4 forwards

## 4:09

1 a D b A c E d C e F f B

2 a E b A, B, F c C d D

3 a A, C, F b A, B, C, F

4 a rectangle, triangle b triangle, triangle
c hexagon, square
d circle, rhombus (or diamond)

5 a 2 b 23 c 9 d 5

## 4:10

1 a 3, 3 b 4, 4 c 3 d 4

2 a 3
b any quadrilateral: square, rectangle, rhombus, trapezium, parallelogram, kite
c 5 d hexagon

## 4:11

1 J

2 L

3 a H, I b E, G c C d D e yes

## 4:12

1 a

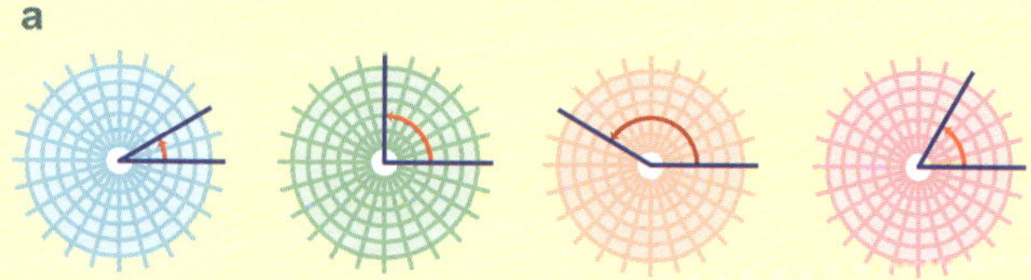

1 3 4 2

b

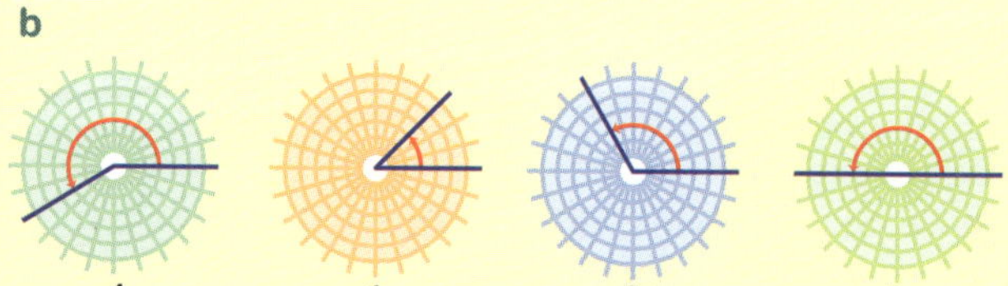

4 1 2 3


 ISBN 978 1 4886 2184 0

2 a 

b 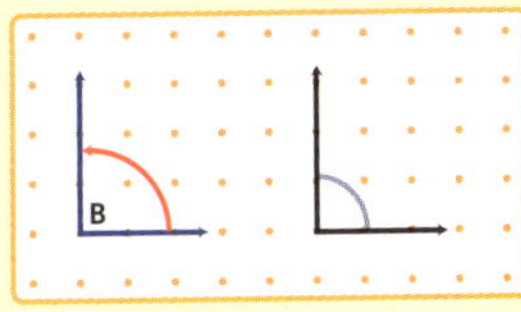


c B is a right angle because the two arms makes a square corner.

## 4:13

1 a C, E, F (Note: F is a parallelogram, which is a kind of trapezium.) b F

2 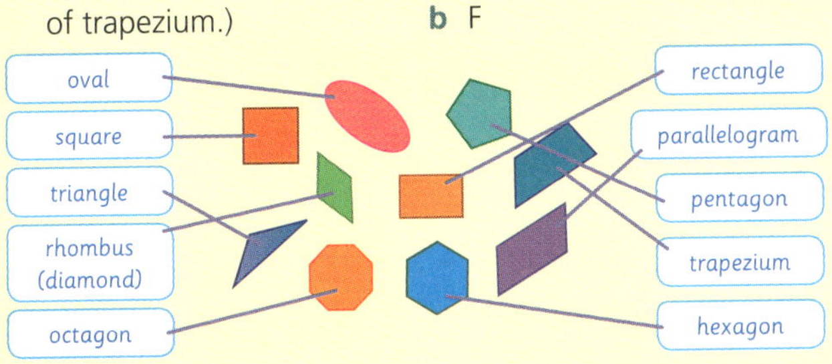


3 Answers will vary.

## 4:14

1 a C, G b A, B, H, J c E, I d A, C, D, E, F, J e H, I
2 a A, J b F
3 triangle (C, G)
4 a Answers will vary.
b 8 (including 4 squares leaning at an angle)
c 9 d 12
5 Each shape has a curved side.

## 4:15

1 A, D, E, G
2 B
3 yes

## 4:16

1 a clockwise, quarter turn b anticlockwise, half turn
c anticlockwise, quarter turn d clockwise, half turn
e clockwise, three-quarter turn
f anticlockwise, three-quarter turn

2 a  b 

c yes

## 4:17

1 a pencil b duck c doll d ball
e right side of top shelf f left side of top shelf
g middle of middle shelf
2 a saucers b plates c mugs d pots
e left (or fourth from the right) on the top
f second from the right (or third from the left) on the bottom
g right (or fourth from the left) on the bottom
h right (or fourth from the left) on the top
3 a Tran b Sophie c Mark d Felicity
e second from the left (or fourth from the right) on the top
f fourth from the left (or second from the right) on the bottom

## 4:18

1 a

| Moves |
|---|
| 3 right |
| 4 up |
| 1 left |
| 2 up |
| 3 left |
| 3 up |
| 4 right |
| 2 down |

b

| Moves |
|---|
| 5 left |
| 9 up |
| 2 right |
| 7 down |
| 2 right |
| 4 up |
| 1 right |
| 3 up |

2 a 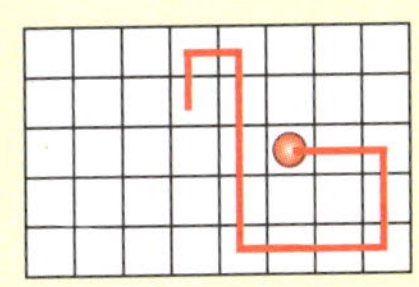 b 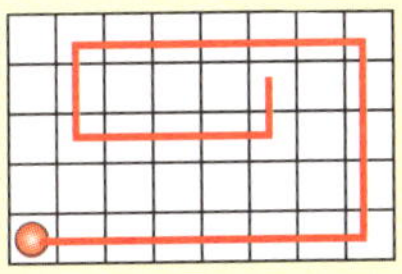

3 H C K B J F E A I D G O

## 4:19

1 a cylinder b neither c prism d prism e neither
f prism g cylinder h neither i neither j neither
2 a Y b X c Z d Y
3 a 8 b 18 c 12
4

| Prism | Shape of Each Face |
|---|---|

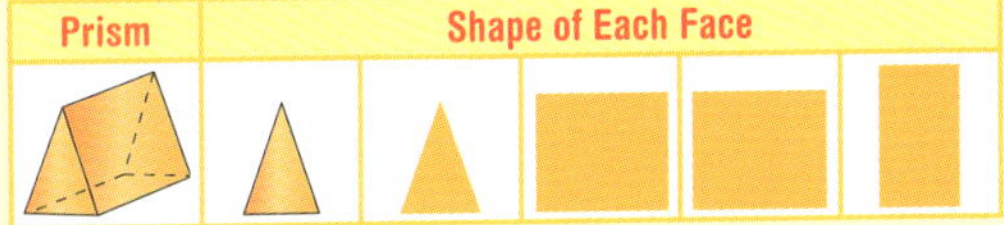

## 4:20

1 b, d, e, f, g, i
2 a A b C c A
3 a 5 b 8 c 5 d 4
4

| Pyramid | Shape of Each Face |
|---|---|

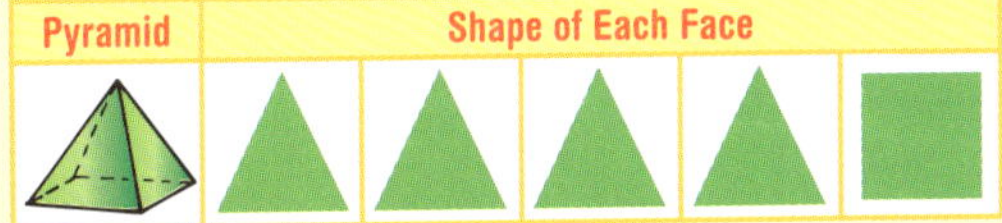

## 4:21

1 a 37 b 3 c cupboard d on the back wall
2 Answers will vary.

## 4:22

1 Koala
2 Echidna
3 Platypus
4 Wallaby

## 4:23

1 a B, E, I b D, F, H c B, D, E, F, H, I
2 a false b true c false
d false e true f false

## 4:24

1 a pyramid b cone c cylinder d sphere
e prism f cylinder g prism h pyramid

2 a 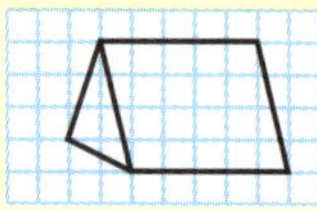 b 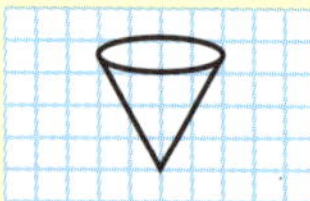 c 
d 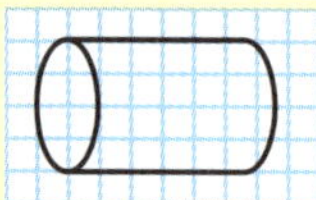 e 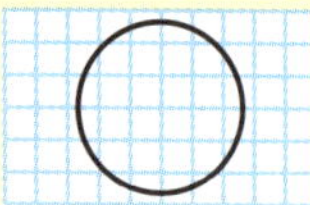 f 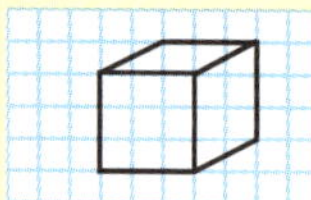

## 4:25

1 a flip b slide c flip d turn e slide
f flip g turn h flip i slide j turn
k a, c, f and h are symmetrical. They are all made by flips.

## 4:26

1 a flip  b slide 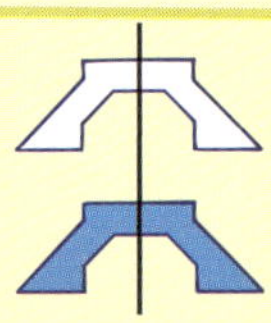
c turn (no lines of symmetry) 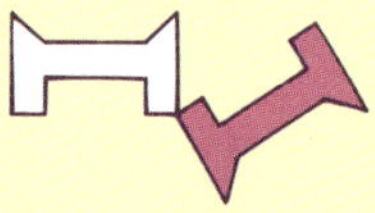

2 a  b 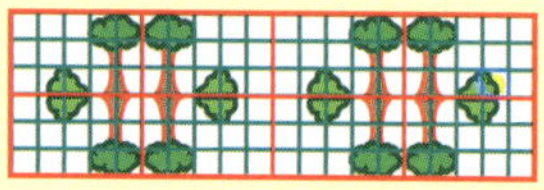
c yes

3 a 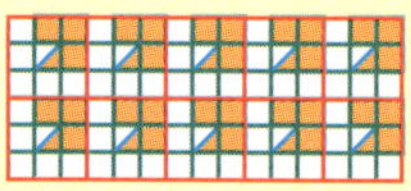 b 
c no

4 a 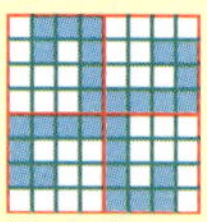 b 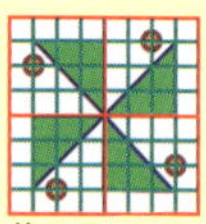 c no

5 no (the shapes in a jigsaw are not the same shapes repeated)

6 a yes b yes c no d yes

## 4:27

1 6

2  

3 B

## 5:01

1 a 2 b 7 c 4 d 20
e Answers will vary. f Answers will vary.

2 a 13 b Silver c 30 d 5
e 48 f Answers will vary. g Answers will vary.

3

| | |
|---|---|
| Birds | 7 |
| Lions | 4 |
| Monkeys | 5 |

## 5:02

1 a blue b orange c 8 d 30

2 a 825 b 46 c 166 d 72 e 88
f 201 g 405 h 193 i 388

3 a 87 b 27 c 23 d 0
e Answers will vary.

## 5:03

1 a 1 b 3 c Fish and Mice d Birds e 16

2

| Pets Kept in the Classroom | | | |
|---|---|---|---|
| Birds | Fish | Mice | Turtles |
| 3 | 6 | 6 | 1 |

| Pets Kept in the Classroom | |
|---|---|
| Birds | 3 |
| Fish | 6 |
| Mice | 6 |
| Turtles | 1 |

a 3 b 16 c Answers will vary.

d 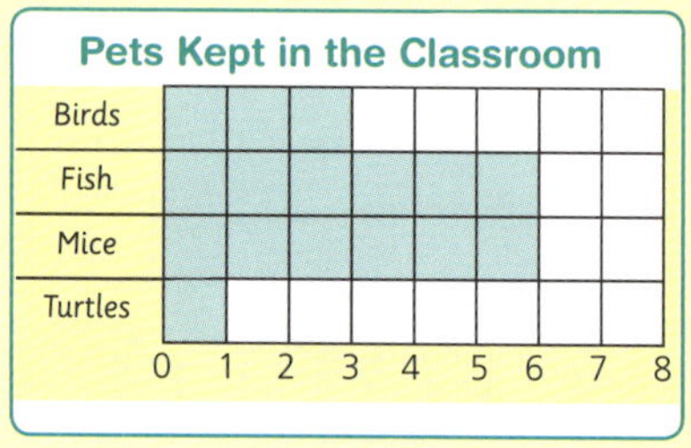


3 a 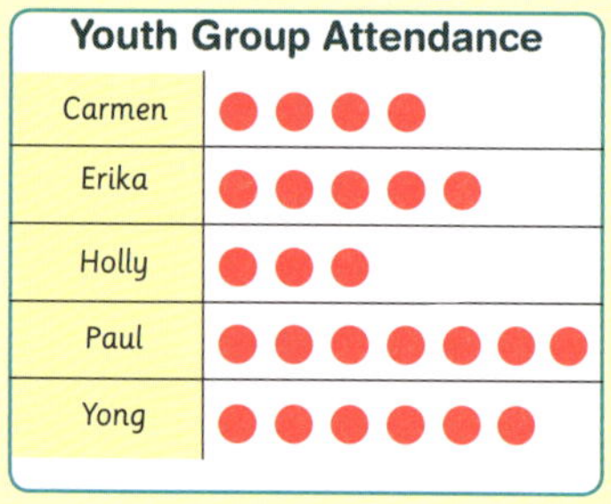


b Holly c yes d 8 times

4 4

5 6

6 39

## 5:04

1 a no b yes

2 a yes b yes
c no (but there is an equal chance)

3 a blue b no c no

4 a C b B c A, D, E d yes
e no f no g yes

## 5:05

1 a red b yellow c no d yes
e blue or yellow f blue, red or yellow

2 a red b yellow

3 a Yes. There is an even chance of spinning blue or red.
b 5

4 a yes b 2 c yes

## 5:06

1. a 6 b Tuesday c 20 d 10
2. a 7 b 6 c 8 d Chinese
3. a 21 b 15 c 10 d 47

## 5:07

1.

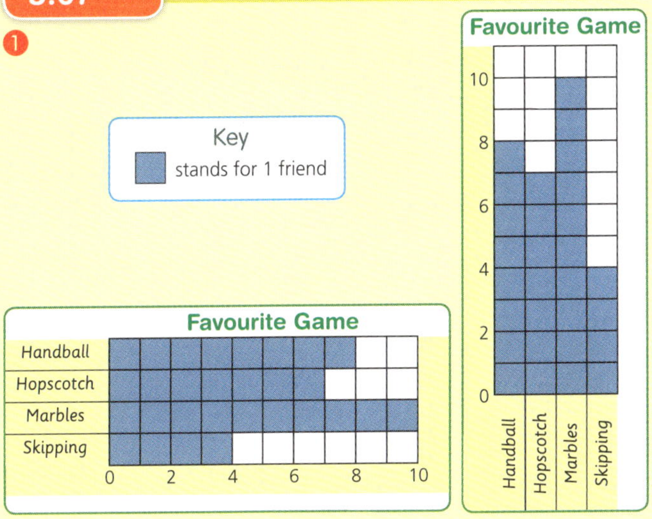


2. Answers will vary.
3.

| Dice | Tally | Number |
|---|---|---|
| 1 | \|\|\| | 3 |
| 2 | \|\|\| | 3 |
| 3 | \|\| | 2 |
| 4 | \|\|\|\| | 4 |
| 5 | \|\|\|\|/ | 5 |
| 6 | \|\|\| | 3 |

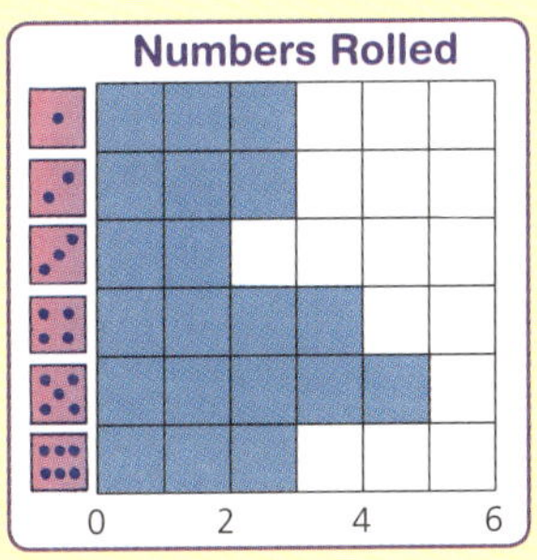


## 5:08

1. a 6 b 100 c 70 d 10 e 184 f 380
2. a Amanda b Juan c 10
   d 8 e 33 f 24
3. a 14 b 18 c Tuesday and Thursday
   d 26 e 8
   f No, because some students may have practised more than once during this week.

## 5:09

1. Indigenous peoples
2. hunting, fishing, gathering
3. Dreamings
4. 9
5. 4
6. 7
7. 3
8. 1
9.

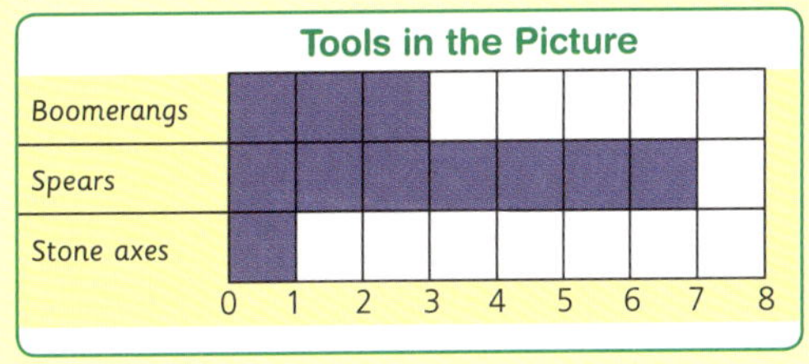


## 5:10

1. a 400 b 200 c 2008 d 1400 e 50 pears
2. a 16 b 12 c Thursday d Monday and Saturday
   e 12 f 74 g The shop was closed.
3. a 20 b 2 pm to 3 pm c 4 pm to 5 pm d 125
   e That is the time when most people are driving home from work.

## 5:11

1.

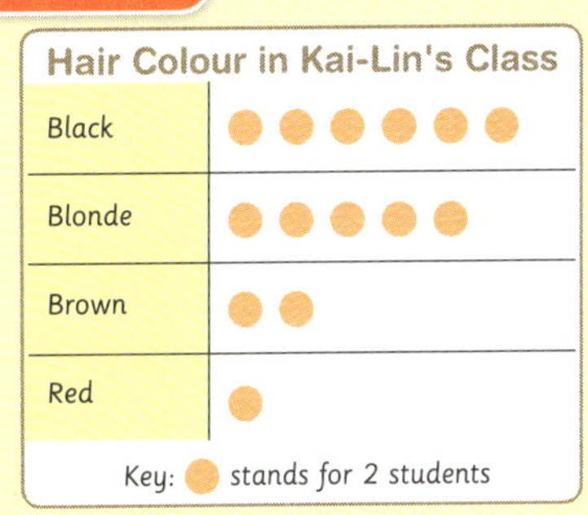


   a Black b none c 6 d 28
   e To tell the value of each picture on the graph.
   f Answers will vary. Example: They are easy to draw and read and are attractive.
2. Answers will vary. Examples:
   a It is useful for recording information as it is observed.
   b It looks impressive and interesting.
   c It is easy to draw and to read.

## 5:12

1. a unlikely b impossible c certain d very likely
   e very likely f very likely g unlikely h impossible
   i certain j even chance
2.

| Will always happen | Will sometimes happen | Will never happen |
|---|---|---|
| c i | a d e f g j | b h |

3. Answers will vary. Example: E A C D B

## 5:13

Answers will vary.

# Diagnostic Test 1

## Number

see page

❶ **a** Circle the even numbers. 2

36 77 89 64 22
51 18 90 47

**b** How many odd numbers are between 27 and 54? ☐

❷ **a** What number is modelled? 4

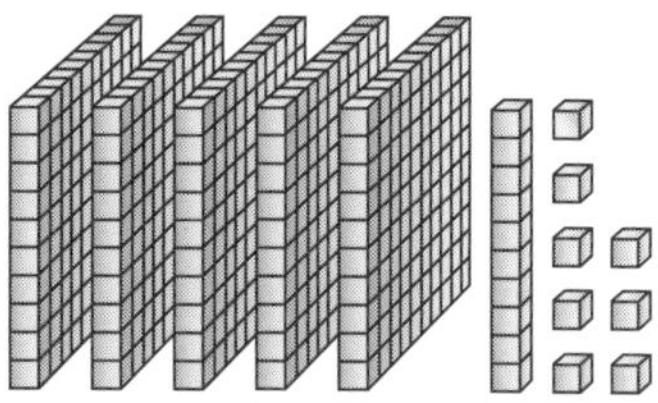

☐

**b** Write the number in words and complete the numeral expander.

☐

☐

❸ What number is shown on the abacus? 4

☐

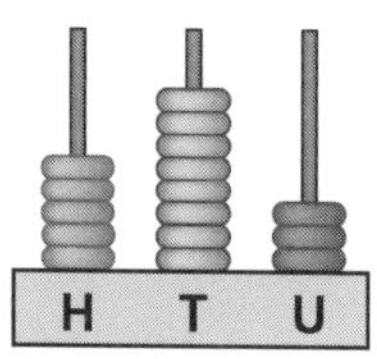


❹ Circle the smaller number. 4

**a** 796 or 439

**b** 916 or 961

Order these from smallest to largest.

**c** 631, 943, 425 ☐

**d** 849, 884, 852 ☐

❺ Skip count to complete these. 1

7 | 14 | ☐ | ☐ | ☐

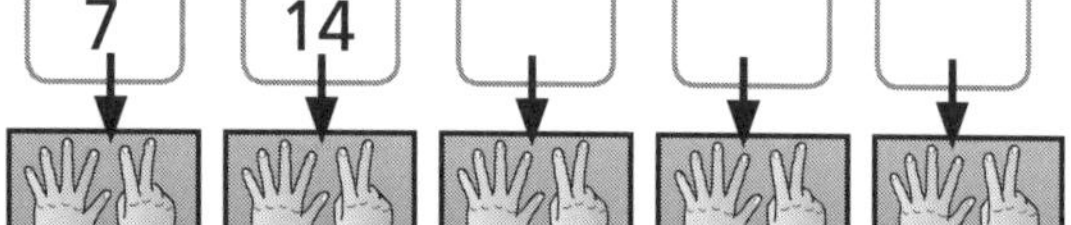

## Operations

see page

❻ 38

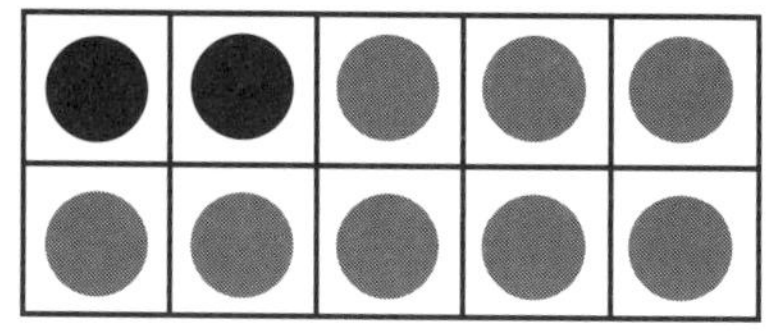

**a** 2 + 8 = ☐

**b** 8 + 2 = ☐

**c** 10 – 2 = ☐

**d** 10 – 8 = ☐

❼ Use the best strategy to complete. 39

**a** 5 + 8 = ☐

**b** 7 + 8 = ☐

**c** 6 – 4 = ☐

**d** 9 – 8 = ☐

❽ 40

| | 3 | 10 | 2 | 1 | 5 | 4 | 7 | 9 | 8 | 6 |
|---|---|---|---|---|---|---|---|---|---|---|
| × 2 | 6 | 20 | | | | | | | | |

❾ Write the value of each coin in the rectangle. 37

**a** ☐ **b** ☐ **c** ☐

❿ Write the value of our bank notes in order from smallest to largest. 37

$5.00 | ☐ | ☐
☐ | ☐



 • *AUSTRALIAN SIGNPOST MATHS 3* • ISBN 978 1 4886 2184 0

# Diagnostic Test 1 (continued)

## Measurement

see page

**11** Which is shorter: 87

**a** G or R?

**b** B or Y?

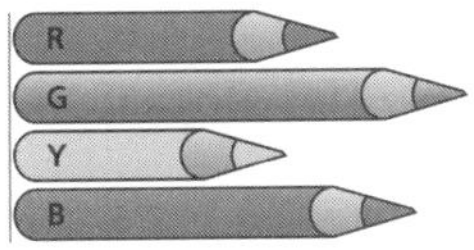


**c** Which pencil is the longest?

**12** Write the time using 'to' or 'past'. 88

**a**

**b**

Show the time given.

**c**

Quarter to 11

**d**

Quarter past 8

**e**

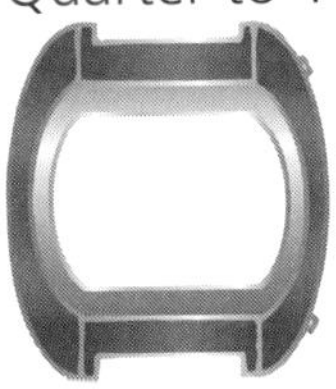

Quarter to 1

**f**

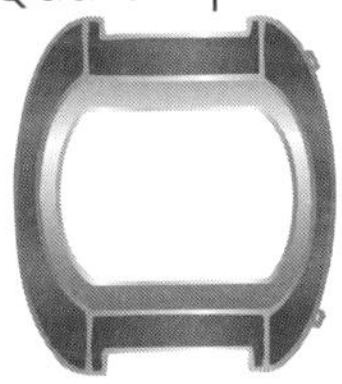

Quarter past 12

**13** Record the time shown. 89

**a**

past

**b**

past

Show the time given.

**c**

to

**d**

to

## Geometry

see page

**14** 120

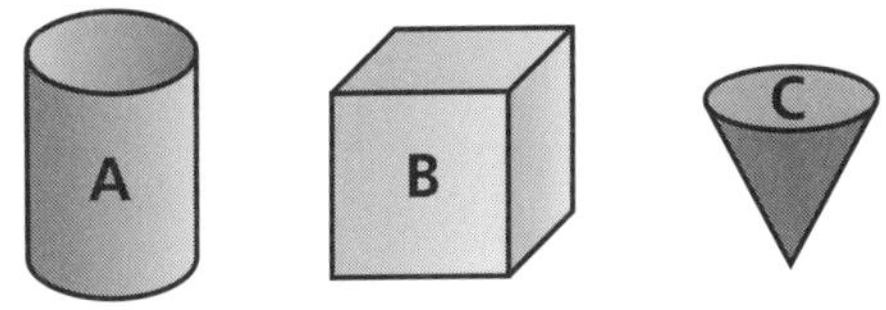


Which objects can be easily stacked?

**15** 122

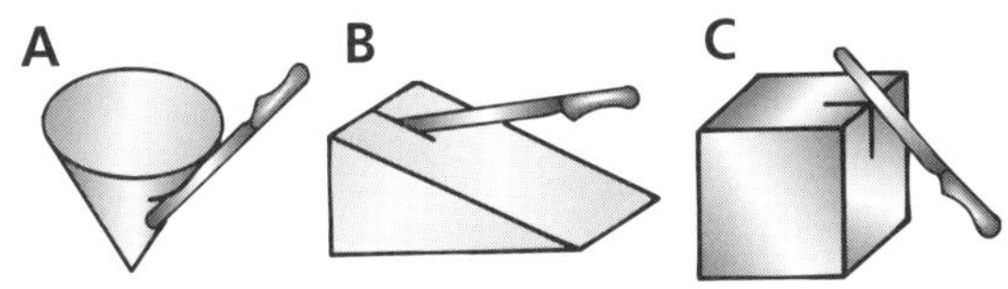


How many surfaces has:

**a** A? **b** B? **c** C?

What is the shape of the cross-section of:

**d** A?

**e** B?

**16** Draw the lines of symmetry. 121

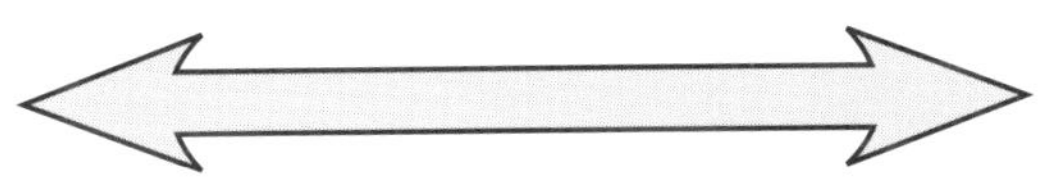

## Statistics

**17** 147

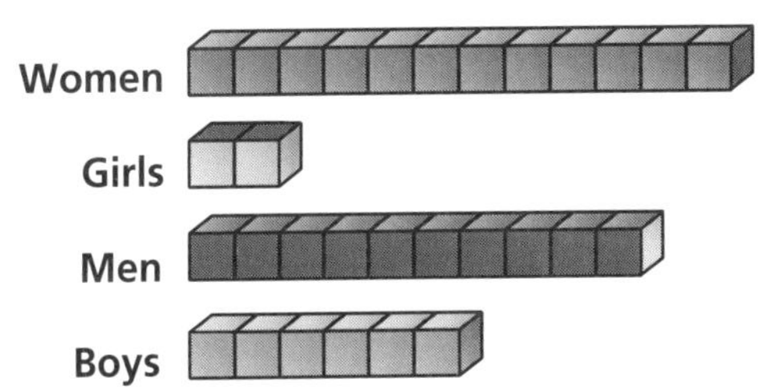


How many:

**a** men? **b** women?

**c** girls? **d** boys?

**e** altogether?

 • *AUSTRALIAN SIGNPOST MATHS 3* • ISBN 978 1 4886 2184 0

# Diagnostic Test 2

## Number

see page

**1** Write the next three numbers. 6

**a** 35, 30, 25, ☐, ☐, ☐

**b** 67, 77, 87, ☐, ☐, ☐

**2** Describe this number pattern. 6

(900) (800) (700) (600)

**3** Complete the pattern on the number line. 7

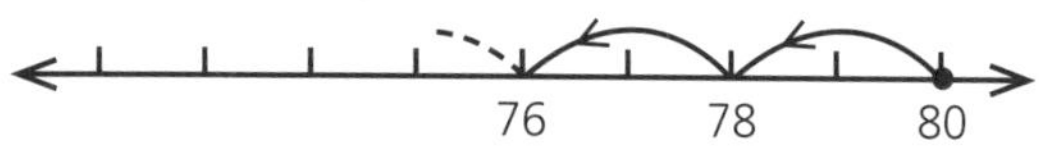


The rule is ☐

**4** 5

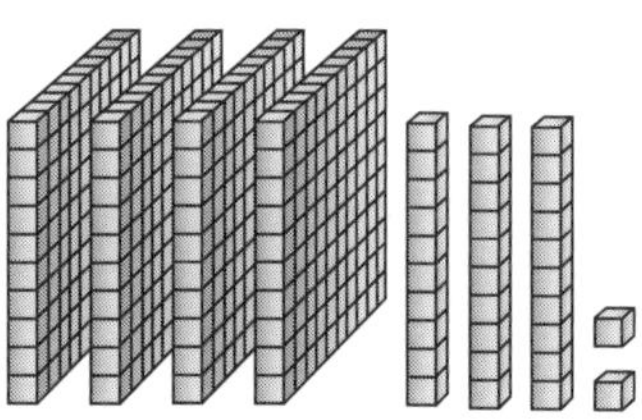

**a** Write the numeral. ☐

**b** Complete the numeral expander.

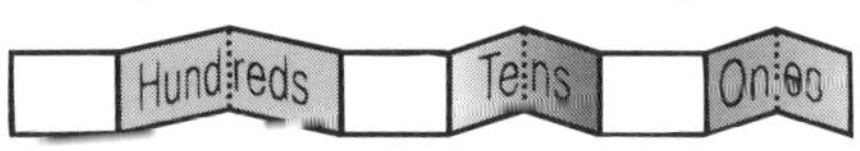


**c** Use words to write 863.

**5** How many digits in: 5

**a** 109? ☐ **b** 65? ☐

Write the number that follows:

**c** 645 ☐ **d** 849 ☐

## Operations

see page

**6** 41

**a**
$$\begin{array}{r} 6 \\ 3 \\ +\ 7 \\ \hline \end{array}$$

**b**
$$\begin{array}{r} 5 \\ 6 \\ 5 \\ +\ 3 \\ \hline \end{array}$$

**c**
$$\begin{array}{r} 7 \\ 4 \\ 3 \\ 5 \\ +\ 6 \\ \hline \end{array}$$

**7** What happens to the tens and the ones when we add 9? 42

| | |
|---|---|
| | 9 |
| 9 + 9 = | 18 |
| 18 + 9 = | 27 |
| 27 + 9 = | 36 |
| 36 + 9 = | 45 |
| 45 + 9 = | 54 |

The tens: ☐

The ones: ☐

Conclusion: to add 9 we can add ☐, then subtract ☐.

**8** Complete the pattern in each part. 43

14 – 8 = 6

24 – 8 = 16

34 – 8 = ☐

☐ – 8 = 36

54 – ☐ = 46

**9** Complete the table. 44

| × | 3 | 10 | 2 | 1 | 5 | 4 | 7 | 9 | 8 | 6 |
|---|---|---|---|---|---|---|---|---|---|---|
| 5 | | | | | | | | | | |
| 10 | | | | | | | | | | |
| 1 | | | | | | | | | | |
| 0 | | | | | | | | | | |



 • *AUSTRALIAN SIGNPOST MATHS 3* • ISBN 978 1 4886 2184 0

## Measurement

see page 92

**10** **a** Is the height of a door closest to:

- **A** 1 metre
- **B** 2 metres or
- **C** 4 metres? ☐

**b** Is the length of your foot:

- **A** less than 1 metre
- **B** about 1 metre or
- **C** more than 1 metre? ☐

93

**11** Write the abbreviation of metre. ☐

90

**12** Write the time shown by completing the labels.

**a**

8: ☐

☐ past ☐

**b**

4: ☐

☐ past ☐

Show the time given.

**c**

5: ☐

☐ past ☐

**d**

2: ☐

☐ past ☐

90

**13** Complete the labels.

**a**

☐ past ☐

**b**

☐ past ☐

## Geometry

see page 124

**14** Which lines are parallel?

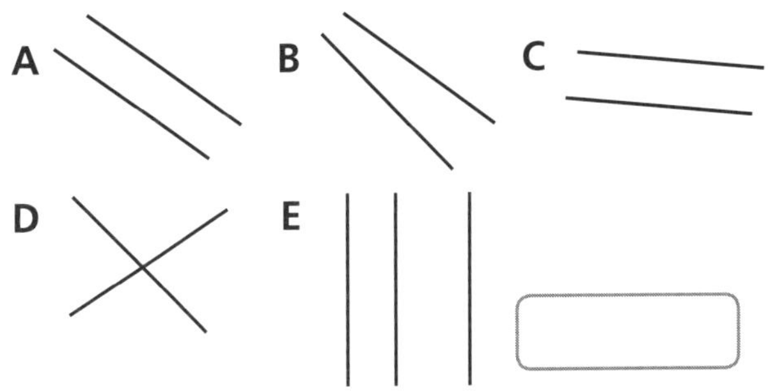


☐

123

**15** Make a symmetrical pattern by completing the picture.

## Statistics

148

**Balloons**

| Red | Green | Blue | Orange |
|---|---|---|---|
| 6 | 10 | 4 | 13 |

**16** How many balloons are:

**a** orange? ☐ **b** red? ☐

**c** blue? ☐ **d** green? ☐

**e** either red or green? ☐

**f** either red or orange? ☐

**g** either green or orange? ☐

**h** How many balloons altogether? ☐

 • *AUSTRALIAN SIGNPOST MATHS 3* • ISBN 978 1 4886 2184 0

# Diagnostic Test 3

## Number

**1** see page 12

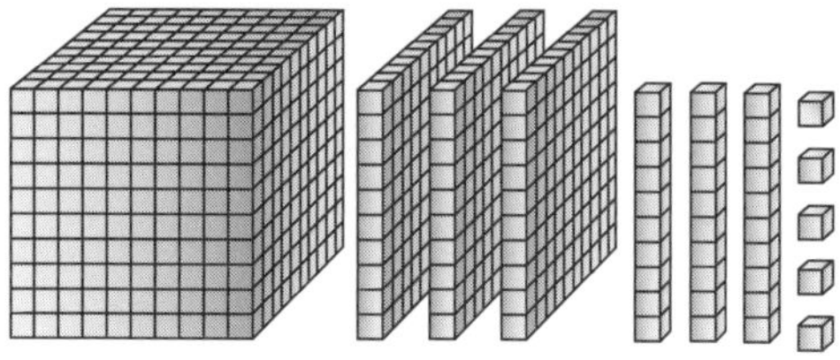

**a** What number is modelled? ☐

**b** Write this number as a numeral. ☐

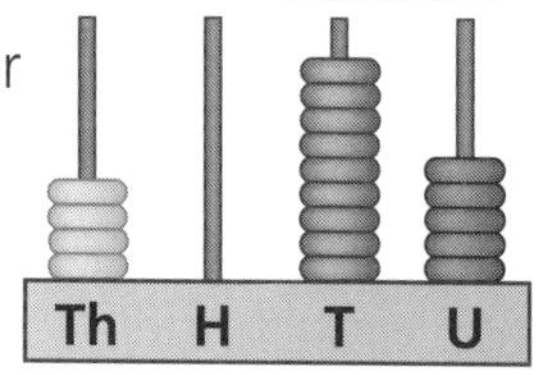


**2** see page 8

Complete the numeral expanders.

479

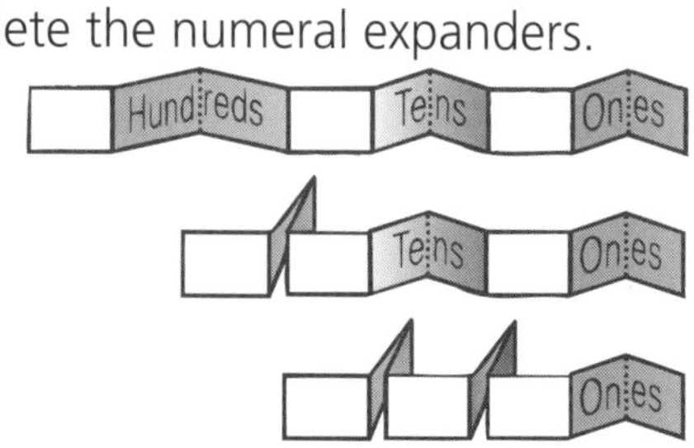


**3** see page 8

**a** Write the numeral for:

six hundred and thirty-two ☐

**b** Write this number in words: 607

☐

**4** see page 9

Round each number to the nearest hundred.

**a** 296 ☐ **b** 324 ☐

**5** see page 9

Use < or > to show the bigger number in each pair.

**a** 743 ☐ 816 **b** 963 ☐ 575

**6** see page 10

What part is shaded?

**a**  ☐ **b** 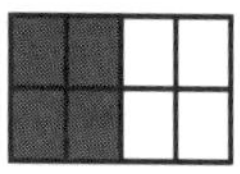 ☐

**c** Colour $\frac{2}{6}$ of this group.

## Operations

**7** see page 50, 51

**a** 52 + 26 = ☐ **b** 28 + 50 = ☐

**c** 33 + 42 = ☐ **d** 37 + 61 = ☐

**e**

| Tens | Ones |
|---|---|
| 3 | 5 |
| +2 | 1 |
| | |

**f**

| Tens | Ones |
|---|---|
| 1 | 6 |
| +2 | 3 |
| | |

**8** see page 51

**a** 4 tens and 6 ones
\+ 2 tens and 2 ones

**b** 5 tens and 4 ones
\+ 3 tens and 3 ones

**c** 1 ten and 5 ones
\+ 6 ten and 2 ones

**9** see page 51

13 horses

32 horses

How many horses altogether?

☐ + ☐ = ☐

**10** see page 46

Use any strategy to find:

**a** 46 + 19 = ☐

**b** 17 + 9 + 3 = ☐

**c** 58 + 7 = ☐ **d** 35 + 41 = ☐

**e** 34 – 6 = ☐ **f** 48 – 23 = ☐

**11** see page 48

**a** 3 × 2 = ☐ **b** 7 × 5 = ☐

**c** 7 × 2 = ☐ **d** 5 × 5 = ☐

**e** 6 × 3 = ☐ **f** 9 × 3 = ☐

**g** 8 × 10 = ☐ **h** 4 × 10 = ☐



 • *AUSTRALIAN SIGNPOST MATHS 3* • ISBN 978 1 4886 2184 0

## Measurement

see page 94

**12** What is the abbreviation for centimetre? ☐

94

**13** What is the length of each bar?

**a**  ☐ cm

**b** 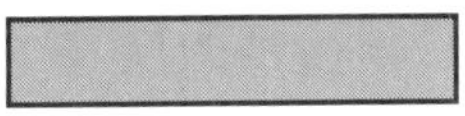 ☐ cm

95

**14** Write the length of each side.

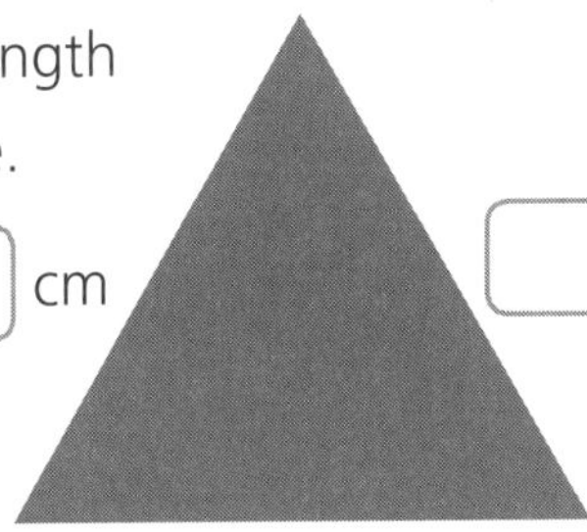

☐ cm ☐ cm ☐ cm

96

**15** Circle the larger area.

## Geometry

126

**16**

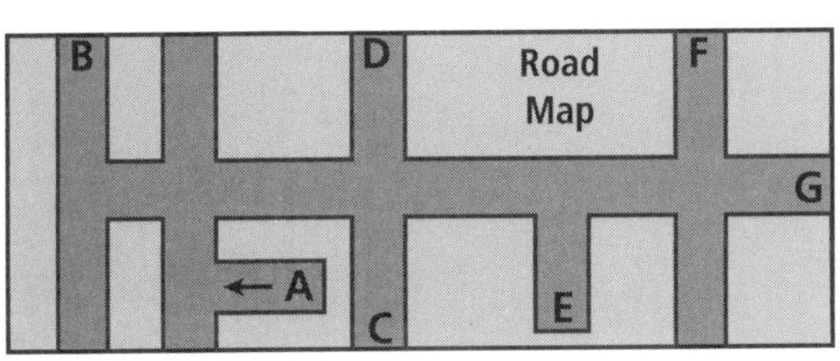


How would I get from **A** to **F**?

☐

see page 127

**17**

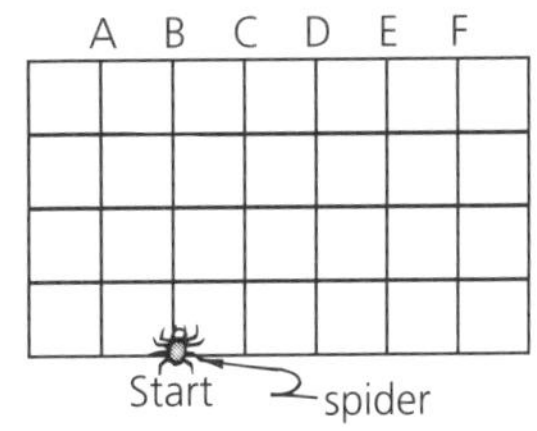


Write the letter where the spider stops if it goes:

**a** 2 up, 4 right, 1 up, 3 left then 1 up. ☐

**b** 1 up, 2 left, 2 up, 4 right then 1 up. ☐

## Probability

151

**18** Without looking, Alan drew a ball from this jar.

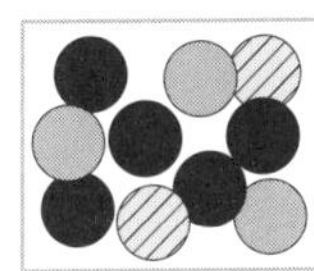

**a** Is black more likely than grey? ☐

**b** Is striped less likely than black? ☐

**c** Is striped less likely than grey? ☐

**d** What colour is least likely? ☐

**e** What colour is most likely? ☐

150

**19** Which spinner is most likely to:

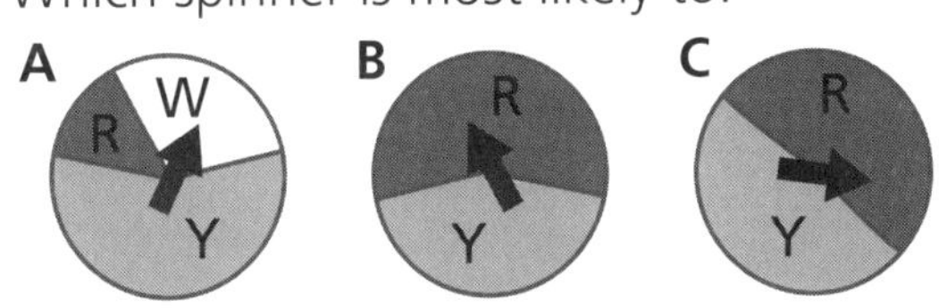


**a** stop on **R**? ☐

**b** stop on **Y**? ☐

**c** stop on **W**? ☐

 • *AUSTRALIAN SIGNPOST MATHS 3* • ISBN 978 1 4886 2184 0

# Diagnostic Test 4

## Number

see page

❶ 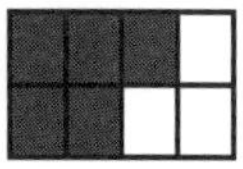

a Part that is shaded? ☐ 14 15

b Part not shaded? ☐

c Is $\frac{1}{8}$ bigger than $\frac{3}{8}$? ☐

d Is $\frac{4}{4}$ equal to 1? ☐

What part is shaded?

e 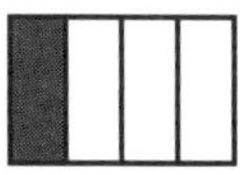 ☐  f 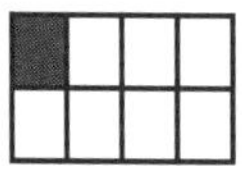 ☐

❷ Use words to describe each number pattern. 16

a 4, 8, 16, 32

☐

b 48, 24, 12, 6

☐

❸ 16

| | | | | | | | | | |
|---|---|---|---|---|---|---|---|---|---|
| 1 | 2 | 3 | 4 | 5 | 6 | 7 | 8 | 9 | 10 |
| 11 | 12 | 13 | 14 | 15 | 16 | 17 | 18 | 19 | 20 |
| 21 | 22 | 23 | 24 | 25 | 26 | 27 | 28 | 29 | 30 |

In your head, count by fives and circle each number counted. Count by tens and put a cross on each number counted. Explain what you noticed.

☐

## Operations

see page

❹ Write a number sentence for '5 rows of 10 packets'. 52

☐ ☐ ☐ ☐ ☐

❺ 53

a $4 \times 10 =$ ☐  b $7 \times 5 =$ ☐

c $3 \times 4 =$ ☐  d $6 \times 3 =$ ☐

❻ 53

a $\begin{array}{r} 4 \\ \times \quad 6 \\ \hline \end{array}$  b $\begin{array}{r} 5 \\ \times \quad 8 \\ \hline \end{array}$  c $\begin{array}{r} 10 \\ \times \quad 9 \\ \hline \end{array}$

❼ 55

a $4 \times 3 =$ ☐

$12 \div 3 =$ ☐

☐ $\times 3 = 12$

b $3 \times 5 =$ ☐

$15 \div 5 =$ ☐

☐ $\times 5 = 15$

❽ Write the value of: 56

a 

☐

b 

☐

❾ Fill in the boxes by using four coins to make: 57

a $2.75 ☐ ☐ ☐ ☐

b $3.15 ☐ ☐ ☐ ☐



 • *AUSTRALIAN SIGNPOST MATHS 3* • ISBN 978 1 4886 2184 0

# Diagnostic Test 4 (continued)

## Measurement

see page

**10** Record the area of each shape. 97

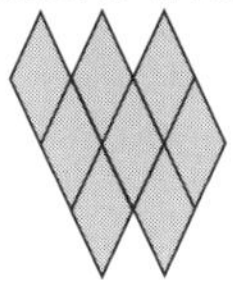
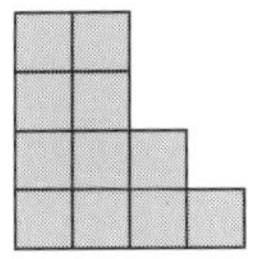
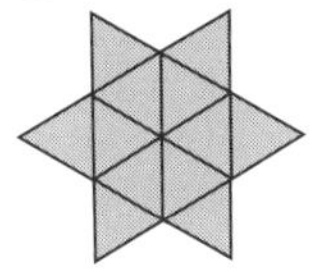

**a** [ ] diamonds **b** [ ] squares **c** [ ] triangles

**d** Is it easy to compare areas using different units? [ ]

**11** Match to show what each holds. 98

| | | |
|---|---|---|
| can of drink | holds less than 1 L | bath |
| bucket | holds more than 1 L | glass |

## Geometry

**12** Name each shape. 128

**a**  [ ]

**b**  [ ]

**c** 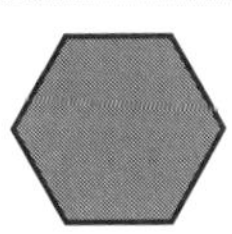 [ ]

**d**  [ ]

**e** 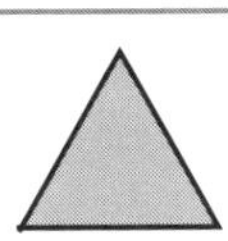 [ ]

**f**  [ ]

**13** How many sides has a: 129

**a** square? [ ] **b** triangle? [ ]

How many corners has a:

**c** triangle? [ ] **d** rectangle? [ ]

see page

**14** 

Which angle is smaller? [ ]

## Statistics

**15** 152

Flags Seen

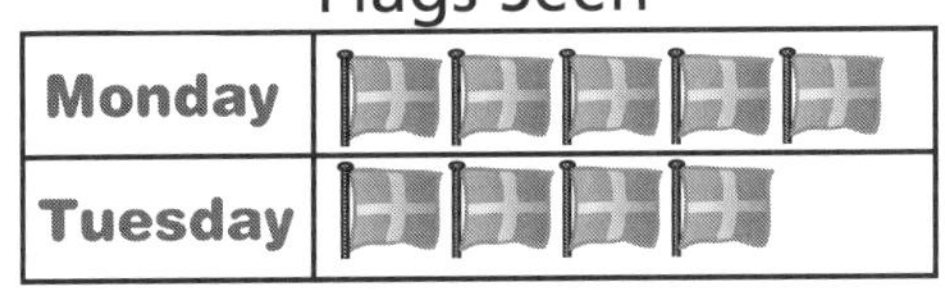


How many flags were seen:

**a** on Tuesday? [ ]

**b** altogether? [ ]

**16** 153

Fruit Chosen

| Apple | Peach | Pear | Plum |
|---|---|---|---|
| 卌 卌 卌 III | 卌 卌 II | 卌 卌 I | IIII |

How many:

**a** apples were chosen? [ ]

**b** pears were chosen? [ ]

**c** choices altogether? [ ]

**17** What is the total of each set of tally marks?

**a** 卌 卌 卌 卌 卌 I [ ]

**b** 卌 卌 卌 卌 卌 卌 III [ ]

**c** 卌 卌 卌 卌 [ ]

**18** Votes for captain: 153

John 卌 III Mark 卌 Chi 卌 II

Complete the column graph.

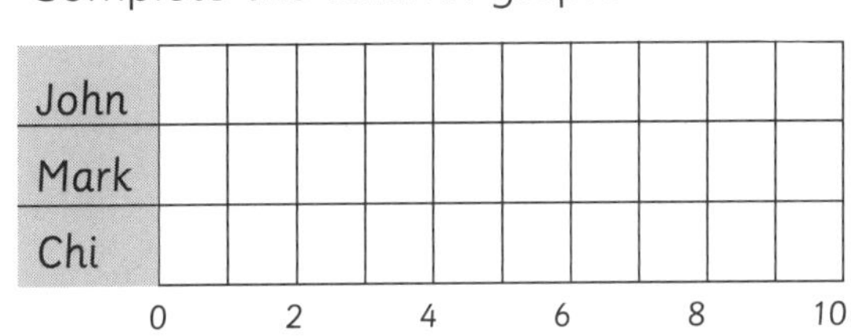


 • *AUSTRALIAN SIGNPOST MATHS 3* • ISBN 978 1 4886 2184 0

# Diagnostic Test 5

## Number

see page

❶ **a** What number is modelled? 17

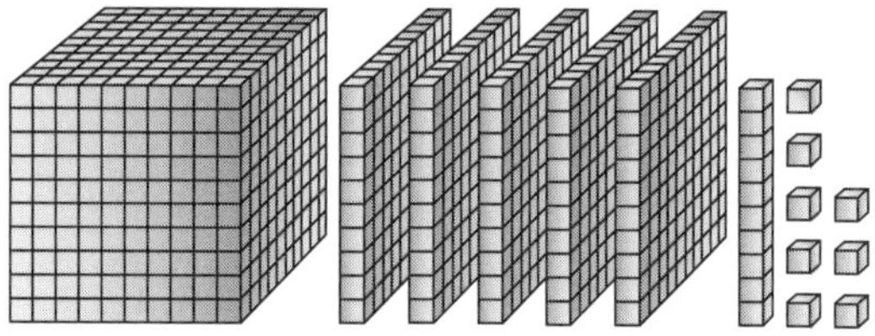

**b** Write the number in words and complete the numeral expander.

❷ Write the numeral for: 17

**a** five hundred and twenty-one

**b** one thousand two hundred and sixteen

❸ How many digits in: 17

**a** 438? **b** 1 076?

❹ Order from the smallest to largest. 18

3 745, 3 457, 3 574

❺ **a** Is 439 > 427 true? 18

**b** Is 2 136 < 2 036 true?

❻ Is 2 469 closer to 2 000 or 3 000? 19

❼ Circle the larger fraction. 20

**a** $\frac{1}{8}$ or $\frac{1}{4}$ **b** $\frac{3}{8}$ or $\frac{3}{4}$

see page

❽ Name the part shaded. 20

Colour to match the fraction.

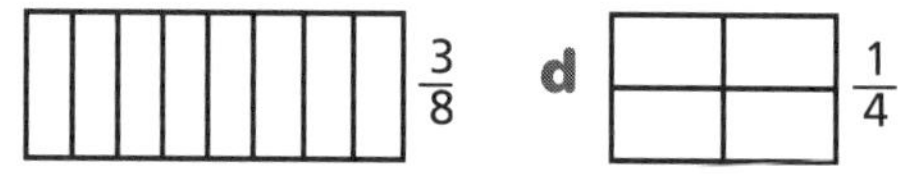

❾ Round to the nearest hundred. 19

**a** 437 **b** 2 583

**c** 5 139 **d** 7 648

## Operations

❿ Find the difference in length between these ribbons. Complete both algorithms. 58

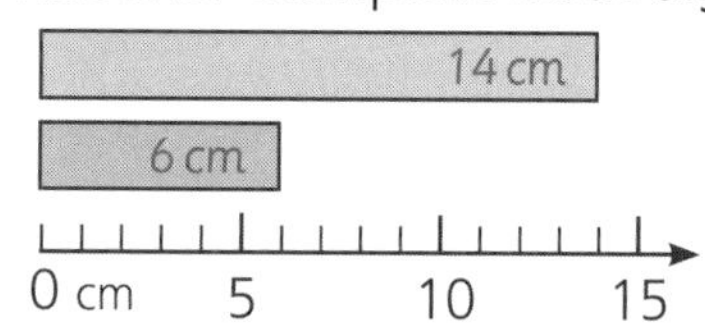


14 – 6 = ☐ so 6 + ☐ = 14

Difference = ☐ cm

⓫ **a** 5 + 3 – 3 = 59

**b** 10 + 8 – 8 =

⓬ 20 shared by 4 = 60

Groups of 4 in 20 =

20 ÷ 4 =

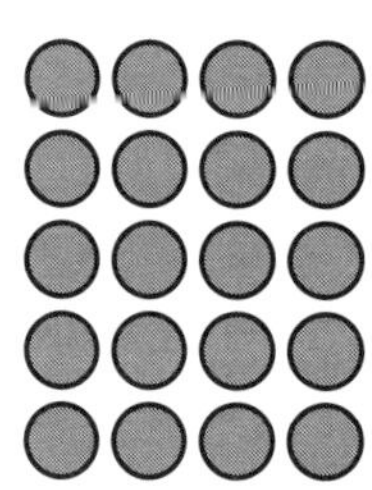

⓭ 62

**a** Colour the change from $2.00 if you spend $1.35. Change =



 • *AUSTRALIAN SIGNPOST MATHS 3* • ISBN 978 1 4886 2184 0

see page 63

b What is the total value of these notes? ______

## Measurement

see page 100

**14** Would these things have a mass of less than one kilogram?

a your maths book ______

b a table ______

see page 101

**15** Use the short form to write:

a 5 kilograms ______

b 9 kilograms ______

see page 104

**16** Use the short form to write:

a 1 metre 25 centimetres ______

b 3 metres 60 centimetres ______

Would you use centimetres (cm) or metres (m) to measure:

c the height of a tree? ______

d the length of a pen? ______

see page 102

**17** Write the analogue time for:

a

______

b

______

## Geometry

see page 132

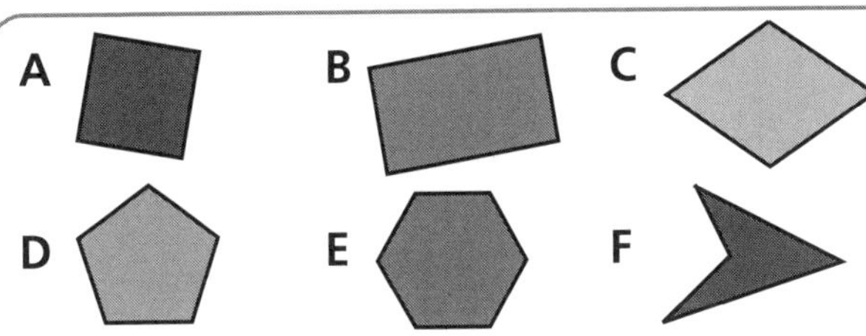


**18** Which of these shapes is a:

a hexagon? ______ b square? ______

c rhombus? ______ d pentagon? ______

e Which shapes are quadrilaterals? ______

f Is **B** a regular shape? ______

g Which shapes have parallel sides? ______

see page 130

**19**

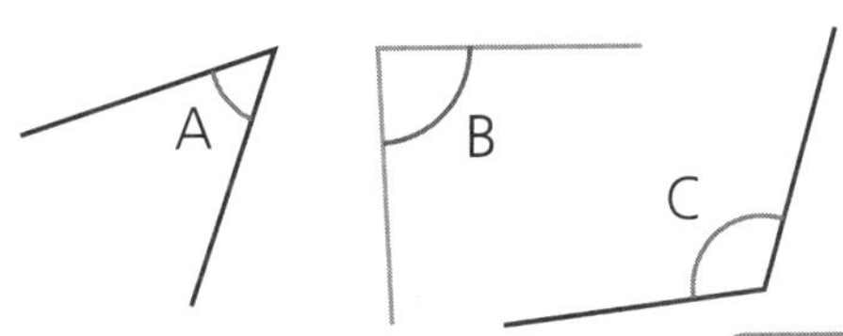


a Which is the largest angle? ______

b Which is the smallest? ______

c Order the angles from smallest to largest. ______

## Statistics

see page 154

**20** **Pet Show**

| | Birds | Cats | Dogs |
|---|---|---|---|
| **Male** | 12 | 9 | 12 |
| **Female** | 5 | 7 | 19 |

How many animals at the pet show were:

a dogs? ______ b cats? ______

c birds? ______ d male cats? ______

e male dogs? ______ f female? ______

g there altogether? ______

# Diagnostic Test 6

## Number

see page

**1** Write this number as a numeral and in words. ☐

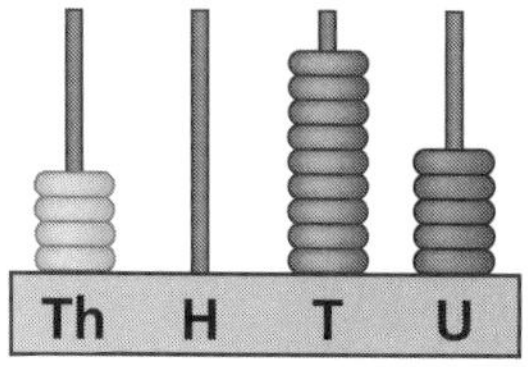


☐
☐

22

**2** How many digits in:

**a** 349? ☐ **b** 4725? ☐

Write the number that follows:

**c** 3796 ☐ **d** 2019 ☐

23

**3** Write the numeral for:

**a** 6000 + 400 + 60 + 5 ☐

**b** 7000 + 200 + 90 + 6 ☐

What is the value of the 6 in:

**c** 3657 ☐ **d** 6193 ☐

26

**4** Write the first four numbers.

**a** Start at 29 and count by tens.

☐, ☐, ☐, ☐

**b** Start at 30 and count by fives.

☐, ☐, ☐, ☐

24

**5** Write the next three numbers.

**a** Add 6: 11, ☐, ☐, ☐

**b** Subtract 2: 24, ☐, ☐, ☐

**c** Multiply by 2: 3, ☐, ☐, ☐

**d** Write the rule for: 20, 16, 12, …

☐

25

## Operations

see page

**6**

**a**

| | Tens | Ones |
|---|---|---|
| | 3 | 4 |
| | 2 | 1 |
| + | 4 | 3 |

**b**

| | Tens | Ones |
|---|---|---|
| | $4 | 2 |
| | $1 | 6 |
| + | $2 | 3 |

**c** 35 buttons, 23 used.
How many buttons left?

☐ ☐ ☐ ☐ ☐

**d**

| | Tens | Ones |
|---|---|---|
| | 3 | 7 |
| − | 2 | 4 |

**e**

| | Tens | Ones |
|---|---|---|
| | 7 | 8 |
| − | 5 | 1 |

**f**

| | Tens | Ones |
|---|---|---|
| | $7 | 8 |
| − | $5 | 6 |

**g**

| | Tens | Ones |
|---|---|---|
| | $5 | 9 |
| − | $3 | 4 |

64

**7** Use the jump strategy to find:

**a** 65 + 18 = ☐ **b** 74 − 2 = ☐

**c** 36 + 28 = ☐ **d** 57 + 29 = ☐

70

**8** Do what is in brackets first.

**a** 5 × 2 = (4 × 2) + (1 × 2)

= ☐ + ☐ = ☐

66

**9** Write the first five multiples of:

**a** 6: ☐, ☐, ☐, ☐, ☐

**b** 9: ☐, ☐, ☐, ☐, ☐

68

**10** **a** 6 × 4 = 24

24 ÷ 6 = ☐

24 ÷ 4 = ☐

**b** 7 × 5 = 36

35 ÷ 7 = ☐

35 ÷ 5 = ☐

69



 • *AUSTRALIAN SIGNPOST MATHS 3* • ISBN 978 1 4886 2184 0

## Measurement

see page

**11** Write these in centimetres: 105

**a** 3 m 50 cm

**b** 5 m 75 cm

**c** Find the perimeter of:

**12** Record the length of each interval. 108

**a** mm

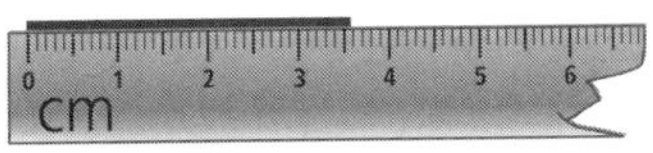

**b** mm

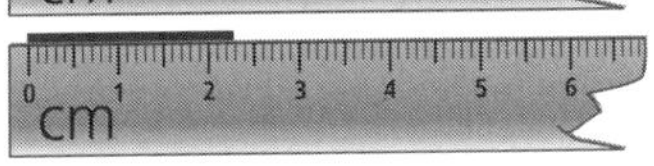

Write these as millimetres.

**c** 4 cm 1 mm

**d** 8 cm 7 mm

Write these as centimetres and millimetres.

**e** 95 millimetres

**f** 31 millimetres

**13** Complete by filling in the boxes. 107

52 minutes past 7 is the same as ___ to ___, so it is nearly ___ o'clock.

**14** Use this calendar to answer the following questions. 106

March

| S | M | T | W | T | F | S |
|---|---|---|---|---|---|---|
| 30 | 31 | | | | | 1 |
| 2 | 3 | 4 | 5 | 6 | 7 | 8 |
| 9 | 10 | 11 | 12 | 13 | 14 | 15 |
| 16 | 17 | 18 | 19 | 20 | 21 | 22 |
| 23 | 24 | 25 | 26 | 27 | 28 | 29 |

**a** What month is shown?

On what day is:

**b** 1st

**c** 24th

## Geometry

see page

**15** 131

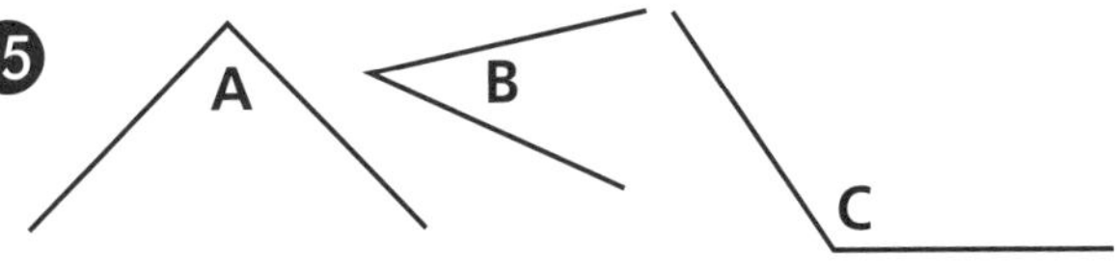


**a** Which angle is the right angle?

**b** Which is the largest angle?

**16** What is kept: 136

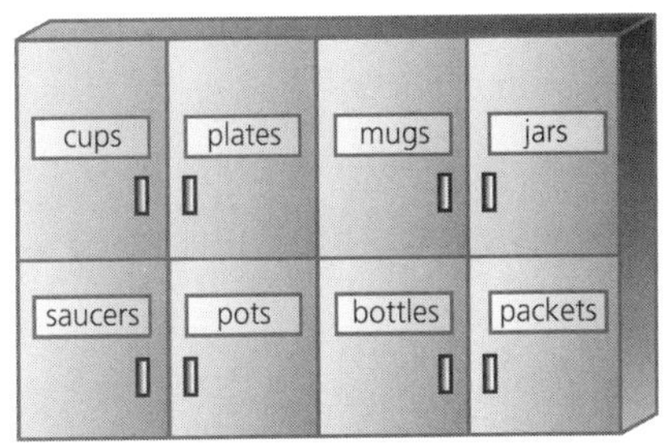


**a** second from left, top row?

**b** third from left, bottom row?

**c** Where are the mugs kept?

**17** Describe the turn using **clockwise** or **anticlockwise** and **a quarter turn**, **half turn** or **three-quarter turn**. 135

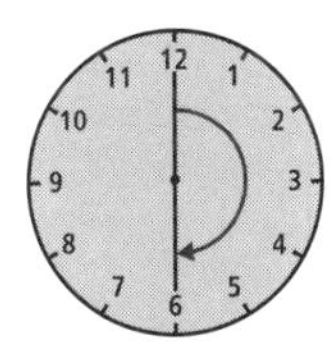


**18** Move the counter along the path. 137

Record each move.

Move 4 left,

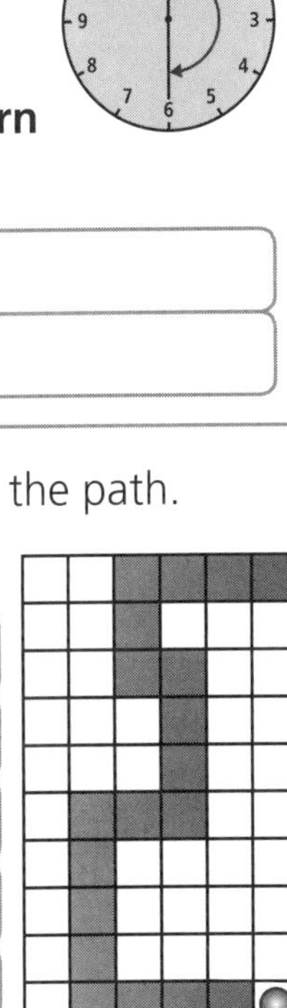

 • *AUSTRALIAN SIGNPOST MATHS 3* • ISBN 978 1 4886 2184 0

# Diagnostic Test 7

## Number

see page

30

**1** Write this number as a numeral and in words. [ ]

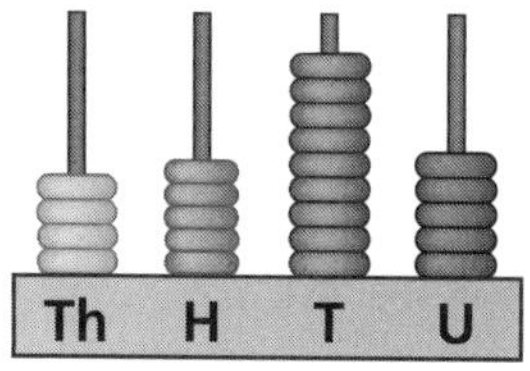


[ ]

[ ]

27

**2** Write the numeral for nine thousand seven hundred and two. [ ]

27

**3** Write **T** (true) of **F** (false) for:

**a** 2459 < 2160 [ ]

**b** 4560 > 4460 [ ]

Circle the smaller number.

**c** 399; 2000 **d** 2581; 2851

**e** 6923; 6723 **f** 3735; 3753

28

**4** Complete each number pattern.

**a** 20 40 60 ( ) ( ) ( )

**b** 95 90 85 ( ) ( ) ( )

29

**5** Complete the table.

**a**

| First number | 1 | 2 | 3 | 4 | 5 | 6 |
|---|---|---|---|---|---|---|
| Second number | 8 | 9 | 10 | | | |

Rule: [ ]

[ ]

**b**

| First number | 20 | 18 | 16 | 14 | 12 | 10 |
|---|---|---|---|---|---|---|
| Second number | 10 | 9 | 8 | | | |

Rule: [ ]

[ ]

## Operations

see page

71

**6** 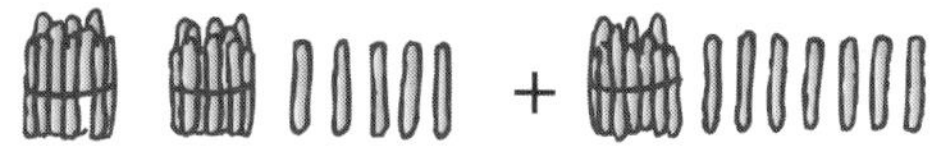

[ ] + [ ] = [ ]

72

**7**

**a**

| | Tens | Ones |
|---|---|---|
| | 3 | 6 |
| + | | 9 |
| | | |

**b**

| | Tens | Ones |
|---|---|---|
| | $5 | 7 |
| + | $3 | 8 |
| | | |

**c**

| Hund | Tens | Ones |
|---|---|---|
| | 7 | 2 |
| + | 5 | 7 |
| | | |

**d**

| Hund | Tens | Ones |
|---|---|---|
| | 9 | 4 |
| + | 6 | 1 |
| | | |

74

**8** In a library, one shelf holds 24 books. Another holds 55. How many books do the shelves hold altogether?

Find: [ ]

Number sentence: [ ]

Answer: [ ]

**9** Joanne planted seeds in rows of 5. She planted 6 full rows and had 4 seeds left. How many seeds were there altogether?

Find. [ ]

Number sentence: [ ]

Answer: [ ]

# Diagnostic Test 7 (continued)

## Measurement

see page

**10** 110

A 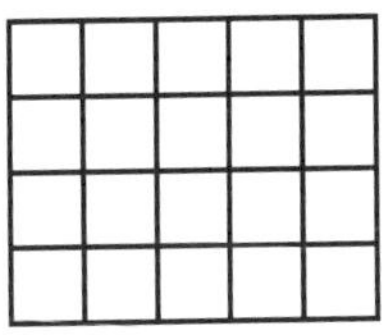 B 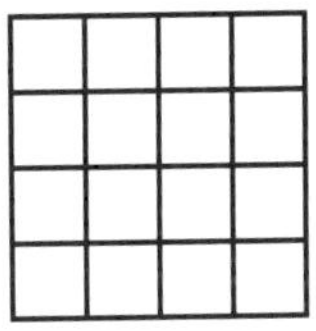

How many small squares has:

**a** A ☐ **b** B ☐

**c** Which has the larger area? ☐

**11** 112 Use the short form to write:

**a** forty grams ☐

**b** six hundred grams ☐

**c** how many grams in 1 kg? ☐

**12** 113 Use the short form to write:

**a** 45 kilograms ☐

**b** 90 kilograms ☐

**c** 750 kilograms ☐

## Geometry

**13** 138

**a** What is this shape? ☐

**b** is this a prism? ☐

**14** 139

A 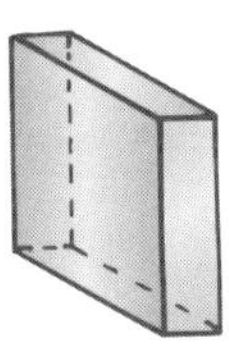

B 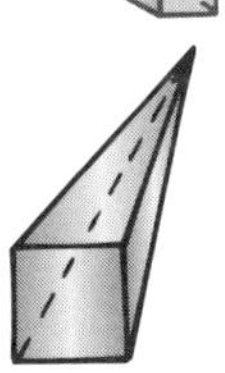

C 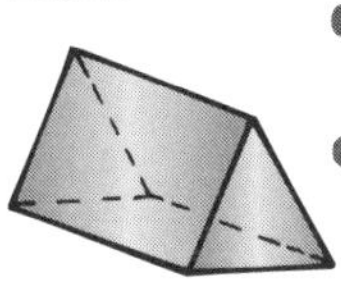

Is it a prism or a pyramid?

**a** A ☐

**b** B ☐

**c** C ☐

Am I **A**, **B** or **C**?

**d** I have 5 vertices. ☐

**e** I have 12 edges. ☐

## Statistics

see page

**15** 157

Birds
Fish
Mice
Turtles

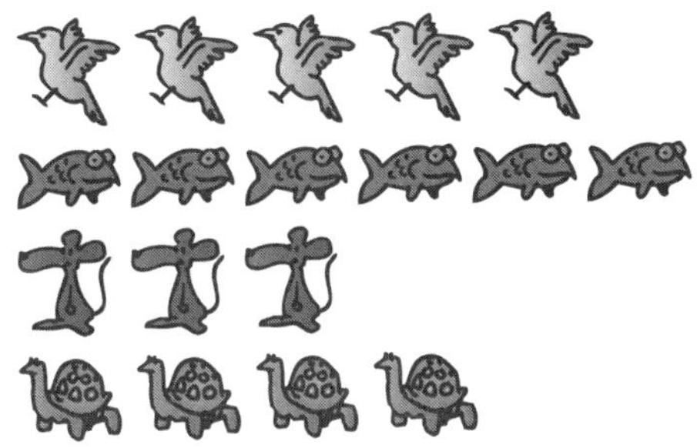

Each picture stands for one creature.

**a** Use the data to complete the table.

| Birds | Fish | Mice | Turtles |
|---|---|---|---|
| | | | |

**b** Use the data to complete the graph.

| | | | | | | | |
|---|---|---|---|---|---|---|---|
| Birds | | | | | | | |
| Fish | | | | | | | |
| Mice | | | | | | | |
| Turtles | | | | | | | |

**c** Which group has the most? ☐

**d** Which group has the least? ☐

**e** How many birds? ☐

**16** If each picture in Question 15 stands for 3 creatures, how many:

**a** fish would there be? ☐

**b** turtles would there be? ☐

**17** 157 **a** Name one advantage of using the table in Question 15a.

☐

**b** Name one advantage of using the graph in Question 15b.

☐

# Diagnostic Test 8

## Number

see page

❶ Write the numeral for:

a 6000 + 700 + 80 + 5 ☐

b 9000 + 200 + 7 ☐

Use expanded notation to write:

c 5623 ☐

d 7058 ☐

e Write the number shown on the abacus. ☐

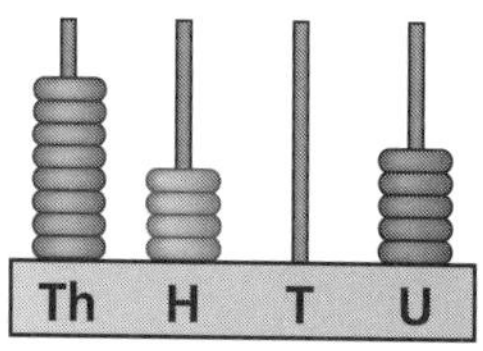


33

❷

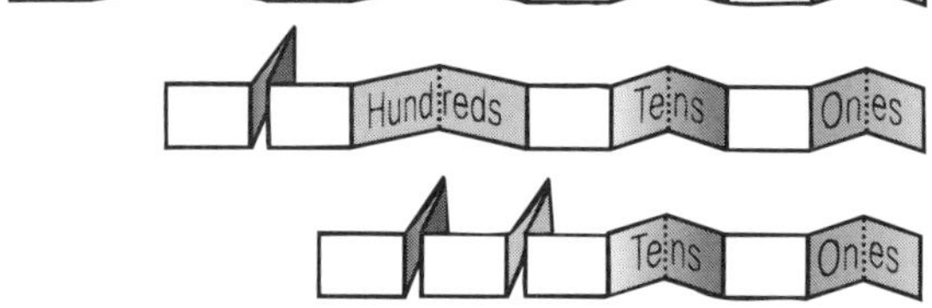


a Fill in the numeral expanders for 1436.

Write the numeral for:

b two thousand and forty ☐

c one thousand nine hundred ☐

34

❸ Find the value of the 7 in:

a 4793 ☐ b 7852 ☐

c 567 ☐ d 1679 ☐

e 6478 ☐

Use > or < to link these numbers.

f 960 ☐ 845

g 1307 ☐ 1695

h 4852 ☐ 4825

i 3628 ☐ 3862

## Operations

see page

77

❹ Complete these problems.

a

| Hund | Tens | Ones |
|---|---|---|
| 2 | 5 | 4 |
| +1 | 8 | 3 |
| | | |

b 65 tens and 7 ones
− 6 tens and 5 ones

c 427 − 47 = ☐

d 833 − 53 = ☐

80

❺ Check subtraction by addition.

Dianne has 57 books. She gave 31 to the library. How many did she have left?

| Tens | Ones |
|---|---|
| 5 | 7 |
| − 3 | 1 |
| | |

(Check)

| Tens | Ones |
|---|---|
| | |
| + 3 | 1 |
| | |

❻

a 5 × 2 = ☐

10 ÷ 2 = ☐

so 5 × 2 ÷ 2 = ☐

b 7 × 5 ÷ 5 = ☐

c 6 × 10 ÷ 10 = ☐

82

❼ Find the change from $20 if $14.40 is the cost. ☐



© PEARSON AUSTRALIA 2018 • *AUSTRALIAN SIGNPOST MATHS 3* • ISBN 978 1 4886 2184 0

# Diagnostic Test 8 (continued)

## Measurement

**8** How many: see page 114

**a** centimetres in 1 metre? ☐

**b** millimetres in 1 centimetre? ☐

**c** millimetres in 1 metre? ☐

**9** Use the short form to write: see page 114

**a** 60 grams ☐

**b** 350 grams ☐

**c** 5 kilograms ☐

**d** 10 kilograms ☐

**10** **a** Circle the items that would have a mass greater than 1 kilogram.

a car    a ruler    a refrigerator

**b** Circle the object that would be measured in kilograms.

a packet of chips    a case of tomatoes

**11** How many millilitres in: see page 116

**a** 3 L ☐    **b** 6 L ☐

How many litres in:

**c** 4000 mL ☐    **d** 5000 mL ☐

**12**

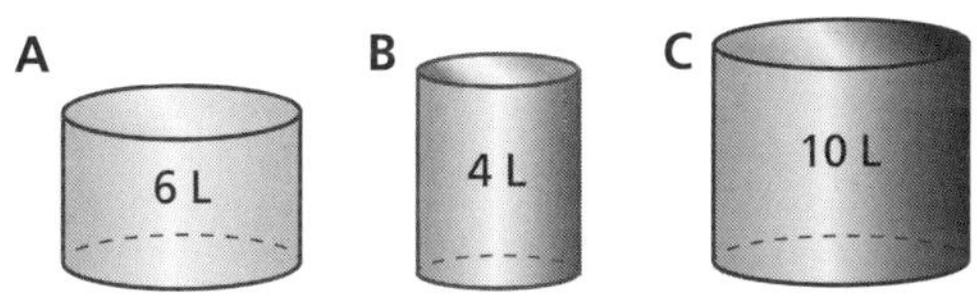


Of these containers, which one would hold:

**a** the most water? ☐

**b** the least water? ☐

**c** more than **A**? ☐

## Geometry

**13** see page 142

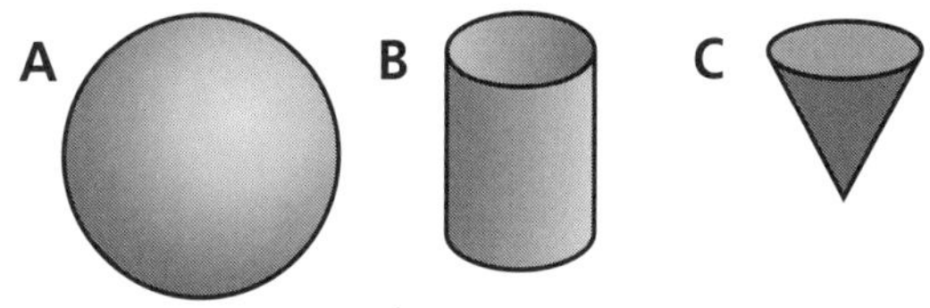


Name these 3D objects.

**a** A ☐

**b** B ☐

**c** C ☐

**14** **A** flip    **B** slide    **C** turn see page 144

Match a label to each coloured shape.

**a** ☐    **b** ☐    **c** ☐

**15**

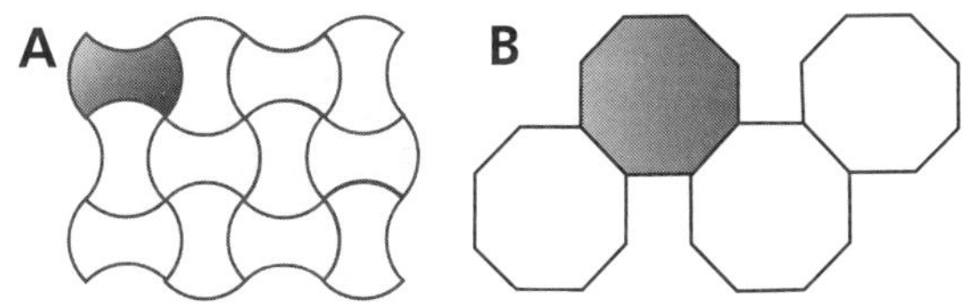


Which of these patterns is a tessellation? ☐

## Probability

**16** Choose the best label for each statement. see page 158

**A** impossible    **B** unlikely

**C** even chance    **D** likely    **E** certain

**a** The sun will set this evening. ☐

**b** I will roll zero on a dice. ☐

**c** The coin will land on heads. ☐

**d** The bus will be on time. ☐

**e** You will see a bird tomorrow. ☐

 • *AUSTRALIAN SIGNPOST MATHS 3* • ISBN 978 1 4886 2184 0